D1377695

IMS

THE

IMS

IP Multimedia Concepts and Services in the Mobile Domain

Miikka Poikselkä, Georg Mayer, Hisham Khartabil and Aki Niemi

John Wiley & Sons, Ltd

Other Wiley Editorial Offices

John Wiley & Sons, Inc., 111 River Street, Hoboken, NJ 07030, USA

Jossey-Bass, 989 Market Street, San Francisco, CA 94103-1741, USA

Wiley-VCH Verlag GmbH, Boschstr. 12, D-69469 Weinheim, Germany

John Wiley & Sons Australia Ltd, 33 Park Road, Milton, Queensland 4064, Australia

John Wiley & Sons (Asia) Pte Ltd, 2 Clementi Loop #02-01, Jin Xing Distripark, Singapore 129809

John Wiley & Sons Canada Ltd, 22 Worcester Road, Etobicoke, Ontario, Canada M9W 1L1

British Library Cataloguing in Publication Data

A catalogue record for this book is available from the British Library

ISBN 0-470-87113-X

Project management by Originator, Gt Yarmouth, Norfolk (typeset in 10/13pt Times)
Printed and bound in Great Britain by TJ International, Padstow, Cornwall
This book is printed on acid-free paper responsibly manufactured from sustainable forestry
in which at least two trees are planted for each one used for paper production.

Contents

Foreword   xvii

Preface   xix

Acknowledgements   xxi

List of Figures   xxiii

List of Tables   xxvii

PART I: ARCHITECTURE   1

1 **Introduction**   3
    1.1 Why the Internet Protocol Multimedia Subsystem was developed   3
    1.2 Where did it come from?   5
        1.2.1 From GSM to 3GPP Release 6   5
        1.2.2 3GPP Release 99 (3GPP R99)   5
        1.2.3 3GPP Release 4   6
        1.2.4 3GPP Release 5 and Release 6   6
    1.3 Other relevant standardization bodies   8
        1.3.1 Internet Engineering Task Force   8
        1.3.2 Open Mobile Alliance   9
        1.3.3 Third Generation Partnership Project 2   9

2 **IP Multimedia Subsystem Architecture**   11
    2.1 Architectural requirements   11
        2.1.1 IP connectivity   11
        2.1.2 Access independence   12
        2.1.3 Ensuring quality of service for IP multimedia services 13

    2.1.4   IP policy control for ensuring correct usage of media
           resources   13
    2.1.5   Secure communication   14
    2.1.6   Charging arrangements   14
    2.1.7   Support of roaming   15
    2.1.8   Interworking with other networks   16
    2.1.9   Service control model   16
    2.1.10 Service development   17
    2.1.11 Layered design   17
2.2   Description of IMS-related entities and functionalities   18
    2.2.1   Proxy-CSCF   19
    2.2.2   Policy Decision Function   20
    2.2.3   Interrogating-CSCF   21
    2.2.4   Serving-CSCF   22
    2.2.5   Home Subscriber Server   23
    2.2.6   Subscription Locator Function   24
    2.2.7   Multimedia Resource Function Controller   24
    2.2.8   Multimedia Resource Function Processor   24
    2.2.9   Application server   25
    2.2.10 Breakout Gateway Control Function   26
    2.2.11 Media Gateway Control Function   27
    2.2.12 IP Multimedia Subsystem-Media Gateway Function   27
    2.2.13 Signalling gateway   27
    2.2.14 Security gateway   27
    2.2.15 Charging entities   28
    2.2.16 GPRS Service entities   28
2.3   IMS reference points   29
    2.3.1   Gm reference point   30
    2.3.2   Mw reference point   31
    2.3.3   IMS Service Control reference point   32
    2.3.4   Cx reference point   32
    2.3.5   Dx reference point   38
    2.3.6   Sh reference point   39
    2.3.7   Si reference point   42
    2.3.8   Dh reference point   42
    2.3.9   Mm reference point   43
    2.3.10 Mg reference point   43
    2.3.11 Mi reference point   43
    2.3.12 Mj reference point   43
    2.3.13 Mk reference point   44
    2.3.14 Ut reference point   44
    2.3.15 Mr reference point   44

2.3.16   Mp reference point   44
2.3.17   Go reference point   45
2.3.18   Gq reference point   45

3   IMS Concepts   49
3.1   Overview   49
3.2   Registration   49
3.3   Session initiation   51
3.4   Identification   53
       3.4.1   Identification of users   53
       3.4.2   Identification of services (public service identities)   58
       3.4.3   Identification of network entities   58
3.5   Identity modules   59
       3.5.1   IP Multimedia Services Identity Module   59
       3.5.2   Universal Subscriber Identity Module   60
3.6   Security services in the IMS   60
       3.6.1   IMS Security Model   61
       3.6.2   Authentication and Key Agreement   62
       3.6.3   Network domain security   62
       3.6.4   IMS access security for SIP-based services   66
       3.6.5   IMS access security for HTTP-based services   70
3.7   Discovering the IMS entry point   72
3.8   S-CSCF assignment   73
       3.8.1   S-CSCF assignment during registration   73
       3.8.2   S-CSCF assignment for an unregistered user   75
       3.8.3   S-CSCF assignment in error cases   75
       3.8.4   S-CSCF de-assignment   75
       3.8.5   Maintaining S-CSCF assignment   75
3.9   Mechanism for controlling bearer traffic   75
       3.9.1   SBLP functions   77
3.10   Charging   91
       3.10.1   Charging architecture   91
       3.10.2   Charging information correlation   99
       3.10.3   Charging information distribution   101
3.11   User profile   101
       3.11.1   Service profile   101
3.12   Service provision   105
       3.12.1   Introduction   105
       3.12.2   Creation of the filter criteria   105
       3.12.3   Selection of AS   107
       3.12.4   AS behaviour   108

3.13   Connectivity between traditional Circuit-Switched users and
        IMS users   109
        3.13.1   IMS-originated session toward a user in the CS core
                 network   109
        3.13.2   CS-originated session toward a user in IMS   111
3.14   Mechanism to register multiple user identities at once   111
3.15   Sharing a single user identity between multiple terminals   113
3.16   SIP compression   114

PART II: DETAILED PROCEDURES   117

4   Introduction   119
    4.1   The example scenario   119
    4.2   Base standards   121

5   An example IMS registration   123
    5.1   Overview   123
    5.2   Signalling PDP context establishment   125
    5.3   P-CSCF discovery   126
    5.4   Transport protocols   126
    5.5   SIP registration and registration routing aspects   126
           5.5.1   Overview   126
           5.5.2   Constructing the REGISTER request   129
           5.5.3   From the UE to the P-CSCF   131
           5.5.4   From the P-CSCF to the I-CSCF   131
           5.5.5   From the I-CSCF to the S-CSCF   132
           5.5.6   Registration at the S-CSCF   132
           5.5.7   The 200 (OK) response   133
           5.5.8   The Service-Route header   134
           5.5.9   The Path header   135
           5.5.10   Third-party registration to application servers   135
           5.5.11   Related standards   137
    5.6   Authentication   137
           5.6.1   Overview   137
           5.6.2   HTTP digest and 3GPP AKA   139
           5.6.3   Authentication information in the initial REGISTER
                    request   140
           5.6.4   S-CSCF challenges the UE   141
           5.6.5   UE's response to the challenge   142
           5.6.6   Integrity protection and successful authentication   142
           5.6.7   Related standards   143

5.7	Access security—IPsec SAs	143
	5.7.1 Overview	143
	5.7.2 Establishing an SA during initial registration	144
	5.7.3 Handling of multiple sets of SAs in case of re-authentication	146
	5.7.4 SA lifetime	149
	5.7.5 Port setting and routing	150
	5.7.6 Related standards	155
5.8	SIP Security Mechanism Agreement	155
	5.8.1 Why the SIP Security Mechanism Agreement is needed	155
	5.8.2 Overview	155
	5.8.3 SIP Security Mechanism Agreement-related headers in the initial REGISTER request	157
	5.8.4 The Security-Server header in the 401 (Unauthorized) response	158
	5.8.5 SIP Security Mechanism Agreement headers in the second REGISTER	159
	5.8.6 SIP Security Mechanism Agreement and re-registration	159
	5.8.7 Related standards	161
5.9	Compression negotiation	162
	5.9.1 Overview	162
	5.9.2 Indicating willingness to use SigComp	163
	5.9.3 comp=SigComp parameter during registration	164
	5.9.4 comp=SigComp parameter in other requests	165
	5.9.5 Related standards	165
5.10	Access and location information	165
	5.10.1 P-Access-Network-Info	165
	5.10.2 P-Visited-Network-ID	166
	5.10.3 Related standards	166
5.11	Charging-related information during registration	167
5.12	User identities	167
	5.12.1 Overview	167
	5.12.2 Public and private user identities for registration	168
	5.12.3 Identity derivation without ISIM	169
	5.12.4 Default public user identity/P-Associated-URI header	170
	5.12.5 UE's subscription to registration-state information	171
	5.12.6 P-CSCF's subscription to registration-state information	173
	5.12.7 Elements of registration-state information	175
	5.12.8 Registration-state information in the body of the NOTIFY request	175
	5.12.9 Example registration-state information	177
	5.12.10 Multiple terminals and registration-state information	179

		5.12.11 Related standards	180
5.13		Re-registration and re-authentication	181
	5.13.1	User-initiated re-registration	181
	5.13.2	Network-initiated re-authentication	181
	5.13.3	Network-initiated re-authentication notification	182
	5.13.4	Related standards	183
5.14		De-registration	184
	5.14.1	Overview	184
	5.14.2	User-initiated de-registration	185
	5.14.3	Network-initiated de-registration	188
	5.14.4	Related standards	188

6 An Example IMS Session — 191

6.1		Overview	191
6.2		Caller and callee identities	192
	6.2.1	Overview	192
	6.2.2	From and To headers	194
	6.2.3	Identification of the calling user: P-Preferred-Identity and P-Asserted-Identity	194
	6.2.4	Identification of the called user	196
	6.2.5	Related standards	198
6.3		Routing	198
	6.3.1	Overview	198
	6.3.2	Session, dialog, transactions and branch	201
	6.3.3	Routing of the INVITE request	202
	6.3.4	Routing of the first response	207
	6.3.5	Re-transmission of the INVITE request and the 100 (Trying) response	209
	6.3.6	Routing of subsequent requests in a dialog	210
	6.3.7	Stand-alone transactions from one UE to another	212
	6.3.8	Routing to and from ASs	212
	6.3.9	Related standards	216
6.4		Compression negotiation	216
	6.4.1	Overview	216
	6.4.2	Compression of the initial request	216
	6.4.3	Compression of responses	217
	6.4.4	Compression of subsequent requests	218
	6.4.5	Related standards	219
6.5		Media negotiation	219
	6.5.1	Overview	219
	6.5.2	Reliability of provisional responses	221
	6.5.3	SDP offer/answer in IMS	223

	6.5.4	Related standards	229
6.6	Resource reservation		229
	6.6.1	Overview	229
	6.6.2	The 183 (Session in Progress) response	230
	6.6.3	Are preconditions mandatorily supported?	230
	6.6.4	Preconditions	233
	6.6.5	Related standards	238
6.7	Controlling the media		239
	6.7.1	Overview	239
	6.7.2	Media authorization	240
	6.7.3	Grouping of media lines	241
	6.7.4	A single reservation flow	241
	6.7.5	Separated flows	242
	6.7.6	Media policing	242
	6.7.7	Related standards	244
6.8	Charging-related information for sessions		245
	6.8.1	Overview	245
	6.8.2	Exchange of ICID for a media session	245
	6.8.3	Correlation of GCID and ICID	246
	6.8.4	Distribution of charging function addresses	248
	6.8.5	Related standards	249
6.9	Release of a session		250
	6.9.1	User-initiated session release	250
	6.9.2	P-CSCF performing network-initiated session release	250
	6.9.3	S-CSCF performing network-initiated session release	253
7	**Routing of PSIs**		255
7.1	Scenario 1: routing from a user to a PSI		255
7.2	Scenario 2: routing from a PSI to a user		255
7.3	Scenario 3: routing from a PSI to another PSI		256
PART III: PROTOCOLS			259
8	**SIP**		261
8.1	Background		261
8.2	Design principles		262
8.3	SIP architecture		263
8.4	Message format		265
	8.4.1	Requests	265
	8.4.2	Response	266
	8.4.3	Header fields	267
	8.4.4	Body	268

8.5	The SIP URI	268
8.6	The tel URI	269
8.7	SIP structure	270
	8.7.1 Syntax and encoding layer	270
	8.7.2 Transport layer	270
	8.7.3 Transaction layer	270
	8.7.4 TU layer	271
8.8	Registration	273
8.9	Dialogs	274
8.10	Sessions	276
	8.10.1 The SDP offer/answer model with SIP	277
8.11	Security	278
	8.11.1 Threat models	278
	8.11.2 Security framework	278
	8.11.3 Mechanisms and protocols	279
8.12	Routing requests and responses	283
	8.12.1 Server discovery	283
	8.12.2 The loose routing concept	284
	8.12.3 Proxy behaviour	284
	8.12.4 Populating the request-URI	286
	8.12.5 Sending requests and receiving responses	286
	8.12.6 Receiving requests and sending responses	287
8.13	SIP extensions	287
	8.13.1 Event notification framework	287
	8.13.2 State publication (the PUBLISH method)	289
	8.13.3 SIP for instant messaging	289
	8.13.4 Reliability of provisional responses	290
	8.13.5 The UPDATE method	291
	8.13.6 Integration of resource management and SIP (preconditions)	292
	8.13.7 The SIP REFER method	293
	8.13.8 The "message/sipfrag" MIME type	294
	8.13.9 SIP extension header for registering non-adjacent contacts (the Path header)	294
	8.13.10 Private SIP extensions for asserted identity within trusted networks	295
	8.13.11 Security mechanism agreement for SIP	296
	8.13.12 Private SIP extensions for media authorization	298
	8.13.13 SIP extension header for service route discovery during registration	299
	8.13.14 Private header extensions to SIP for 3GPP	299
	8.13.15 Compressing SIP	300

9 SDP 301
  9.1 SDP message contents 301
     9.1.1 Session description 302
     9.1.2 Time description 302
     9.1.3 Media description 302
  9.2 SDP message format 303
  9.3 Selected SDP lines 303
     9.3.1 Protocol version line 303
     9.3.2 Connection information line 303
     9.3.3 Media line 304
     9.3.4 Attribute line 304
     9.3.5 The rtpmap attribute 305

10 The Offer/Answer Model with SDP 307
  10.1 The offer 307
  10.2 The answer 307
  10.3 Offer/Answer processing 308
     10.3.1 Modifying a session description 308
     10.3.2 Putting the media stream on hold 309

11 RTP 311
  11.1 RTP for real-time data delivery 311
     11.1.1 RTP fixed header fields 311
     11.1.2 What is jitter? 312
  11.2 RTCP 313
     11.2.1 RTCP packet types 313
     11.2.2 RTCP report transmission interval 313
  11.3 RTP profile and payload format specifications 314
     11.3.1 Profile specification 314
     11.3.2 Payload format specification 314
  11.4 RTP profile and payload format specification for audio and video (RTP/AVP) 314
     11.4.1 Static and dynamic payload types 314

12 DNS 317
  12.1 DNS resource records 317
  12.2 The naming authority pointer (NAPTR) DNS RR 317
     12.2.1 NAPTR example 319
  12.3 ENUM – the E.164 to URI Dynamic Delegation Discovery System (DDD) application 319
     12.3.1 ENUM service registration for SIP addresses-of-record 320
  12.4 Service records (SRVs) 321
     12.4.1 SRV example 321

13 GPRS   323
  13.1   Overview   323
  13.2   Packet Data Protocol (PDP)   323
         13.2.1   Primary PDP context activation   323
         13.2.2   Secondary PDP context activation   324
         13.2.3   PDP context modification   324
         13.2.4   PDP context deactivation   324
  13.3   Access points   324
  13.4   PDP context types   325

14 TLS   327
  14.1   Introduction   327
  14.2   TLS Record Protocol   327
  14.3   TLS Handshake Protocol   328
  14.4   Summary   330

15 Diameter   331
  15.1   Introduction   331
  15.2   Protocol components   332
  15.3   Message processing   332
  15.4   Diameter clients and servers   334
  15.5   Diameter agents   334
  15.6   Message structure   335
  15.7   Error handling   337
  15.8   Diameter services   338
         15.8.1   Authentication and authorization   338
         15.8.2   Accounting   339
  15.9   Specific Diameter applications used in 3GPP   339
  15.10 Diameter SIP application   340
  15.11 Diameter credit control application   340
  15.12 Summary   343

16 MEGACO   345
  16.1   Introduction   345
  16.2   Connection model   345
  16.3   Protocol operation   346

17 COPS   349
  17.1   Introduction   349
  17.2   Message structure   350
  17.3   COPS usage for policy provisioning (COPS-PR)   351
  17.4   The PIB for the Go interface   354
  17.5   Summary   354

18 IPsec   355
  18.1 Introduction   355
  18.2 Security associations   356
  18.3 Internet Security Association and Key Management Protocol
       (ISAKMP)   356
  18.4 Internet Key Exchange (IKE)   357
  18.5 Encapsulated Security Payload (ESP)   357
  18.6 Summary   359

19 Signalling Compression   361
  19.1 SigComp architecture   361
  19.2 Compartments   362
  19.3 Compressing a SIP message in IMS   363
      19.3.1 Initialization of SIP compression   363
      19.3.2 Compressing a SIP message   363
      19.3.3 Decompressing a compressed SIP message 363

20 DHCPv6   365
  20.1 DHCP options   366
  20.2 DHCP options for SIP servers   366

21 XCAP   369
  21.1 XCAP application usage   369

22 CPCP   371

PART IV: SERVICES   373

23 Presence   375
  23.1 SIP for presence   376
  23.2 Presence service architecture in IMS   377
  23.3 Resource (presentity) list   378
  23.4 XCAP usage for resource (presentity) lists   378
  23.5 Setting presence authorization   379
  23.6 Publishing presence   379
  23.7 Watcher information event template package   379
  23.8 Example signalling flows of presence service operation 380
      23.8.1 Successful subscription to presence 380
      23.8.2 Successful publication of presence information 381
      23.8.3 Subscribing to a resource list   381
      23.8.4 Subscribing to watcher information   382

24 Messaging   383
   24.1 Overview of IMS messaging   383
   24.2 IMS messaging architecture   384
   24.3 Immediate messaging   384
   24.4 Session-based messaging   385
   24.5 Deferred delivery messaging   385

25 Conferencing   387
   25.1 Conferencing architecture   387
   25.2 SIP event package for conference state   388
   25.3 Example signalling flows of conferencing service operation   389
      25.3.1 Creating a conference with a conference factory URI   389
      25.3.2 Referring a user to a conference using the REFER
             request   389
      25.3.3 Subscribing to a conference state   390
      25.3.4 Conference creation using CPCP   391

References   393

Abbreviations   401

Index   409

Foreword

We have telephony to talk to each other. We have messaging to dispatch mail or instant notes. We have browsing to read published content on known sites. We even have search engines to located content sites, which may have content relevant to. This may look as if we have a lot on our plate; so, do we need Internet Protocol (IP) Multimedia?

The problem is that we have no practical mechanism to engage another application-rich terminal in a peer-to-peer session. Enormously successful mobile telephony shows that there is immense value in sharing with peers. With increasingly attractive terminals, the sharing experience will be something more than just exchanging voice.

We will be sharing real time video (see what I see), an MP3-coded music stream,* a white board to present objects and we will be exchanging real time game data. Many of these will be exercised simultaneously. No doubt, we want to break into this completely new ground of communication.

Telephony is sufficient for telephones. Multimedia terminals need IP Multimedia networks.

Session Initiation Protocol (SIP) enables clients to invite others into a session and negotiate control information about the media channels needed for the session. IP Multimedia builds on top of this and provides a full suite of network operator capabilities enabling authentication of clients, network-to-network interfaces and administration capabilities like charging. All this is essential in order to build interoperating networks that, when combined, can provide truly global service coverage, in the spirit of good old telephony. This enables a global market of multimedia terminals.

As IP Multimedia is now emerging as the key driver of renewal of maturing mass-market communication services, several technical audiences have an urgent need to understand how it works. Georg Mayer, Aki Niemi, Hisham Khartabil and Miikka Poikselkä are major contributors to IP Multimedia industry develop-

---

*MP3 is the voice compression method developed by the Moving Picture Experts Group (MPEG), by means of which the size of a voice-containing file can be reduced to one-tenth of the original without significantly affecting the quality of voice.

ment through their work in the standardization arena. This book provides the essential insight into the architecture and structure of these new networks.

Petri Pöyhönen
Vice President
Nokia Networks

Preface

The Internet Protocol (IP) Multimedia Subsystem, better known as "The IMS", is based on the specification of the Session Initiation Protocol (SIP) as standardized by the Internet Engineering Task Force (IETF). But SIP as a Protocol is only one part of it; the IMS is more than just a protocol. It is an architecture for the convergence of data, speech and mobile networks and is based on a wide range of protocols, of which most have been developed by the IETF. It combines and enhances them to allow real time services on top of the Universal Mobile Telecommunications System (UMTS) packet-switched domain.

This book was written to provide detailed insights about what the IMS is, its concepts, architecture, protocols and services. Its intended audience ranges from marketing managers, research and development engineers, test engineers to university students. The book is written in a manner that allows readers to choose the level of knowledge they need and the depth in understanding they desire to achieve about the IMS. The book is also very well suited as a reference.

The first few chapters in Part I provide a detailed overview of the system architecture and the entities that, when combined, provide the IMS. The chapters also present the reference points (interfaces) between those entitites and introduces the protocols assigned to those interfaces.

As with every communication system, the IMS is built on concepts that offer basic and advanced services to its users. Security is a concept that is required by any communication architecture. In this book we describe the security threats and the models used to secure communications in the IMS. IMS security, along with concepts such as registration, session establishment, charging and service provisioning, are explained in Chapter 3.

SIP and SDP are two of the main building blocks within IMS and their usage therein gets complemented by a large number of vital extensions. Chapters 4–7 in Part II go step by step through an example IMS registration and session establishment at the protocol level, detailing the procedures taken at every entity.

Chapters 8–22 in Part III describe the protocols used within the IMS in more detail, paying special attention to signalling as well as security protocols. This part of the book shows how different protocols are built up, how they work and why they are applied within the IMS.

Finally, the last part gives an introduction to some of the advanced services in IMS with call flows. This part proves that the convergence of services and networks is not a myth, but a real added value for the user.

The Third Generation Partnership Project (3GPP) and the IETF have worked together during recent years in an amazing way to achieve the IMS and the protocols used by it. We, the authors, have had the chance to participate in many technical discussions regarding architecture and protocols and are still very active in further discussions on the ever-improving protocols and communication systems. Some of those discussions, which often can be described as debates or negotiations, frequently take a long time to conclude and even more frequently do not result in an agreement or consensus on the technical solutions. We want to thank all the people in these standardization bodies as well as in our company who have had ideas as well as patience and who have worked hard to standardize this communication system of the future called IMS.

Miikka Poikselkä
Georg Mayer
Hisham Khartabil
Aki Niemi

(April 2004)

Acknowledgements

The authors of this book would like to extend their thanks to colleagues working in the 3GPP and the IETF for their great efforts in creating the IMS specifications and related protocols. The authors would also like to give special thanks to the following who helped in the writing of this book by providing excellent review comments and suggestions:

Erkki Koivusalo
Hannu Hietalahti
Tao Haukka
Risto Mononen
Kalle Tammi
Risto Kauppinen
Marco Stura
Ralitsa Gateva
Juha-Pekka Koskinen
Markku Tuohino
Juha Räsänen
Peter Vestergaard

The authors welcome any comments and suggestions for improvements or changes that could be used to improve future editions of this book. Our email addresses are:

miikka.poikselka@nokia.com
georg.mayer@nokia.com
hisham.khartabil@nokia.com
aki.niemi@nokia.com

Figures

1.1 The key ingredient to new, enriching user experiences is peer-to-peer
         IP connections of applications   4
1.2 The IMS and its relationship with existing communication systems   4
1.3 Main 3GPP working groups doing IMS work   8

2.1 IMS connectivity options when a user is roaming   12
2.2 IMS/CS roaming alternatives   16
2.3 IMS and layering architecture   18
2.4 Structure of HSS   24
2.5 Relationship between different AS types   26
2.6 Signalling conversion in the SGW   28
2.7 IMS architecture   30
2.8 HSS resolution using the SLF   39
3.1 A high-level IMS registration flow   50
3.2 A high-level IMS session establishment flow   52
3.3 Relationship between user identities   58
3.4 IP Multimedia Services Identity Module   59
3.5 Security architecture of the IMS   61
3.6 Security domains underlining the IMS   64
3.7 NDS/IP and SEGs   66
3.8 GBA   71
3.9 GPRS-specific mechanism for discovering P-CSCF   72
3.10 Generic mechanism for discovering P-CSCF   73
3.11 Example of an S-CSCF assignment   74
3.12 SBLP entities   76
3.13 Bearer authorization using SBLP   79
3.14 IMS offline charging architecture   92
3.15 IMS online charging architecture   95
3.16 IMS charging correlation   100
3.17 Distribution of charging information   102

3.18 Structure of IMS user profile   103
3.19 Media authorization in the S-CSCF   103
3.20 Structure of Initial Filter Criteria   104
3.21 Structure of service point trigger   106
3.22 IMS–CS interworking configuration when an IMS user calls a CS user   110
3.23 IMS–CS interworking configuration when a CS user calls an IMS user   111
3.24 Example of implicit registration sets   112
3.25 Multiple terminals   113

4.1   The example scenario   120

5.1   Initial IMS registration flow   124
5.2   Routing during registration   128
5.3   Third-party registration by S-CSCF   136
5.4   Authentication information flows during IMS registration   138
5.5   SA establishment during initial registration   145
5.6   Two sets of SAs during re-authentication   147
5.7   Taking a new set of SAs into use and dropping an old set of SAs   149
5.8   Request and response routing between the UE and the P-CSCF over
          UDP   154
5.9   Request and response routing between the UE and the P-CSCF over
          TCP   154
5.10 SIP Security Mechanism Agreement during initial registration   162
5.11 Tobias's subscription to his registration-state information   173
5.12 P-CSCF subscription to Tobias's registration-state information   174
5.13 User-initiated re-registration (without re-authentication)   181
5.14 Network-initiated re-authentication   183
5.15 User-initiated de-registration   184
5.16 Network-initiated de-registration   185

6.1   IMS session establishment call flow   193
6.2   Routing an initial INVITE request and its responses   200
6.3   Routing of subsequent requests and their responses   211
6.4   Routing to an AS   214
6.5   SDP offer/answer in IMS   220
6.6   SIP, SDP offer/answer and preconditions during session
          establishment   231
6.7   SIP session establishment without preconditions   232
6.8   Transport of media authorization information   239
6.9   Media streams and transport in the example scenario   243
6.10 Worst case scenario for media policing   244
6.11 Theresa releases the session   251

6.12 P-CSCF terminates a session   252
6.13 S-CSCF terminates a session   254

7.1   Routing from a user to a PSI   256
7.2   Routing from a PSI to a user   256
7.3   Routing from an AS to a PSI   257

8.1   Protocol stack   262
8.2   SIP trapezoids   263
8.3   SIP message format   265
8.4   SIP protocol layers   270
8.5   Normal digest AKA message flow   282
8.6   Digest AKA message flow at the time of a synchronization failure   282
8.7   Security agreement handshake message flow   297

11.1 RTP packet format   312
11.2 Packet jitter   312

12.1 CS to IP cell example   320

13.1 PDP context types   325

14.1 The TLS handshake   329

15.1 Diameter header   336
15.2 Diameter AVP header   337
15.3 Diameter SIP application architecture   342

17.1 COPS model   350
17.2 COPS header   351
17.3 COPS-specific objects   353

18.1 ESP packet format   358

19.1 SigComp architecture   362

20.1 Client–Server DHCP message format   366
20.2 DHCP options format   366

23.1 Dynamic presence   376
23.2 Reference architecture to support a presence service in the IMS   377
23.3 Successful subscription to presence   380

23.4   Successful publication   381
23.5   Subscription to a resource list   381
23.6   Subscription to watcher information   382

24.1   Immediate messaging flow   384
24.2   IMS session-based messaging flow   386

25.1   Conferencing architecture   388
25.2   Creating a conference using a conference factory URI   390
25.3   Referring a user to a conference using the REFER request   390
25.4   Subscribing to a conference state   391
25.5   Conference creation using CPCP   391

Tables

1.1   IMS features   7

2.1   Cx commands   37
2.2   Sh commands   42
2.3   Summary of reference points   46

3.1   Information storage before, during and after the registration process   51
3.2   The high-level content of a SIP INVITE request during session
          establishment   53
3.3   AKA parameters   63
3.4   Flow identifier information in PDF #1   81
3.5   The maximum data rates per media type   82
3.6   The maximum data rates and QoS class per flow identifier in PDF #1   82
3.7   Requested QoS parameters   85
3.8   The maximum authorized traffic class per media type in the UE   86
3.9   The values of the maximum authorized UMTS QoS parameters per
          flow identifier as calculated by UE #1 (Tobias) from the example   87
3.10 The values of the maximum authorized UMTS QoS parameters per
          PDP context as calculated by UE #1 from the example   87
3.11 Offline charging messages reference table   94
3.12 Online charging messages reference table   98

4.1   Location of CSCFs and GPRS access for the example scenario   121

5.1   Routing-related headers   127
5.2   Filter criteria in Tobias's S-CSCF   136
5.3   Tobias's public user identities   167

6.1   Filter criteria in Tobias's S-CSCF   213

9.1   Session-level description SDP lines   302
9.2   Time-level description SDP lines   302
9.3   Media-level description SDP lines   303
9.4   Most common SDP attribute lines   306

11.1 RTP/AVP-specific profile   315
11.2 Sample payload formats for audio and video   315

12.1 NAPTR RR fields   318
12.2 SRV RR fields   321

15.1 Diameter local action entries   333
15.2 Diameter result codes   338
15.3 Mapping Cx parameters to the Diameter SIP application   341
15.4 Diameter SIP application Command-Codes   342
15.5 Diameter credit control application Command-Codes   343

16.1 MEGACO descriptors   347

17.1 COPS operation codes   352
17.2 COPS-specific object description   353

Part I

Architecture

1

Introduction

1.1 Why the Internet Protocol Multimedia Subsystem (IMS) was developed

The new communication paradigm is about networking Internet Protocol (IP)-based mobile devices. These terminals have large, high-precision displays, they have built-in cameras and a lot of resources for applications. They are always-on-always-connected application devices. This redefines applications. Applications are no longer isolated entities exchanging information only with the user interface. The next generation of more exciting applications are peer-to-peer entities, which facilitate sharing: shared browsing, shared whiteboard, shared game experience, shared two-way radio session (i.e., push to talk). The concept of being connected will be redefined. Dialing a number and talking will soon be seen as a narrow subset of networking. The ability to establish a peer-to-peer connection between the new IP-enabled mobile devices is the key ingredient required (Figure 1.1).

This new paradigm of communications reaches far beyond the capabilities of good old telephony. It can be built on current General Packet Radio Service (GPRS) networks.

In order to communicate, the IP-based applications must have a mechanism to reach the correspondent. The telephone network currently provides this critical task of establishing a connection. By dialing the B number, the network can establish an *ad hoc* connection between any two terminals. This critical IP connectivity capability is offered only in isolated and single-service provider environments in the Internet. We need a global system—the IMS. It enables applications in mobile devices to establish peer-to-peer connections.

True integration of voice and data services increases productivity and overall effectiveness, while the development of innovative applications integrating voice, data and multimedia will create demands for new services, such as presence, multimedia chat, conferencing, push to talk and conferencing. The skill to combine mobility and the IP network will be crucial to service success in the future.

The IMS. Miikka Poikselkä, Georg Mayer, Hisham Khartabil and Aki Niemi
Copyright 2004 by John Wiley & Sons, Ltd. ISBN 0-470-87113-X

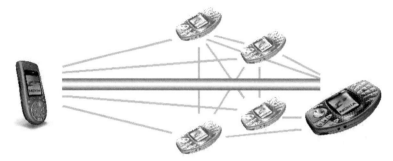

Figure 1.1 The key ingredient to new, enriching user experiences is peer-to-peer IP connections of applications.

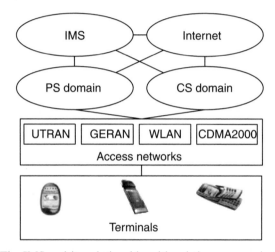

Figure 1.2 The IMS and its relationship with existing communication systems.

Figure 1.2 shows a consolidated network where the IMS introduces multimedia session control in the packet-switched domain and at the same time brings circuit-switched functionality in the packet-switched domain. The IMS is a key technology for network consolidation.

Traditionally, the mobile communication system has been divided in three parts: terminals, the radio access network (RAN) and the core network. This approach needs one change when we are talking about an IMS-based system. The term "radio access network" should be replaced by "access network" because an IMS system can be deployed over non-RANs as well.

It is important to remember that each of these parts can be further split into smaller functional parts along different interfaces. It is important that these interfaces are open and standardized. This book splits IMS into smaller parts and describes how it works as defined in the Third Generation Partnership Project (3GPP).

1.2 Where did it come from?

1.2.1 From GSM to 3GPP Release 6

The European Telecommunications Standards Institute (ETSI) was the standardization organization that defined the Global System for Mobile Communications (GSM) during the late 1980s and 1990s. ETSI also defined the GPRS network architecture. The last GSM-only standard was produced in 1998, and in the same year the 3GPP was founded by standardization bodies from Europe, Japan, South Korea, the USA and China to specify a third-generation mobile system comprising Wideband Code Division Multiple Access (WCDMA) and Time Division/Code Division Multiple Access (TD-CDMA) radio access and an evolved GSM core network (http://www.3gpp.org/About/3gppagre.pdf). Most of the work and cornerstone specifications were inherited from the ETSI Special Mobile Group (SMG). The 3GPP originally decided to prepare specifications on a yearly basis, the first specification release being Release 99 [3GPP R99].

1.2.2 3GPP Release 99 (3GPP R99)

It took barely a year to produce the first release—Release 1999. The functionality of the release was frozen in December 1999 although some base specifications were frozen afterward—in March 2001. Fast completion was possible because the actual work was divided between two organizations: 3GPP and ETSI SMG. 3GPP developed the services, system architecture, WCDMA and TD-CDMA radio accesses, and the common core network. ETSI SMG developed the GSM/Enhanced Data Rates for Global Evolution (EDGE) radio access.

WCDMA radio access was the most significant enhancement to the GSM-based 3G system in Release 1999. In addition to WCDMA, UTRAN (UMTS terrestrial radio access network) introduced the Iu interface as well. Compared with the A and Gb interfaces, there are two significant differences. First, speech transcoding for Iu is performed in the core network. In the GSM it was logically a BTS (base transceiver station) functionality. Secondly, encryption and cell-level mobility management for Iu are done in the radio network controller (RNC). In the GSM they were done in the Serving GPRS Support Node (SGSN) for GPRS services.

The Open Service Architecture (OSA) was introduced for service creation. On the service side the target was to stop standardizing new services and to concentrate on service capabilities, such as toolkits (CAMEL, SIM Application Toolkit and OSA). This principle was followed quite well, even though the virtual home environment (VHE), an umbrella concept that covers all service creation, still lacks a good definition.

1.2.3 3GPP Release 4

After Release 1999, 3GPP started to specify Release 2000, including the so-called All-IP that was later renamed as the IMS. During 2000 it was realized that the development of IMS could not be completed during the year. Therefore, Release 2000 was split into Release 4 and Release 5.

It was decided that Release 4 would be completed without the IMS. The most significant new functionalities in 3GPP Release 4 were: the MSC Server-MGW concept, IP transport of core network protocols, LCS enhancements for UTRAN and multimedia messaging and IP transport for the Gb user plane.

3GPP Release 4 was functionally frozen and officially completed in March 2001. The backward compatibility requirement for changes, essential for the radio interface, was enforced as late as September 2002.

1.2.4 3GPP Release 5 and Release 6

Release 5 finally introduced the IMS as part of 3GPP standards. The IMS is supposed to be a standardized access-independent IP-based architecture that interworks with existing voice and data networks for both fixed (e.g., PSTN, ISDN, Internet) and mobile users (e.g., GSM, CDMA). The IMS architecture makes it possible to establish peer-to-peer IP communications with all types of clients with the requisite quality of services. In addition to session management the IMS architecture also addresses functionalities that are necessary for complete service delivery (e.g., registration, security, charging, bearer control, roaming). All in all, the IMS will form the heart of the IP core network.

The content of Release 5 was heavily discussed, and finally the functional content of 3GPP Release 5 was frozen in March 2002. The consequence of this decision was that many features were postponed to the next release—Release 6. After freezing the content the work continued and 21 months later there are still a number of changes to be made in Release 5 IMS.

Release 6 IMS is going to fix the shortcomings in Release 5 IMS and also contains novel features. Release 6 is to be completed in 2004. Table 1.1 shows the most important features of Release 5 and the items postponed to Release 6.

From Table 1.1 you can see that 3GPP has defined a finite architecture for SIP-based IP multimedia service machinery. It contains a functionality of logical elements, a description of how elements are connected, selected protocols and procedures. It is important to realize that optimization for the mobile communication environment has been designed in the form of user authentication and authorization based on mobile identities, definite rules at the user network interface for compressing SIP messages and security and policy control mechanisms that allow radio loss

Table 1.1 IMS features.

Release 5	Release 6
Architecture: network entities and reference points including charging functions.	*Architecture:* interworking (CS, other IP networks, WLAN) and a few, new entities and reference points.
Signalling: general routing principles, registration, session initiation, session modification, session tear-down, network-initiated session release/ deregistration flows: <ul><li>SIP compression between UE and IMS network;</li><li>data transfer between user information storage (HSS) and session control entities (CSCF);</li><li>data transfer between user information storage (HSS) and application server (AS).</li></ul>	*Signalling:* routing of group identities, multiple registration, emergency sessions
Security: IMS AKA for authenticating users and network, integrity protection of SIP messages between UE and IMS network, network domain security.	*Security:* confidentiality protection of SIP messages, usage of public key infrastructure, subscriber certificates.
Quality of service: policy control between IMS and GPRS access network, preconditions and authorization token.	
Service provisioning: usage of applications servers and IMS service control reference point.	*Services:* presence, messaging, conferencing, group management, local services.
General: ISIM	

and recovery detection. Moreover, important aspects from the operator point of view are addressed while developing the architecture, such as charging framework and policy and service control. This book explains how these aspects have been defined.

The development of IMS is distributed to multiple working groups in 3GPP. 3GPP follows a working method in which the work has three different stages. In stage 1 a service description from a service user and operator point of view are evaluated. In stage 2 problems are broken down into functional elements and the interactions between the elements are identified. In stage 3 all the protocols and procedures are defined in detail. Figure 1.3 shows the most important working groups and responsibility areas that are involved in the development of IMS.

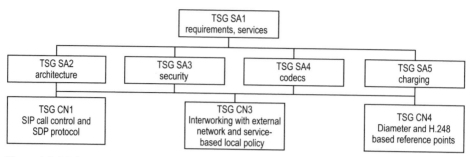

Figure 1.3 Main 3GPP working groups doing IMS work.

1.3 Other relevant standardization bodies

1.3.1 Internet Engineering Task Force

The Internet Engineering Task Force (IETF) is a standardization body that assumes the task of developing and evolving the Internet and its architecture, as well as ensuring the smooth and secure operation of it. The IETF is made up of network designers, academics, engineers and researchers from many companies, volunteering their time and effort to achieve the common goal. IETF participation does not require membership and is open to any individuals who share the same interests.

The IETF is divided into areas that are managed by area directors. Each area has a specific topic to work on. Each area has a number of working groups each tasked to complete a specific charter, concentrating on a specific topic within the area. The areas are: applications, general, Internet, operations and management, routing, security, sub-IP and transport. Each working group produces Internet Drafts that, after many reviews, become standards and are labelled as Requests For Comment (RFC) which get assigned a number.

The area directors are members of the Internet Engineering Steering Group (IESG). The IESG makes sure that the solutions have sufficient security considerations and follow Internet methodologies. The Internet Architecture Board (IAB) provides architectural guidance. The Internet Assigned Numbers Authority (IANA) is where protocol designers can request the assignment of unique parameter names and values.

3GPP and IETF work closely together. 3GPP adopts protocols developed at the IETF as needed (e.g., SIP, SDP, RTP, DIAMETER). 3GPP generates requirements for a specific problem and then contacts the IETF for a possible solution to its requirements. The IETF evaluates the 3GPP requirements and provides 3GPP with a protocol that satisfies those requirements. If no suitable protocol is found, then the IETF assumes responsibility and begins to design a solution to suit the requirements, documenting it in the form of an Internet Draft. The solution gets reviewed and modified time and time again until a satisfactory one has been agreed. 3GPP then adopts that solution. In some situations a partial solution is only available or the 3GPP community feels that the solution provided is not satisfactory. In this case an extension to the available protocol is needed.

1.3.2 Open Mobile Alliance

In June 2002 the mobile industry set up a new, global organization called the Open Mobile Alliance (OMA). OMA has taken its place as the leading standardization organization for doing mobile service specification work. OMA's role is to specify different service enablers, such as digital rights management and push to talk over the cellular service (PoC).

OMA has recognized that it is not beneficial for each service enabler to have its own mechanism for security, quality of service, charging, session management, etc. On the contrary, service enablers should be able to use an infrastructure that provides these basic capabilities. This is where the IMS steps into the OMA landscape. Different service enablers developed in OMA can interface to the IMS, can utilize IMS capabilities and the resources of their underlying network infrastructure via the IMS. Usage of the IMS infrastructure would greatly shorten the specification time of service enablers and would bring modularity to the system, which is definitely a common interest in the industry. Therefore, co-operation between the OMA and 3GPP will increase in the future. It is very likely that the OMA will gradually take overall responsibility for the invention and design of various applications and services on top of the IMS architecture, while 3GPP will continue to develop the core IMS.

1.3.3 Third Generation Partnership Project 2

The Third Generation Partnership Project 2 (3GPP2) is a collaborative project for developing a third-generation mobile system for the ANSI (American National Standards Institute) community. 3GPP2 comprises organizational partners (ARIB, CCSA, TIA, TTA and TTC) and market representation partners (the CDMA Development Group and the IPv6 Forum).

3GPP2's role in IMS standardization lies in specifying IMS as part of the Multimedia Domain solution that further contains the Packet Data Subsystem. The Multimedia Domain and the CDMA2000 Access Network together form the third-generation All IP Core Network in 3GPP2. 3GPP2 has adopted core Release 5 IMS specifications as a baseline from its sister project, 3GPP. However, there are differences between 3GPP2 IMS and 3GPP IMS Release 5 solutions due to different underlying packet and radio technology. Additionally, in some areas 3GPP2 has defined further additions or limitations. Here are some of the main issues that relate to the first IMS releases:

- IP Policy Control between IMS and the Packet Data Subsystem is not supported in 3GPP2.

- The IMS entry point P-CSCF may be located in a different network than the Packet Data Subsystem. In 3GPP the P-CSCF and the Gateway GPRS Support Node are always located in the same network.

- IP version 4 is also supported in 3GPP2 IMS, whereas 3GPP IMS exclusively supports IP version 6.

- No default codec is specified in 3GPP2.

- Differences in charging solutions.

- No support for a Universal Integrated Circuit Card that could contain an IP Multimedia Services Identity Module for storing, say, IMS access parameters.

- Customized Applications for Mobile network Enhanced Logic (CAMEL)-related functions are not supported.

- The architecture does not contain the Subscription Locator Functional entity nor a reference point for discovering a database that holds the user's subscription.

2

IP Multimedia Subsystem Architecture

This chapter introduces the reader to the Internet Protocol (IP) Multimedia Subsystem (IMS). Section 2.1 explains basic architectural concepts: for instance, we explain why bearers are separated and why the home control model was selected. Section 2.2 gives a wide overview of IMS architecture, including an introduction to different network entities and main functionalities. Section 2.3 goes deeper and shows how the entities are connected and what protocols are used between them; it also describes their relationships to other domains: IP networks, UMTS and CS CN.

2.1 Architectural requirements

There is a set of basic requirements which guides the way in which the IMS architecture has been created and how it should evolve in the future. This section covers the most significant requirements. Third Generation Partnership Project (3GPP) stage 1 IMS requirements are documented in [3GPP TS 22.228].

2.1.1 IP connectivity

A fundamental requirement is that a client has to have IP connectivity to access IMS services. In addition, it is required that IPv6 is used [3GPP TS 23.221].

IP connectivity can be obtained either from the home network or the visited network. The leftmost part of Figure 2.1 presents an option in which user equipment (UE) has obtained an IP address from a visited network. In the Universal Mobile Telecommunications System (UMTS) network this means that the radio access

The IMS. Miikka Poikselkä, Georg Mayer, Hisham Khartabil and Aki Niemi
Copyright 2004 by John Wiley & Sons, Ltd. ISBN 0-470-87113-X

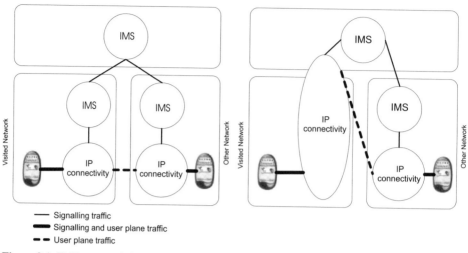

Figure 2.1 IMS connectivity options when a user is roaming.

network (RAN), Serving GPRS Support Node (SGSN) and Gateway GPRS Support Node (GGSN) are located in the visited network when a user is roaming in the visited network. The rightmost part of Figure 2.1 presents an option in which a UE has obtained an IP address from the home network. In the UMTS network this means that the RAN and SGSN are located in the visited network when a user is roaming in the visited network. Obviously, when a user is located in the home network all necessary elements are in the home network and IP connectivity is obtained in that network.

It is important to note that a user can roam and obtain IP connectivity from the home network as shown in the figure. This would allow users to use new, fancy IMS services even when they are roaming in an area that does not have an IMS network but provides IP connectivity. In theory, it is possible to deploy an IMS network in a single area/country and use, say, General Packet Radio Service (GPRS) roaming to connect customers to the home network. In practice this would not happen because routing efficiency would not be high enough. Consider routing real time transport protocol (RTP) voice packets from the USA to Europe and then back to the USA. However, this deployment model is important when operators are ramping up IMS networks or, in an initial phase, when they are offering non or near-real time multimedia services.

2.1.2 Access independence

The IMS is designed to be access-independent so that IMS services can be provided over any IP connectivity networks (e.g., GPRS, WLAN, broadband access x-Digital

Subscriber Line). Unfortunately, Release 5 IMS specifications contain some GPRS-specific features. In Release 6 (e.g., GPRS) access-specific issues will be separated from the core IMS description. 3GPP uses the term "IP connectivity access network" to refer to the collection of network entities and interfaces that provides the underlying IP transport connectivity between the UE and the IMS entities. In this book we use GPRS as an example.

2.1.3 Ensuring quality of service for IP multimedia services

On the public Internet, delays tend to be high and variable, packets arrive out of order and some packets are lost or discarded. This will no longer be the case with the IMS. The underlying access and transport networks together with the IMS provide end-to-end quality of service (QoS).

Via the IMS, UE negotiates its capabilities and expresses its QoS requirements during a Session Initiation Protocol (SIP) session set-up or session modification procedure. The UE is able to negotiate such parameters as:

- Media type, direction of traffic.

- Media type bit rate, packet size, packet transport frequency.

- Usage of RTP payload for media types.

- Bandwidth adaptation.

After negotiating the parameters at the application level, UEs reserve suitable resources from the access network. When end-to-end QoS is created, the UEs encode and packetize individual media types with an appropriate protocol (e.g., RTP) and send these media packets to the access and transport network by using a transport layer protocol (e.g., TCP or UDP) over IP. It is assumed that operators negotiate service-level agreements for guaranteeing the required QoS in the interconnection backbone. In the case of UTMS, operators could utilize the GPRS Roaming Exchange backbone.

2.1.4 IP policy control for ensuring correct usage of media resources

IP policy control means the capability to authorize and control the usage of bearer traffic intended for IMS media, based on the signalling parameters at the IMS session. This requires interaction between the IP connectivity access network and the IMS. The means of setting up interaction can be divided into three different categories [3GPP TS 22.228, 23.207, 23.228]:

- The policy control element is able to verify that values negotiated in SIP signalling are used when activating bearers for media traffic. This allows an operator to verify that its bearer resources are not misused (e.g., the source and destination IP address and bandwidth in the bearer level are exactly the same as used in SIP session establishment).

- The policy control element is able to enforce when media traffic between end points of a SIP session start or stop. This makes it possible to prevent the use of the bearer until the session establishment is completed and allows traffic to start/ stop in synchronization with the start/stop of charging for a session in IMS.

- The policy control element is able to receive notifications when the IP connectivity access network service has either modified, suspended or released the bearer(s) of a user associated with a session. This allows IMS to release ongoing session because, for instance, the user is no longer in the coverage area.

Policy control is further described in Section 3.9.

2.1.5 Secure communication

Security is a fundamental requirement in every telecommunication system and the IMS is not an exception. The IMS provides at least a similar level of security as the corresponding GPRS and circuit-switched networks: for example, the IMS ensures that users are authenticated before they can start using services, and users are able to request privacy when engaged in a session. Section 3.6 will discuss security features in more detail.

2.1.6 Charging arrangements

From an operator or service provider perspective the ability to charge users is a must in any network. The IMS architecture allows different charging models to be used. This includes, say, the capability to charge just the calling party or to charge both the calling party and the called party based on used resources in the transport level. In the latter case the calling party could be charged entirely on IMS-level session: that is, it is possible to use different charging schemes at the transport and IMS level. However, an operator might be interested to correlate charging information generated at transport and IMS (service and content) charging levels. This capability is provided if an operator utilizes a policy control reference point. The charging correlation mechanism is further described in Section 3.10.2 and policy control is explained in Section 3.9.

As IMS sessions may include multiple media components (e.g., audio and video), it is required that the IMS provides a means for charging per media component. This would allow a possibility to charge the called party if she adds a new media component in a session. It is also required that different IMS networks are able to exchange information on the charging to be applied to a current session [3GPP TS 22.101, TR 23.815].

The IMS architecture supports both online and offline charging capabilities. Online charging is a charging process in which the charging information can affect in real time the service rendered and therefore directly interacts with session/service control. In practice, an operator could check the user's account before allowing the user to engage a session and to stop a session when all credits are consumed. Prepaid services are applications that need online charging capabilities. Offline charging is a charging process in which the charging information does not affect in real time the service rendered. This is the traditional model in which the charging information is collected over a particular period and, at the end of the period, the operator posts a bill to the customer.

2.1.7 Support of roaming

From a user point of view it is important to get access to her services regardless of her geographical location. The roaming feature makes it possible to use services even though the user is not geographically located in the service area of the home network. Section 2.1.1 has already described two instances of roaming: namely, GPRS roaming and IMS roaming. In addition to these two there exists an IMS circuit-switched (CS) roaming case. GPRS roaming means the capability to access the IMS when the visited network provides the RAN and SGSN and the home network provides the GGSN and IMS. The IMS roaming model refers to a network configuration in which the visited network provides IP connectivity (e.g., RAN, SGSN, GGSN) and the IMS entry point (i.e., P-CSCF) and the home network provides the rest of the IMS functionalities. The main benefit of this roaming model compared with the GPRS roaming model is optimum usage of user-plane resources. Roaming between the IMS and the CS CN domain refers to inter-domain roaming between IMS and CS. When a user is not registered or reachable in one domain a session can be routed to the other domain. It is important to note that both the CS CN domain and the IMS domain have their own services and cannot be used from another domain. Some services are similar and available in both domains (e.g., Voice over IP in IMS and speech telephony in CSCN). Figure 2.2 shows different IMS/CS roaming cases.

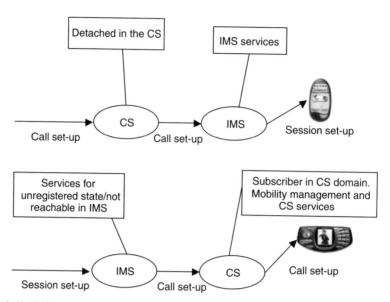

Figure 2.2 IMS/CS roaming alternatives.

2.1.8 Interworking with other networks

It is evident that the IMS is not deployed over the world at the same time. Moreover, people may not be able to switch terminals or subscriptions very rapidly. This will raise the issue of being able to reach people regardless of what kind of terminals they have or where they live. To be a new, successful communication network technology and architecture the IMS has to be able to connect to as many users as possible. Therefore, the IMS supports communication with PSTN, ISDN, mobile and Internet users. Additionally, it will be possible to support sessions with Internet applications that have been developed outside the 3GPP community [3GPP TS 22.228].

2.1.9 Service control model

In 2G mobile networks the visited service control is in use. This means that, when a user is roaming, an entity in the visited network provides services and controls the traffic for the user. This entity in 2G is called a visited mobile service switching centre. In the early days of Release 5 both visited and home service control models were supported. Supporting two models would have required that every problem have more than one solution; moreover, it would reduce the number of optimal architecture solutions, as simple solutions may not fit both models. Supporting both models would have meant additional extensions for Internet Engineering

Task Force (IETF) protocols and increased the work involved in registration and session flows. The visited service control was dropped because it was a complex solution and did not provide any noticeable added value compared with the home service control. On the contrary, the visited service control imposes some limitations. It requires a multiple relationship and roaming models between operators. Service development is slower as both the visited and home network would need to support similar services, otherwise roaming users would experience service degradations. In addition, the number of interoperator reference points increase, which requires complicated solutions (e.g., in terms of security and charging). Therefore, the home service control was selected; this means that the entity that has access to the subscriber database and interacts directly with service platforms is always located at the user's home network.

2.1.10 Service development

The importance of having a scalable service platform and the possibility to launch new services rapidly has meant that the old way of standardizing complete sets of teleservices, applications and supplementary services is no longer acceptable. Therefore, 3GPP is standardizing service capabilities and not the services themselves [3GPP TS 22.101]. The IMS architecture should actually include a service framework that provides the necessary capabilities to support speech, video, multimedia, messaging, file sharing, data transfer, gaming amd basic supplementary services within the IMS. Section 3.12 further describes how the IMS service control works and Chapters 23–25 explain in more detail how presence, messaging and conferencing services are offered.

2.1.11 Layered design

3GPP has decided to use a layered approach to architectural design. This means that transport and bearer services are separated from the IMS signalling network and session management services. Further services are run on top of the IMS signalling network. Figure 2.3 shows the design.

In some cases it may be impossible to distinguish between functionality at the upper and lower layers. The layered approach aims at a minimum dependency between layers. A benefit is that it facilitates the addition of new access networks to the system later on. Wireless Local Area Network (WLAN) access to the IMS, in 3GPP Release 6, will test how well the layering has been done. Other accesses may follow (e.g., fixed broadband).

The layered approach increases the importance of the application layer. When applications are isolated and common functionalities can be provided by the underlying IMS network the same applications can run on UE using diverse access types.

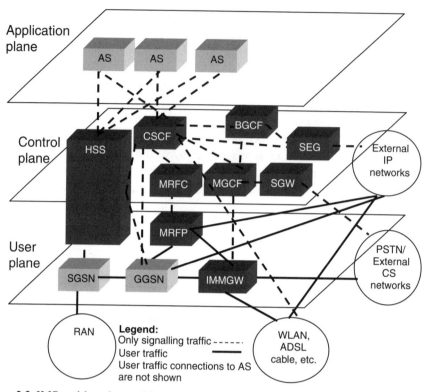

Figure 2.3 IMS and layering architecture.

2.2 Description of IMS-related entities and functionalities

This section discusses IMS entities and key functionalities. These entities can be roughly classified in six main categories: session management and routing family (CSCFs), databases (HSS, SLF), interworking elements (BGCF, MGCF, IM-MGW, SGW), services (application server, MRFC, MRFP), support entities (THIG, SEG, PDF) and charging. It is important to understand that IMS standards are set up so that the internal functionality of network entities is not specified in detail. For instance, the Home Subscriber Server (HSS) contains three internal functions: IMS functionality, necessary functions for the CS domain and necessary functions for the PS domain. 3GPP standards do not describe how IMS functionality interacts with functions designed for Packet Switched (PS); instead, they describe reference points between entities and functionalities supported at the reference points (e.g., how does CSCF obtain user data from HSS). Different reference points will be described in Section 2.3. Additionally, General Packet Radio Service (GPRS) functions are described at the end of this section.

2.2.1 Proxy-CSCF

The Proxy-Call Session Control Function (P-CSCF) is the first contact point for users within the IMS. All SIP signalling traffic from or to the UE go via the P-CSCF. As the name of the entity indicates the P-CSCF behaves like a proxy as defined in [RFC3261]. It means that the P-CSCF validates the request, forwards it to selected destinations and processes and forwards the response. In addition, the P-CSCF may behave as a user agent (UA) as defined in [RFC3261]. The UA role is needed for releasing sessions in abnormal conditions (e.g., when a bearer loss is detected according to service-based local policy—see Section 3.9) and for generating independent SIP transactions, as explained in Section 5.12.6, which deals with registration. There can be one or many P-CSCFs within an operator's network. The functions performed by the P-CSCF are [3GPP TS 23.228, TS 24.229]:

- To forward SIP REGISTER requests to the Interrogating-CSCF (I-CSCF) based on a home domain name provided by the UE in the request. Section 5.5 gives a detailed description of what actions the P-CSCF needs to take before forwarding the SIP REGISTER request (e.g., to resolve an address of the CSCF or to let it be known that a REGISTER request was not received with a security association).

- To forward SIP requests and responses received by the UE to the Serving-CSCF (S-CSCF). Chapter 6 gives a detailed description of what actions the P-CSCF needs to take before forwarding a non-REGISTER request or response (e.g., to check that the user identity used is valid).

- To forward SIP requests and responses to the UE. Chapter 6 gives a detailed description of what actions the P-CSCF needs to take before forwarding SIP messages to the UE (e.g., to compress the message).

- To detect emergency session establishment requests. In Release 5 the P-CSCF returns a SIP error message, 380, indicating that the UE should try the CSCN instead. The work is ongoing in Release 6 and the P-CSCF behaviour is going to change in such a way that the P-CSCF will select an S-CSCF to handle an emergency session. The selection is needed because in IMS roaming cases the assigned S-CSCF is in the home network and the home S-CSCF is unable to route the request to a correct emergency centre.

- To send accounting-related information to the Charging Collection Function (CCF).

- To provide integrity protection of SIP signalling and maintain a security association between the UE and the P-CSCF. Integrity protection is provided by means of Internet Protocol Security (IPsec) Encapsulating Security Payload (ESP).

Release 6 is able to provide confidentiality protection as well. Section 3.6 explains how IMS security is designed and the security protocols are discussed in Chapter 18.

- To decompress and compress SIP messages from the UE. The P-CSCF supports compression based on three RFCs: [RFC3320], [RFC3485] and [RFC3486]. Sections 3.16 and 6.4 and Chapter 19 describe the usage of SIP compression in more detail [3GPP TS 24.229].

- To subscribe a registration event package at the user's registrar (S-CSCF). This is needed for downloading implicitly registered public user identities and for getting notifications on network-initiated de-registration events. Section 5.12.6 describes a registration event package and Section 3.14 shows how implicit registration works and Section 5.14.3 tells us more about network-initiated de-registrations.

- To execute media policing. The P-CSCF is able to check the content of the Session Description Protocol (SDP) payload and to check whether it contains media types or codecs, which are not allowed for a user. When the proposed SDP does not fit the operator's policy, the P-CSCF rejects the request and sends a SIP error message, 488, to the UE. An operator may want to use this feature for roaming users due to bandwidth restrictions.

- To maintain session timers. Release 5 does not provide a means for a statefull proxy to know the status of sessions. Release 6 corrects this deficiency by introducing session timers. It allows the P-CSCF to detect and free resources used up by hanging sessions.

- To interact with the Policy Decision Function (PDF). The PDF is responsible for implementing the Service Based Local Policy (SBLP). In Release 5 the PDF is a logical entity of the P-CSCF, and in Release 6 the PDF is a stand-alone function.

2.2.2 Policy Decision Function

The Policy Decision Function (PDF) is responsible for making policy decisions based on session and media-related information obtained from the P-CSCF. It acts as a policy decision point for SBLP control. The following policy decision point functionalities for SBLP are identified:

- To store session and media-related information (IP addresses, port numbers, bandwidths, etc.).

- To generate an authorization token that identifies the PDF and the session.

- To provide an authorization decision according to the stored session and media-related information on receiving a bearer authorization request from the GGSN.

- To update the authorization decision at session modifications which changes session and media-related information.

- The capability to revoke the authorization decision at any time.

- The capability to enable the usage of an authorized bearer (e.g., Packet Data Protocol, or PDP, context).

- The capability to prevent the usage of an authorized bearer (e.g., PDP context) while maintaining the authorization.

- To inform the P-CSCF when the bearer (e.g., PDP context) is lost or modified. A modification indication is only given when the bearer is upgraded or downgraded from or to 0 kbit/s.

- To pass an IMS-charging identifier to the GGSN and to pass a GPRS-charging identifier to the P-CSCF.

2.2.3 Interrogating-CSCF

Interrogating-CSCF (I-CSCF) is a contact point within an operator's network for all connections destined to a subscriber of that network operator. There may be multiple I-CSCFs within an operator's network. The functions performed by the I-CSCF are:

- To contact the HSS to obtain the name of the S-CSCF that is serving a user.

- To assign an S-CSCF based on received capabilities from the HSS. An S-CSCF is assigned if there is no S-CSCF allocated. This procedure is described in more detail in Section 3.8.

- To forward SIP requests or responses to the S-CSCF.

- To send accounting-related information to the CCF.

- To provide a hiding functionality. The I-CSCF may contain a functionality called the Topology Hiding Inter-network Gateway (THIG). THIG could be used to hide the configuration, capacity and topology of the network from outside an operator's network.

2.2.4 Serving-CSCF

The Serving-CSCF (S-CSCF) is the brain of the IMS; it is located in the home network. It performs session control and registration services for UEs. While UE is engaged in a session the S-CSCF maintains a session state and interacts with service platforms and charging functions as needed by the network operator for support of the services. There may be multiple S-CSCFs, and S-CSCFs may have different functionalities within an operator's network. More specifically, the functions performed by the S-CSCF are:

- To handle registration requests by acting as a registrar as defined in [RFC3261]. The S-CSCF knows the UE's IP address and which P-CSCF the UE is using as an IMS entry point.

- To authenticate users by means of the IMS Authentication and Key Agreement (AKA) schema. The IMS AKA achieves mutual authentication between the UE and the home network.

- To download user information and service-related data from the HSS during registration or when handling a request to an unregistered user.

- To route mobile-terminating traffic to the P-CSCF and to route mobile-originated traffic to the I-CSCF, the Breakout Gateway Control Function (BGCF) or the application server (AS).

- To perform session control. The S-CSCF can act as a proxy server and UA as defined in [RFC3261].

- To interact with service platforms. Interaction means the capability to decide when a request or response needs to be routed to a specific AS for further processing.

- To translate an E.164 number to a SIP universal resource identifier (URI) using a domain name system (DNS) translation mechanism with the format as specified in [Draft-ietf-enum-rfc2916bis]. This translation is needed because routing of SIP signalling in IMS uses only SIP URIs.

- To supervise registration timers and to be able to de-register users when needed.

- To select an emergency centre when the operator supports IMS emergency sessions. This is a Release 6 feature.

- To execute media policing. The S-CSCF is able to check the content of the SDP payload and check whether it contains media types or codecs, which are not allowed for a user. When the proposed SDP does not fit the operator's policy or user's subscription, the S-CSCF rejects the request and sends a SIP error message, 488. Section 3.11 shows how media policy information can be included as part of the user profile.

- To maintain session timers. Release 5 does not provide the means for a statefull proxy to know the status of sessions. Release 6 corrects this deficiency by introducing session timers. It allows the S-CSCF to detect and free resources used up by hanging sessions.

- To send accounting-related information to the CCF for offline charging purposes and to the Online Charging System (OCS) for online charging purposes.

2.2.5 Home Subscriber Server

The Home Subscriber Server (HSS) is the main data storage for all subscriber and service-related data of the IMS. The main data stored in the HSS include user identities, registration information, access parameters and service-triggering information [3GPP TS 23.002].

User identities consist of two types: private and public user identities. The private user identity is a user identity that is assigned by the home network operator and is used for such purposes as registration and authorization, while the public user identity is the identity that other users can use for requesting communication with the end user. IMS access parameters are used to set up sessions and include parameters like user authentication, roaming authorization and allocated S-CSCF names. Service-triggering information enables SIP service execution. The HSS also provides user-specific requirements for S-CSCF capabilities. This information is used by the I-CSCF to select the most suitable S-CSCF for a user.

In addition to functions related to IMS functionality, the HSS contains the subset of Home Location Register and Authentication Center (HLR/AUC) functionality required by the PS domain and the CS domain. The structure of the HSS is shown in Figure 2.4. Communication between different HSS functions is not standardized.

HLR functionality is required to provide support to PS domain entities, such as SGSN and GGSN. This enables subscriber access to PS domain services. In similar fashion the HLR provides support for CS domain entities like MSC/MSC servers. This enables subscriber access to CS domain services and supports roaming to GSM/UMTS CS domain networks.

The AUC stores a secret key for each mobile subscriber, which is used to generate dynamic security data for each mobile subscriber. Data are used for mutual authentication of the International Mobile Subscriber Identity (IMSI) and the network. Security data are also used to provide integrity protection and ciphering of the communication over the radio path between the UE and the network.

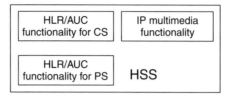

Figure 2.4 Structure of HSS.

There may be more than one HSS in a home network depending on the number of mobile subscribers, the capacity of the equipment and the organization of the network. There are multiple reference points between the HSS and other network entities.

2.2.6 Subscription Locator Function

The Subscription Locator Function (SLF) is used as a resolution mechanism that enables the I-CSCF, the S-CSCF and the AS to find the address of the HSS that holds the subscriber data for a given user identity when multiple and separately addressable HSSs have been deployed by the network operator.

2.2.7 Multimedia Resource Function Controller

The Multimedia Resource Function Controller (MRFC) is needed to support bearer-related services, such as conferencing, announcements to a user or bearer transcoding. The MRFC interprets SIP signalling received via S-CSCF and uses Media Gateway Control Protocol (MEGACO) instructions to control the Multimedia Resource Function Processor (MRFP). The MRFC is able to send accounting information to the CCF and OCS. Chapter 25 shows how the MRFC is used in conferencing services.

2.2.8 Multimedia Resource Function Processor

The Multimedia Resource Function Processor (MRFP) provides user-plane resources that are requested and instructed by the MRFC. The MRFP performs the following functions:

- Mixing of incoming media streams (e.g., for multiple parties).

- Media stream source (for multimedia announcements).

- Media stream processing (e.g., audio transcoding, media analysis) [3GPP TS 23.228, TS 23.002].

2.2.9 Application server

Keeping in mind the layered design, application servers (ASs) are not pure IMS entities; rather, they are functions on top of IMS. However, ASs are described here as part of IMS functions because ASs are entities that provide value-added multimedia services in the IMS.

An AS resides in the user's home network or in a third-party location. The third party here means a network or a stand-alone AS. The main functions of the AS are:

- The possibility to process and impact an incoming SIP session received from the IMS.

- The capability to originate SIP requests.

- The capability to send accounting information to the CCF and the OCS.

Offered services are not limited purely to SIP-based services since an operator is able to offer access to services based on the Customized Applications for Mobile network Enhanced Logic (CAMEL) Service Environment (CSE) and the Open Service Architecture (OSA) for its IMS subscribers [3GPP TS 23.228]. Therefore, "AS" is the term used generically to capture the behaviour of the SIP AS, OSA Service Capability Server (SCS) and CAMEL IP Multimedia Service Switching Function (IM-SSF).

Using the OSA an operator may utilize such service capability features as call control, user interaction, user status, data session control, terminal capabilities, account management, charging and policy management for developing services [3GPP TS 29.198]. An additional benefit of the OSA framework is that it can be used as a standardized mechanism for providing third-party ASs in a secure manner to the IMS, as the OSA itself contains initial access, authentication, authorization, registration and discovery features (the S-CSCF does not provide authentication and security functionality for secure direct third-party access to the IMS). As the support of OSA services is down to operator choice, it is not architecturally sound to support OSA protocols and features in multiple entities. Therefore, OSA SCS is used to terminate SIP signalling from the S-CSCF. The OSA SCS uses an OSA application program interface (API) to communicate with an actual OSA application server.

The IM-SSF function was introduced in the IMS architecture to support legacy services that are developed in the CAMEL Service Environment (CSE). It hosts CAMEL network features (trigger detection points, CAMEL Service Switching

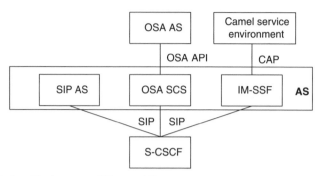

Figure 2.5 Relationship between different AS types.

Finite State Machine, etc.) and interworks with the CAMEL Application Part (CAP) interface.

A SIP AS is a SIP-based server that hosts a wide range of value-added multimedia services. A SIP AS could be used to provide presence, messaging and conferencing services. The different functions of SIP servers are described in more detail in Sections 8.3 and 3.12.4, as part of service provisioning.

Figure 2.5 shows how different functions are connected. From the perspective of the S-CSCF SIP AS, the OSA service capability server and the IM-SSF exhibit the same reference point behaviour.

An AS may be dedicated to a single service and a user may have more than one service, therefore there may be one or more ASs per subscriber. Additionally, there may be one or more ASs involved in a single session. For example, an operator could have one AS to control terminating traffic to a user based on user preferences (e.g., redirecting all incoming multimedia sessions to an answer machine between 5 p.m. and 7 a.m.) and another AS to adapt the content of instant messages according to the capabilities of the UE (screen size, number of colours, etc.).

2.2.10 Breakout Gateway Control Function

The Breakout Gateway Control Function (BGCF) is responsible for choosing where a breakout to the CS domain occurs. The outcome of a selection process can be either a breakout in the same network in which the BGCF is located or another network. If the breakout happens in the same network, then the BGCF selects a Media Gateway Control Function (MGCF) to handle a session further. If the breakout takes place in another network, then the BGCF forwards a session to another BGCF in a selected network [3GPP TS 23.228]. The actual selection rules are not specified. In addition, the BGCF is able to report account information to the CCF and collect statistical information. IMS and CS interworking is described in Section 3.13.

2.2.11 Media Gateway Control Function

The Media Gateway Control Function (MGCF) is a gateway that enables communication between IMS and CS users. All incoming call control signalling from CS users is destined to the MGCF that performs protocol conversion between the ISDN User Part (ISUP), or the Bearer Independent Call Control (BICC), and SIP protocols and forwards the session to IMS. In similar fashion all IMS-originated sessions toward CS users traverses through MGCF. MGCF also controls media channels in the associated user-plane entity, the IMS Media Gateway CIMS-MGW. In addition, MGCF is able to report account information to the CCF. IMS and CS interworking is described in Section 3.13.

2.2.12 IP Multimedia Subsystem-Media Gateway Function

The IMS Multimedia Gateway Function (IMS-MGW) provides the user-plane link between CS networks (PSTN, GSM) and the IMS. It terminates the bearer channels from the CS network and media streams from the backbone network (e.g., RTP streams in an IP network or AAL2/ATM connections in an ATM backbone), executes the conversion between these terminations and performs transcoding and signal processing for the user plane when needed. In addition, the IMS-MGW is able to provide tones and announcements to CS users. The IMS-MGW is controlled by the MGCF.

2.2.13 Signalling gateway

A signalling gateway (SGW) is used to interconnect different signalling networks, such as SCTP/IP-based signalling networks and SS7 signalling networks. The SGW performs signalling conversion (both ways) at the transport level between the Signalling System No. 7 (SS7)-based transport of signalling and the IP-based transport of signalling (i.e., between Sigtran SCTP/IP and SS7 MTP). The SGW does not interpret application layer (e.g., BICC, ISUP) messages. In Figure 2.6 ISUP is shown, but BICC could be shown as well.

2.2.14 Security gateway

To protect control-plane traffic between security domains, traffic will pass through a security gateway (SEG) before entering or leaving the security domain. The security domain refers to a network that is managed by a single administrative authority. Typically, this coincides with operator borders. The SEG is placed at the border of

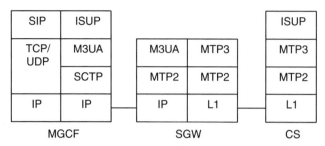

Figure 2.6 Signalling conversion in the SGW.

the security domain and it enforces the security policy of a security domain toward other SEGs in the destination security domain. The network operator may have more than one SEG in its network in order to avoid a single point of failure or for performance reasons. The SEG may be defined for interaction toward all reachable security domain destinations or it may be defined for only a subset of the reachable destinations [3GPP TS 33.203]. The concept behind a security domain is described more thoroughly in Section 3.6.3.

2.2.15 Charging entities

Different charging entities and corresponding reference points will be described separately in Section 3.10.

2.2.16 GPRS entities

2.2.16.1 Serving GPRS Support Node

The Serving GPRS Support Node (SGSN) links the RAN to the packet core network. It is responsible for performing both control and traffic-handling functions for the PS domain. The control part contains two main functions: mobility management and session management. Mobility management deals with the location and state of the UE and authenticates both the subscriber and the UE. The control part of session management deals with connection admission control and any changes in the existing data connections. It also supervises 3G network services and resources. Traffic handling is the part of session management that is executed. The SGSN acts as a gateway for user data tunnelling: in other words, it relays user traffic between the UE and the GGSN. As a part of this function, the SGSN also ensures that connections receive the appropriate QoS. In addition, the SGSN generates charging information.

2.2.16.2 Gateway GPRS Support Node

The Gateway GPRS Support Node (GGSN) provides interworking with external packet data networks. The prime function of the GGSN is to connect the UE to external data networks, where IP-based applications and services reside. The external data network could be the IMS or the Internet, for instance. In other words, the GGSN routes IP packets containing SIP signalling from the UE to the P-CSCF and vice versa. Additionally, the GGSN takes care of routing IMS media IP packets toward the destination network (e.g., to GGSN in the terminating network). The interworking service provided is realized as access points that relate to the different networks the subscriber wants to connect. In most cases the IMS has its own access point. When the UE activates a bearer (PDP context) toward an access point (IMS), the GGSN allocates a dynamic IP address to the UE. This allocated IP address is used in IMS registration and when the UE initiates a session as a contact address of the UE. Additionally, the GGSN polices and supervises the PDP context usage for IMS media traffic and generates charging information.

2.3 IMS reference points

This section explains how the previously described network entities are connected to each other and what protocol is used; moreover, the IMS architecture is depicted (Figure 2.7). You will also find an overview of SIP-based reference points (i.e., where SIP is used and what are the main procedures). However, you will realize that the level of description of SIP-based reference points is not so deep as with Diameter-based reference points. The reason for this division is that several chapters in this book are dedicated for SIP and SDP procedures where such descriptions are given in detail.

For the sake of clarity, it is impossible to include everything in one figure; so, please note the following:

- Figure 2.7 does not show charging-related functions or reference points (see Section 3.10 for more details).

- The figure does not show different types of ASs (see Section 2.2.9 for more details).

- The figure does not show the user-plane connections between different IMS networks and the AS.

- The figure does not show the SEG at the Mm, Mk, Mw reference points.

- The dotted line between the entities indicates a direct link.

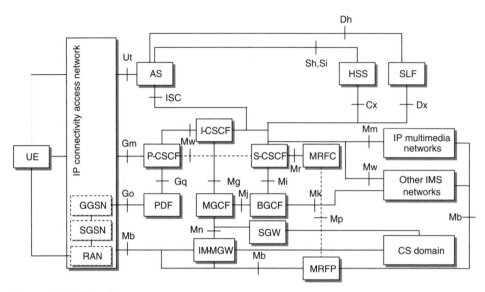

Figure 2.7 IMS architecture.

- ISC, Cx, Dx, Mm, Mw terminate at both the Serving-CSCF (S-CSCF) and the I-CSCF.

2.3.1 Gm reference point

The Gm reference point connects the UE to the IMS. It is used to transport all SIP signalling messages between the UE and the IMS. The IMS counterpart is P-CSCF. Procedures in the Gm reference point can be divided into three main categories: registration, session control and transactions:

- In the registration procedure the UE uses the Gm reference point to send a registration request with an indication of supported security mechanisms to the P-CSCF. During the registration process the UE exchanges the necessary parameters for authenticating both itself and the network, gets implicit registered user identities, negotiates the necessary parameters for a security association with the P-CSCF and possibly starts SIP compression. In addition, the Gm reference point is used to inform the UE if network-initiated de-registration or network-initiated re-authentication occurs.

- Session control procedures contain mechanisms for both mobile-originated sessions and mobile-terminated sessions. In mobile-originated sessions the Gm reference point is used to forward requests from the UE to the P-CSCF. In mobile-terminated sessions the Gm reference point is used to forward request from the P-CSCF to the UE.

- Transaction procedures are used to send stand-alone requests (e.g., MESSAGE) and to receive all responses (e.g., 200 OK) to that request via the Gm reference point. The difference between transaction procedures and session control procedures is that a dialog is not created.

2.3.2 Mw reference point

The Gm reference point links the UE to the IMS (namely, to P-CSCF). Next, a SIP-based reference point between different CSCFs is needed. This reference point is called Mw. The procedures in the Mw reference point can be divided into three main categories: registration, session control and transaction:

- In the registration procedure the P-CSCF uses the Mw reference point to forward a registration request from the UE to the I-CSCF. The I-CSCF then uses the Mw reference point to pass the request to the S-CSCF. Finally, the response from the S-CSCF traverves back via the Mw reference point. In addition, the S-CSCF uses the Mw reference point in network-initiated de-registration procedures to inform the UE about network-initiated de-registration and network-initiated re-authentication to inform the P-CSCF that it should release resources regarding a particular user.

- Session control procedures contain mechanisms for both mobile-originated sessions and mobile-terminated sessions. In mobile-originated sessions the Mw reference point is used to forward requests both from the P-CSCF to the S-CSCF and from the S-CSCF to the I-CSCF. In mobile-terminated sessions the Mw reference point is used to forward requests both from the I-CSCF to the S-CSCF and from the S-CSCF to the P-CSCF. This reference point is also used for network-initiated session releases: for example, the P-CSCF could initiate a session release toward the S-CSCF if it receives an indication from the PDF that media bearer(s) are lost. In addition, charging-related information is conveyed via the Mw reference point.

- Transaction procedures are used to pass a stand-alone request (e.g., MESSAGE) and to receive all responses (e.g., 200 OK) to that request via the Mw reference point. As already stated, the difference between transaction procedures and session control procedures is that a dialog is not created.

2.3.3 IMS Service Control reference point

In the IMS architecture, ASs are entities that host and execute services, such as presence, messaging and session forwarding. Therefore, there has to be a reference point for sending and receiving SIP messages between the CSCF and an AS. This reference point is called the IMS Service Control (ISC) reference point and the selected protocol is SIP. ISC procedures can be divided into two main categories: routing the initial SIP request to an AS and AS-initiated SIP requests:

- When the S-CSCF receives an initial SIP request it will analyse it. Based on the analysis the S-CSCF may decide to route the request to an AS for further processing. The AS may terminate, redirect or proxy the request from the S-CSCF.

- An AS may initiate a request (e.g., on behalf of a user).

- The concept of service control is thoroughly described in Section 3.12.

2.3.4 Cx reference point

Subscriber and service data are permanently stored in the HSS. These centralized data need to be utilized by the I-CSCF and the S-CSCF when the user registers or receives sessions. Therefore, there has to be a reference point between the HSS and the CSCF. This reference point is called the Cx reference point and the selected protocol is Diameter. The procedures can be divided into three main categories: location management, user data handling and user authentication. Generally, descriptions only cover successful cases—unsuccessful ones are not covered here. The result information element could be used to carry information about why a request fails. If an error occurs, an answer message would not contain any further information elements in most cases.

2.3.4.1 Location management procedures

Location management procedures can be further divided in two groups: registration and de-registration, and location retrieval.

Registration and de-registration procedures between I-CSCF and HSS

When the I-CSCF receives a SIP REGISTER request from the P-CSCF via the Mw reference point it will invoke a user registration status query, or as it is known in the standards a User-Authorization-Request (UAR) command. This command contains:

- Private User Identity—the identity to uniquely identify the user from a network perspective. It identifies subscription and correct authentication data (see Section 3.4.1.1 for further details on private user identity).

- Public User Identity—the identity to be registered (see Section 3.4.1.2 for further details on public user identity).

- Visited Network Identifier—identifies the visited IMS network in the case of IMS roaming. Based on this identifier the HSS is able to enforce roaming restrictions.

- Routing Information—contains the address of the HSS if the I-CSCF is aware of it. If the I-CSCF does not know the address of the HSS, then it contains the destination realm (i.e., the SLF is used to resolve a correct HSS).

- Type of Authorization—three possible values for the type of authorization information element are defined:

  o REGISTRATION—it is included when the expires value in the REGISTER request does not equal zero.

  o REGISTRATION_CAPABILITIES—it is included when the expires value in the REGISTER request is not equal to zero and the I-CSCF explicitly queries S-CSCF capabilities (e.g., when a previously given S-CSCF is not responding).

  o DE-REGISTRATION—it is included when the expires value in the REGISTER request is equal to zero.

After receiving the UAR command the HSS sends a User-Authorization-Answer (UAA) command. It contains:

- Result—informs the outcome of the UAR command.

- S-CSCF Name and/or S-CSCF Capabilities (if the UAR command does not fail due, say, to the private and public identities received in the request not belonging to the same user) depending on the user's current registration status.

S-CSCF capabilities are returned if the user does not have an S-CSCF name assigned yet in the HSS or if the I-CSCF explicitly requests S-CSCF capabilities. Otherwise, the S-CSCF name is returned. When capabilities are returned the I-CSCF needs to perform S-CSCF selection as described in Section 3.8.

Registration and de-registration procedures between S-CSCF and HSS

We explained above how I-CSCF finds an S-CSCF that will serve the user. Having done this, the I-CSCF forwards a SIP REGISTER request to the S-CSCF. When the

S-CSCF receives the SIP REGISTER request from the I-CSCF it uses a Server-Assignment-Request (SAR) command to communicate with the HSS. The SAR command is used to inform the HSS about which S-CSCF will be serving the user when the expires value is not equal to zero. Similarly, if the expires value equals zero, then the SAR command is used to inform that the S-CSCF is no longer serving a user. A precondition for sending the SAR command is that the user has been successfully authenticated by the S-CSCF. The SAR command contains:

- Private User Identity—see the UAR command.

- Public User Identity—the identity to be registered/de-registered (see Section 3.4.1.2 for further details on public user identity).

- Routing Information—contains the address of the HSS if the S-CSCF is aware of it. If the S-CSCF does not know the address of the HSS, then it contains the destination realm.

- S-CSCF Name—contains the

- SIP URI of the S-CSCF.

- Server Assignment Type—the server assignment type contains information about why this operation is executed (e.g., due to registration, re-registration, session to unregistered user, de-registration that is user-initiated or S-CSCF-initiated and authentication failure).

- User Data Already Available—indicates to the HSS whether or not the S-CSCF has already the part of the user profile that it needs for serving the user.

- User Data Request Type—tells whether the S-CSCF wants to download a complete, registered or unregistered profile.

After receiving the SAR command the HSS will respond with a Server-Assignment-Answer (SAA) command. It contains:

- Result—informs the outcome of the SAR command.

- User Profile—based on the set values of Server Assignment Type and User Data Already Available in the SAR command the User Profile is sent (the User Profile is explained in Section 3.11).

- Charging Information—contains the addresses of the charging functions. This is an optional information element.

Previous sections have described how user-initiated registration and de-registration (user-initiated or S-CSCF-initiated) procedures are handled over the Cx reference point. There is still the need for additional operations to bring about network-initiated de-registration (e.g., due to stolen UE or when a subscription is terminated). In this case it is the HSS that starts network-initiated de-registration by using a command called Registration-Termination-Request (RTR). The RTR command contains:

- Private User Identity—the identity to uniquely identify the user from a network perspective. It identifies the subscription and the correct authentication data (see Section 3.4.1.1 for further details on private user identity).

- Public User Identity—one or more identities to be deregistered (see Section 3.4.1.2 for further details on public user identity).

- Routing Information—contains the name of the S-CSCF that is serving the user.

- Reason for de-registration—contains a reason code that determines S-CSCF behaviour and optionally includes a textual message to be shown to the user.

The RTR command is acknowledged by a Registration-Termination-Answer (RTA) command, which simply indicates the result of the operation. Note that it is possible to deregister the public user identity in one go by only sending the private user identity.

Location retrieval procedures

Previously, we have described how the I-CSCF uses a user registration status query (UAR command) to find the S-CSCF when it receives a SIP REGISTER request. Correspondingly, there has to be a procedure to find the S-CSCF when a SIP method is different than REGISTER. The required procedure is to make use of a Location-Info-Request (LIR) command. This request contains:

- Public User Identity—contains the identity from the request URI field of a SIP method.

- Routing Information—contains the address of the HSS if the I-CSCF is aware of it. If the I-CSCF does not know the address of the HSS, then it contains the destination realm.

The HSS responds with a Location-Info-Answer (LIA) command. The response contains:

- Result—inform the outcome of the LIR command.

- The S-CSCF Name or S-CSCF Capabilities—the latter are returned if the user does not have the S-CSCF name assigned, otherwise the SIP URI of the S-CSCF is returned.

2.3.4.2 User data-handling procedures

During the registration process, user and service-related data will be downloaded from the HSS to the S-CSCF via the Cx reference point using SAR and SAA commands as described earlier. However, it is possible for these data to be changed later when the S-CSCF is still serving a user. To update the data in the S-CSCF the HSS initiates a Push-Profile-Request (PPR) command. This request contains:

- Private User Identity—the identity to uniquely identify the user from a network perspective (see Section 3.4.1.1 for further details on private user identity).

- Routing Information—contains the name of the S-CSCF that is serving the user.

- User Data—contains the updated user profile (the user profile is explained in Section 3.11).

Update takes place immediately after the change with one exception: when the S-CSCF is serving an unregistered user or the S-CSCF is kept for an unregistered user as described in Section 3.8.5 and there is a change in the registered part of user profile, then the HSS will not send a PPR command. The PPR command is acknowledged by a Push-Profile-Answer (PPA) command, which simply indicates the result of the operation.

2.3.4.3 Authentication procedures

IMS user authentication relies on a pre-configured shared secret. Shared secrets and sequence numbers are stored in the IP Multimedia Services Identity Module (ISIM) in the UE and in the HSS in the network. Because S-CSCF takes care of user authorization, there exists the need to transfer security data over the Cx reference point. When the S-CSCF needs to authenticate a user it sends a Multimedia-Auth-Request (MAR) command to the HSS. This request contains:

- Private User Identity—the identity to uniquely identify the user from a network perspective. It identifies subscription and correct authentication data (see Section 3.4.1.1 for further details on private user identity).

- Public User Identity—the identity to be registered (see Section 3.4.1.2 for further details on public user identity).

- S-CSCF Name—contains the SIP URI of the S-CSCF.

- Routing Information—contains the address of the HSS if the I-CSCF is aware

Table 2.1 Cx commands.

Command-Name	Purpose	Abbreviation	Source	Destination
User-Authorization-Request/Answer	User-Authorization-Request/Answer (UAR/UAA) commands are used between the I-CSCF and the HSS during SIP registration for retrieving S-CSCF name or S-CSCF capabilities for S-CSCF selection and during SIP deregistration for retrieving S-CSCF name when the SIP method is REGISTER	UAR	I-CSCF	HSS
		UAA	HSS	I-CSCF
Server-Assignment-Request/Answer	Server-Assignment-Request/Answer (SAR/SAA) commands are used between the S-CSCF and the HSS to update the S-CSCF name to the HSS and to download the user profile data to the S-CSCF	SAR	S-CSCF	HSS
		SAA	HSS	X-CSCF
Location-Info-Request/Answer	Location-Info-Request/Answer (LIR/LIA) commands are used between the I-CSCF and the HSS during the SIP session set-up to obtain the name of the S-CSCF that is serving the user or S-CSCF capabilities for S-CSCF selection	LIR	I-CSCF	HSS
		LIA	HSS	I-CSCF
Multimedia-Auth-Request/Answer	Multimedia-Auth-Request/Answer (MAR/MAA) commands are used between the S-CSCF and the HSS to exchange information to support the authentication between the end user and the home IMS network	MAR	S-CSCF	HSS
		MAA	HSS	S-CSCF
Registration-Termination-Request/Answer	Registration-Termination-Request/Answer (RTR/RTA) commands are used between the S-CSCF and the HSS when the HSS administratively de-registers one or more of the user's public identities	RTR	HSS	S-CSCF
		RTA	S-CSCF	HSS
Push-Profile-Request/Answer	Push-Profile-Request/Answer (PPR/PPA) commands are used between the HSS and the S-CSCF when user profile data are changed by a management operation in HSS and the data need to be updated to the S-CSCF	PPR	HSS	S-CSCF
		PPA	S-CSCF	HSS

of it. If the I-CSCF does not know the address of the HSS, then it contains the destination realm.

- Number of Authentication Items—information about how many authentication vectors the S-CSCF wants to download at once. Multiple authentication vectors can be downloaded (e.g., if an operator wants to re-authenticate all re-registrations).

- Authentication Data—includes authentication scheme (e.g., Digest-AKAv1-MD5) and authentication information in case of synchronization failure.

The HSS responds with a Multimedia-Auth-Answer (MAA) command. The answer contains:

- Result—informs the outcome of the MAR command.

- Private User Identity—the identity to uniquely identify the user from a network perspective. It identifies subscription and correct authentication data (see Section 3.4.1.1 for further details on private user identity).

- Public User Identity—the identity to be registered (see Section 3.4.1.2 for further details on public user identity).

- Number of Authentication Items—contains the authentication vectors.

- Authentication Data—includes an authentication vector, which is comprised of an Authentication Scheme (e.g., Digest-AKAv1-MD5), Authentication Information (authentication challenge RAND and the token AUTN), Authorization Information (expected response, or XRES), Integrity Key and, optionally, a Confidentiality Key. Additionally, it contains an Item Number, which indicates the order in which the authentication vectors are to be consumed when multiple vectors are returned.

2.3.5 Dx reference point

When multiple and separately addressable HSSs have been deployed in a network, neither the I-CSCF nor the S-CSCF know which HSS they need to contact. However, they need to contact the SLF first. For this purpose the Dx reference point has been introduced. The Dx reference point is always used in conjunction with the Cx reference point. The protocol used in this reference point is based on DIAMETER. Its functionality is implemented by means of the routing mechanism provided by an enhanced DIAMETER redirect agent.

    To get an HSS address the I-CSCF or the S-CSCF sends to the SLF the Cx

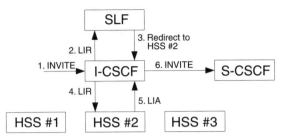

Figure 2.8 HSS resolution using the SLF.

requests aimed for the HSS. On receipt of the HSS address from the SLF, I-CSCF or the S-CSCF will send the Cx requests to the HSS. Figure 2.8 shows how the SLF is used to find a correct HSS when the I-CSCF receives an INVITE request and three HSSs have been deployed.

2.3.6 Sh reference point

An AS (SIP AS or OSA SCS) may need user data or need to know which S-CSCF to send a SIP request. This type of information is stored in the HSS. Therefore, there has to be a reference point between the HSS and the AS. This reference point is called the Sh reference point and the protocol is DIAMETER. Procedures are divided into two main categories: data handling and subscription/notification. The HSS maintains a list of ASs that are allowed to obtain or store data.

2.3.6.1 Data handling

Data handling procedures contain the possibility to retrieve user data from the HSS. Such user data can contain service-related data (transparent or non-transparent), registration information, identities, initial filter criteria, S-CSCF name serving the user, addresses of the charging functions and even location information from the CS and PS domains. Transparent data are understood syntactically but not semantically by the HSS. They are data that an AS may store in the HSS to support its service logic. On the contrary, non-transparent data are understood both syntactically and semantically by the HSS. The AS uses the User-Data-Request (UDR) command to request data. The request contains:

- User identity—includes the public user identity of the user who requires the data (see Section 3.4.1.2 for further details on public user identity).

- AS Identity—identifies the requesting AS. This information is used to check whether the AS has permission to fetch data from the HSS.

- Requested Domain—indicates the access domain for which certain data are requested. Two values are specified: CS domain and PS domain.

- Requested Data—used to indicate what kind of data is requested. The following values are defined:

  - RepositoryData—contains the transparent data stored for the user.

  - PublicIdentifiers—list of public user identities of the user.

  - IMSUserState—information about the user's current state in IMS, defined as REGISTERED, NOT_REGISTERED, AUTHENTICATION_ PENDING and REGISTERED_UNREG_SERVICES.

  - S-CSCFName—name of the S-CSCF that is serving the user.

  - InitialFilterCriteria—contains the relevant triggering information for a service that impacts the requesting AS (see Sections 3.11.1.3 and 3.12 for further information).

  - LocationInformation—consists of location information about the user in the requested domain (e.g., cell global identification).

  - UserState—information about the user's current state in the requested domain.

  - ChargingInformation—contains the addresses of the charging functions.

- Current Location: informs whether the HSS has to perform a location retrieval procedure.

- Service Indication—unique value within the operator's network to identify transparent data.

- AS Name—the identity that is used together with other values to identify the correct InitialFilterCriteria.

The HSS responds with the User-Data-Answer (UDA). The response contains:

- Result informs the outcome of the UDR command.

- The requested data.

The AS can update transparent data in the HSS using the Profile-Update-Request (PUR) command, which contains:

- User Identity—includes the public user identity of the user who required the data (see Section 3.4.1.2 for further details on public user identity).

- AS Identity—identifies the requesting AS. This information is used to check whether the AS has permission to fetch data from the HSS.

- Data—contains the data to be updated.

The PUR command is acknowledged by a Profile-Update-Answer (PUA) command, which simply indicates the result of the operation.

2.3.6.2 Subscription/Notification

Subscription/Notification procedures allow the AS to get a notification when particular data for a specific user is updated in the HSS. The AS sends a Subscribe-Notifications-Request (SNR) command to receive a notification of when a user's data indicated in the SNR command are changed in the HSS:

- User Identity—includes the public user identity of the user who requires the data change.

- Requested Data—contains the reference to the data on which notifications of change are required. Possible values are shown as part of the UDR command (RepositoryData, PublicIdentifiers, etc.).

- Subscription Request Type—informs whether the AS wants to perform a subscribe (initiates notifications) or unsubscribe (stops notifications) operation.

- Service Indication—unique value within the operator's network to identify the transparent data that require the data change.

- Application Server Identity—identifies the requesting AS. This information is used to check whether the AS has permission to fetch data from the HSS.

- Application Server Name—an identity that is used together with other values to identify the correct InitialFilterCriteria that are required for data change.

The HSS acknowledges the subscription request by a Subscribe-Notifications-Answer (SNA) command, which simply indicates the result of the operation.

If the AS has sent the SNR command and requested a notification with subscription request type, then the HSS sends a Push-Notification-Request (PNR) command to the AS when the particular data has changed. It contains the following information:

- User Identity—includes a public user identity of the user for whom the data have changed;

- Requested Data—contains the changed data.

The PNR command is acknowledged by a Push-Notification-Answer (PNA) command, which simply indicates the result of the operation.

Table 2.2 Sh commands.

Command-Name	Purpose	Abbreviation	Source	Destination
User-Data-Request/Answer	User-Data-Request/Answer (UDR/UDA) commands are	UDR	AS	HSS
	used to deliver the user data of a particular user	UDA	HSS	AS
Profile-Update-Request/Answer	Profile-Update-Request/Answer (PUR/PUA) commands are	PUR	AS	HSS
	used to update transparent data in the HSS	PUA	HSS	AS
Subscribe-Notifications-Request/Answer	Subscribe-Notifications-Request/Answer commands are	SNR	AS	HSS
	used to make a subscription/ cancel a subscription to user's data on which notifications of change are required	SNA	HSS	AS
Push-Notification-Request/Answer	Push-Notification-Request/ Answer commands are used to	PNR	HSS	AS
	send the changed data to the AS	PNA	AS	HSS

2.3.7 Si reference point

When the AS is a CAMEL AS (IM-SSF) it uses the Si reference point to communicate to the HSS. The Si reference point is used to transport CAMEL subscription information including triggers from the HSS to the IM-SSF. The used protocol is Mobile Application Part (MAP).

2.3.8 Dh reference point

When multiple and separately addressable HSSs have been deployed in the network, the AS cannot know which HSS it needs to contact. However, the AS needs to contact the SLF first. For this purpose the Dh reference point was introduced in Release 6. In Release 5 the correct HSS is discovered by using proprietary means. The Dh reference point is always used in conjunction with the Sh reference point. The protocol used in this reference point is based on DIAMETER. Its functionality is implemented by means of the routing mechanism provided by an enhanced DIAMETER redirect agent.

To get an HSS address, the AS sends to the SLF the Sh request aimed for the HSS. On receipt of the HSS address from the SLF, the AS will send the Sh request to the HSS.

2.3.9 Mm reference point

For communicating with other multimedia IP networks, a reference point between the IMS and other multimedia IP networks is needed. The Mm reference point allows I-CSCF to receive a session request from another SIP server or terminal. Similarly, the S-CSCF uses the Mm reference point to forward IMS UE-originated requests to other multimedia networks. At the time of writing, a detailed specification of the Mm reference point has not been provided. However, it is very likely that the protocol would be SIP.

2.3.10 Mg reference point

The Mg reference point links the CS edge function, MGCF, to IMS (namely, to the I-CSCF). This reference point allows MGCF to forward incoming session signalling from the CS domain to the I-CSCF. The protocol used for the Mg reference point is SIP. MGCF is responsible for converting incoming ISUP signalling to SIP.

2.3.11 Mi reference point

When the S-CSCF discovers that a session needs to be routed to the CS domain it uses the Mi reference point to forward the session to BGCF. The protocol used for the Mi reference point is SIP. Section 3.13 contains further details about IMS–CS interworking.

2.3.12 Mj reference point

When BGCF receives a session signalling via the Mi reference point it selects the CS domain in which breakout is to occur. If the breakout occurs in the same network, then it forwards the session to MGCF via the Mj reference point. The protocol used for the Mj reference point is SIP. Section 3.13 contains further details about IMS–CS interworking.

2.3.13 Mk reference point

When BGCF receives a session signalling via the Mk reference point it selects the CS domain in which breakout is to occur. If the breakout occurs in another network, then it forwards the session to BGCF in the other network via the Mk reference point. The protocol used for the Mk reference point is SIP. Section 3.13 contains further details about IMS–CS interworking.

2.3.14 Ut reference point

The Ut reference point is the reference point between the UE and the AS. It enables users to securely manage and configure their network services-related information hosted on an AS. Users can use the Ut reference point to create public service identities (PSIs), such as a resource list, and manage the authorization policies that are used by the service. Examples of services that utilize the Ut reference point are presence and conferencing. The AS may need to provide security for the Ut reference point.

HTTP is the chosen data protocol for the Ut reference point. Any protocol chosen for an application that makes use of the Ut reference point needs to be based on HTTP. This reference point is being standardized in Release 6.

2.3.15 Mr reference point

When the S-CSCF needs to activate bearer-related services it passes SIP signalling to the MRFC via the Mr reference point. The functionality of the Mr reference point is not fully standardized: for example, it is not specified how the S-CSCF informs the MRFC to play a specific announcement. The used protocol in the Mr reference point is SIP.

2.3.16 Mp reference point

When the MRFC needs to control media streams (e.g., to create connections for conference media or to stop media in the MRFP) it uses the Mp reference point. This reference point is fully compliant with H.248 standards. However, IMS services may require extensions. This reference point is to be standardized in Release 6.

2.3.17 Go reference point

It is in operators' interests to ensure that the QoS and source and destination addresses of the intended IMS media traffic matches the negotiated values at the IMS level. This requires communication between the IMS (control plane) and the GPRS network (user plane). The Go reference point was originally defined for this purpose. Later on, the charging correlation was added as an additional functionality. The protocol used is the Common Open Policy Service (COPS) protocol. Go procedures can be divided into two main categories:

- Media authorization—as far as access is concerned, the Policy Enforcement Point (PEP) (e.g., GGSN) uses the Go reference point to ask whether a requested bearer activation can be accepted from the PDF that acts as a policy decision point. The PEP also uses the Go reference point to notify the policy decision point about necessary bearer modification and bearer releases (e.g., PDP context). As far as the IMS is concerned, the PDF uses the Go reference point to explicitly indicate when a bearer can or cannot be used; it may also request the PEP to initiate a bearer release. Media authorization is thoroughly explained in the context of the SBLP in Section 3.9.

- Charging correlation—via the Go reference point the IMS is able to pass an IMS charging identifier (ICID) to the GPRS network (user plane). In similar manner, the access network is able to pass a GPRS charging identifier to the IMS. With this procedure it is possible to later merge GPRS charging and IMS charging information in a billing system. This concept is further explained in Section 3.10.

2.3.18 Gq reference point

When a stand-alone PDF is deployed the Gq reference point is used to transport policy set-up information between the application function and the PDF. The term "application function" is used because it is intended that a PDF could authorize other traffic than IMS traffic. In the IMS case the P-CSCF plays the role of an application function. This reference point is being standardized in Release 6.

The P-CSCF sends policy information to the PDF about every SIP message that includes an SDP payload. This ensures that the PDF passes the proper information to perform media authorization for all possible IMS session set-up scenarios. The

Table 2.3 Summary of reference points.

Name of reference point	Involved entities	Purpose	Protocol
Gm	UE, P-CSCF	This reference point is used to exchange messages between UE and CSCFs	SIP
Mw	P-CSCF, I-CSCF, S-CSCF	This reference point is used to exchange messages between CSCFs	SIP
ISC	S-CSCF, I-CSCF, AS	This reference point is used to exchange messages between CSCF and AS	SIP
Cx	I-CSCF, S-CSCF, HSS	This reference point is used to communicate between I-CSCF/S-CSCF and HSS	Diameter
Dx	I-CSCF, S-CSCF, SLF	This reference point is used by I-CSCF/S-CSCF to find a correct HSS in a multi-HSS environment	Diameter
Sh	SIP AS, OSA SCS, HSS	This reference point is used to exchange information between SIP AS/OSA SCS and HSS	Diameter
Si	IM-SSF, HSS	This reference point is used to exchange information between IM-SSF and HSS	MAP
Dh	SIP AS, OSA, SCF, IM-SSF, HSS	This reference point is used by AS to find a correct HSS in a multi-HSS environment	Diameter
Mm	I-CSCF, S-CSCF, external IP network	This reference point will be used for exchanging messages between IMS and external IP networks	Not specified

Reference Point	Description	Protocol
Mg	MGCF → I-CSCF	MGCF converts ISUP signalling to SIP signalling and forwards SIP signalling to I-CSCF	SIP
Mi	S-CSCF → BGCF	This reference point is used to exchange messages between S-CSCF and BGCF	SIP
Mj	BGCF → MGCF	This reference point is used to exchange messages between BGCF and MGCF in the same IMS network	SIP
Mk	BGCF → BGCF	This reference point is used to exchange messages between BGCFs in different IMS networks	SIP
Mr	S-CSCF, MRFC	This reference point is used to exchange messages between S-CSCF and MRFC	SIP
Mp	MRFC, MRFP	This reference point is used to exchange messages between MRFC and MRFP	H.248
Mn	MGCF, IM-MGW	This reference point allows control of user-plane resources	H.248
Ut	UE, AS (SIP AS, OSA SCS, IM-SSF)	This reference point enables UE to manage information related to his services	HTTP
Go	PDF, GGSN	This reference point allows operators to control QoS in a user plane and exchange charging correlation information between IMS and GPRS network	COPS
Gq	P-CSCF, PDF	This reference point is used to exchange policy decisions-related information between P-CSCF and PDF	Diameter

media authorization concept is thoroughly explained in Section 3.9. The P-CSCF provides the following information to the PDF for each media component [3GPP TS 29.207]:

- Destination IP address and destination port number.

- Transport protocol ID (e.g., RTP).

- Media direction information (send, receive, send and receive).

- Direction of the source (originating or terminating side).

- Indication of the group that the media component belongs to.

- Media-type information (audio, video, etc.).

- Bandwidth parameters.

- Indication of forking/non-forking.

Additionally, the P-CSCF passes an ICID to the PDF when the ICID is received in SIP signalling or generated in the P-CSCF.

Similarly, the PDF sends an authorization token and GPRS charging identifier (GCID) to the P-CSCF. Section 3.9 explains further when an authorization token is generated and when the PDF receives the GCID from the GGSN.

At the time of writing, standardization of the Gq reference point was ongoing and therefore it is subject to further changes.

3

IMS Concepts

3.1 Overview

This chapter begins with a first-glance description of IP Multimedia Subsystem (IMS) registration and session establishment. It depicts the IMS entities that are involved. The intention is not to show a full-blown solution; rather, it is to give an overview and help the reader to understand the different IMS concepts explained in this chapter. Detailed registration and session establishment flows will be shown and explained later in the book.

Prior to IMS registration the user equipment (UE) must discover the IMS entity to which it will send a REGISTER request. This concept is called a Proxy-Call Session Control Function (P-CSCF) discovery and is described in Section 3.7. In addition, before a registration process the UE needs to fetch user identities from identity modules. Identity modules are covered in Section 3.5 and identities are presented in Section 3.4. During the registration a Serving-CSCF (S-CSCF) will be assigned (Section 3.8), authentication will be performed and corresponding security associations will be established (Section 3.6), a user profile (Section 3.11) will be downloaded to the assigned S-CSCF, Session Initiation Protocol (SIP) compression is initialized (Section 3.16) and implicitly registered public user identities will be delivered (Section 3.14).

Section 3.9 explains how Internet Protocol (IP) policy control is applied when a user is establishing a session, and Section 3.12 shows how services can be provisioned. Section 3.10 shows how an operator is able to charge a user. Interworking with the Circuit Switched (CS) network is briefly described in Section 3.13. In addition, the concept of sharing a single user identity between multiple terminals is covered in Section 3.15.

3.2 Registration

Prior to IMS registration, which allows the UE to use IMS services, the UE must obtain an IP connectivity bearer and discover an IMS entry point, P-CSCF: for example, in the case of General Packet Radio Service (GPRS) access the UE

The IMS. Miikka Poikselkä, Georg Mayer, Hisham Khartabil and Aki Niemi
Copyright 2004 by John Wiley & Sons, Ltd. ISBN 0-470-87113-X

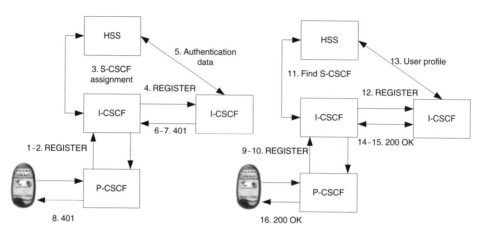

Figure 3.1 A high-level IMS registration flow.

performs the GPRS attach procedure and activates a Packet Data Protocol (PDP) context for SIP signalling. Section 13.2 gives a short overview of the PDP context, and P-CSCF discovery is explained in Section 3.7. This book does not describe the GPRS attach procedure (for further information see [3GPP TS 23.060].

IMS registration contains two phases: the leftmost part of Figure 3.1 shows the first phase—how the network challenges the UE. The rightmost part of Figure 3.1 shows the second phase—how the UE responds to the challenge and completes the registration.

First, the UE sends a SIP REGISTER request to the discovered P-CSCF. This request would contain, say, an identity to be registered and a home domain name (address of the Interrogating-CSCF, or I-CSCF). The P-CSCF processes the REGISTER request and uses the provided home domain name to resolve an IP address of the I-CSCF. The I-CSCF in turn will contact the Home Subscriber Server (HSS) to fetch the required capabilities for S-CSCF selection. After S-CSCF selection the I-CSCF forwards the REGISTER request to the S-CSCF. The S-CSCF realizes that the user is not authorized and therefore retrieves authentication data from the HSS and challenges the user with a 401 Unauthorized response. Second, the UE will calculate a response to the challenge and send another REGISTER request to the P-CSCF. Again the P-CSCF finds the I-CSCF and the I-CSCF in turn will find the S-CSCF. Finally the S-CSCF checks the response and if it is correct downloads a user profile from the HSS and accepts the registration with a 200 OK response. Once the UE is successfully authorized, the UE is able to initiate and receive sessions. During the registration procedure both the UE and the P-CSCF learns which S-CSCF in the network will be serving the UE.

It is the UE's responsibility to keep its registration active by periodically refreshing its registration. If the UE does not refresh its registration, then the S-CSCF will

Table 3.1 Information storage before, during and after the registration process.

Node	Before registration	During registration	After registration
UE	P-CSCF address, home domain name, credentials, public user identity, private user identity	P-CSCF address, home domain name, credentials, public user identity, private user identity, security association	P-CSCF address, home domain name, credentials, public user identity (and implicitly registered public user identities), private user identity, security association, service route information (S-CSCF)
P-CSCF	No state information	Initial network entry point, UE IP address, public and private user IDs, security association	Final network entry point (S-CSCF), UE address, registered public user identity (and implicitly registered public user identities), private user ID, security association, address of CCF
I-CSCF	HSS or SLF address	HSS or SLF entry, P-CSCF address, S-CSCF address	HSS or SLF address
S-CSCF	HSS or SLF address	HSS address/name, user profile (limited—as per network scenario), proxy address/name, public/ private user ID, UE IP address	HSS address/name, user profile (limited—as per network scenario), proxy address/name, public/ private user ID, UE IP address
HSS	User profile authentication data, S-CSCF selection parameters	User profile, P-CSCF, network ID	User profile including updated registration status of public user identities, S-CSCF name

silently remove the registration when the registration timer lapses. When the UE wants to de-register from the IMS it simply sends a REGISTER request including the registration timer (expire) value zero. Sections 5.5 and 5.14 contain a more detailed description of IMS registration and de-registration.

3.3 Session initiation

When user *A* wants to have a session with user *B*, UE *A* generates a SIP INVITE request and sends it via the Gm reference point to the P-CSCF. The P-CSCF

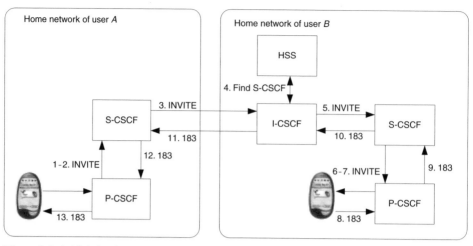

Figure 3.2 A high-level IMS session establishment flow.

processes the request: for example, it decompresses the request and verifies the originating user's identity before forwarding the request via the Mw reference point to the S-CSCF. The S-CSCF processes the request, executes service control which may include interactions with application servers (ASs) and eventually determines an entry point of the home operator of user B based on user B's identity in the SIP INVITE request. The I-CSCF receives the request via the Mw reference point and contacts the HSS over the Cx reference point to find the S-CSCF that is serving user B. The request is passed to the S-CSCF via the Mw reference point. The S-CSCF takes charge of processing the terminating session, which may include interactions with application servers (ASs) and eventually delivers the request to the P-CSCF over the Mw reference point. After further processing (e.g., compression and privacy checking), the P-CSCF uses the Gm reference point to deliver the SIP INVITE request to UE B. UE B generates a response, 183 Session Progress, which traverses back to UE A following the route that was created on the way from UE A (i.e., UE $B \rightarrow$ P-CSCF $\rightarrow$ S-CSCF $\rightarrow$ I-CSCF $\rightarrow$ S-CSCF $\rightarrow$ P-CSCF $\rightarrow$ UE A) (Figure 3.2). After a few more round trips, both sets of UE complete session establishment and are able to start the actual application (e.g., a game of chess). During session establishment an operator may control the usage of bearers intended for media traffic. Section 3.9 explains how this can be done.

Just to give a taste of what is coming in the book, the high-level content of a SIP INVITE request is given in Table 3.2. Each column gives the information elements that are inserted, removed or modified. The meaning of each information element is covered later in the book.

Table 3.2 The high-level content of a SIP INVITE request during session establishment.

UE (*A*)	P-CSCF (*A*)	S-CSCF (*A*)
User *A* identity	*Inserted information*	*Inserted information*
User *B* identity	One piece of routing	Interoperator identifier
Contact address	information (Record-Route	*Removed information*
Access information	header)	One piece of routing
Routing information (Via	IMS charging information	information (Route
and Route headers)	Verified *A* party identity	header)
Support of reliable	*Removed information*	Access information
responses	Security information	*Modified information*
Support of preconditions	Proposed *A* party identity	Routing information
Security information	*Modified information*	(Record-Route and Via)
Privacy indication	Routing information (Route,	Verified *A* party identity,
Compression indication	Via)	also include Tel-URL type
SDP payload reflecting user's		of identity from now on
terminal capabilities and		(if a user has one)
user preferences for the		
session, MIME subtype		
"telephone-event",		
bandwidth information		

I-CSCF (*B*)	S-CSCF (*B*)	P-CSCF (*B*)
Inserted information	*Inserted information*	*Inserted information*
One piece of routing	None	Authorization token
information (Route	*Removed information*	*Removed information*
header)	Interoperator identifier	IMS charging information
Removed information	*Modified information*	One piece of routing
None	Routing information (R-URI,	information (Route
Modified information	Route, Via, Record route)	header)
Routing information (Via)		A party identity is removed
		if privacy is required.
		Modified information
		Routing information (Via,
		Record route)

3.4 Identification

3.4.1 Identification of users

3.4.1.1 Private user identity

The private user identity is a unique global identity defined by the home network operator, which may be used within the home network to uniquely identify the user from a network perspective [3GPP TS 23.228]. It does not identify the user herself; on the contrary, it identifies the user's subscription. Therefore, it is mainly used for authentication purposes. It is possible to utilize private user identities for accounting and administration purposes as well.

The IMS architecture imposes the following requirements for private user identity [3GPP TS 23.228, TS 23.003]:

- The private user identity will take the form of a network access identifier (NAI) defined in [RFC2486].

- The private user identity will be contained in all registration requests passed from the UE to the home network.

- The private user identity will be authenticated only during registration of the user (including re-registration and de-registration).

- The S-CSCF will need to obtain and store the private user identity on registration and on unregistered termination.

- The private user identity will not be used for routing of SIP messages.

- The private user identity will be permanently allocated to a user and securely stored in an IMS Identity Module (ISIM) application. The private user identity will be valid for the duration of the user's subscription within the home network.

- It will not be possible for the UE to modify the private user identity.

- The HSS will need to store the private user identity.

- The private user identity will optionally be present in charging records based on operator policies.

Example of NAI	form_user@realm

3.4.1.2 Public user identity

User identities in IMS networks are called public user identities. They are the identities used for requesting communication with other users. Public identities can be published (e.g., in phone books, Web pages, business cards).

As stated earlier in the book, IMS users will be able to initiate sessions and receive sessions from many different networks, such as GSM networks and the Internet. To be reachable from the CS side, the public user identity must conform to telecom numbering (e.g., +358 50 1234567). In similar manner, requesting communication with Internet clients, the public user identity must conform to Internet naming (e.g., joe.doe@example.com).

The IMS architecture imposes the following requirements for public user identity [3GPP TS 23.228, TS 23.003]:

- The public user identity/identities will take the form of either a SIP uniform resource identifier (URI) or a telephone uniform resource locator (tel URL) format.

- At least one public user identity will be securely stored in an ISIM application.

- It will not be possible for the UE to modify the public user identity.

- A public user identity will be registered before the identity can be used to originate IMS sessions and IMS session-unrelated procedures (e.g., MESSAGE, SUBSCRIBE, NOTIFY).

- A public user identity will be registered before terminating IMS sessions, and terminating IMS session-unrelated procedures will be delivered to the UE of the user that the public user identity belongs to. This does not prevent the execution of services in the network by unregistered users.

- It will be possible to register multiple public user identities through one single UE request. This is described further in Section 3.14.

- The network will not authenticate public user identities during registration.

The tel URL scheme is used to express traditional e.164 numbers in URL syntax. The tel URL is described in [RFC2806], and the SIP URI is described in [RFC3261] and [RFC2396]. Examples of public user identities are given below. A more detailed description of SIP URI and tel URL syntaxes can be found in Sections 8.5 and 8.6.

Example of SIP URI	sip:joe.doe@ims.example.com
Example of tel URL	tel:+358 50 1234567

3.4.1.3 Derived public user identity and private user identity

In Sections 3.4.1.1 and 3.4.1.2 the concepts of private user identity and public user identity have been explained. It was stated that these identities are stored in an ISIM application. When the IMS is deployed there will be a lot of UE in the market place that does not support the ISIM application; therefore, a mechanism to access the IMS without the ISIM was developed.

In this model, private user identity, public user identity and home domain name are derived from an International Mobile Subscriber Identifier (IMSI). This mechanism is suitable for UE that has a Universal Subscriber Identity Module (USIM) application.

Private user identity

The private user identity derived from the IMSI is built according to the following steps [3GPP TS 23.003]:

1. The user part of the private user identity is replaced with the whole string of digits from IMSI.

2. The domain part of the private user identity is composed of the MCC and MNC values of IMSI and has a predefined domain name, IMSI.3gppnetwork.org. These three parts are merged together and separated by dots in the following order: mobile network code (MNC, a digit or a combination of digits uniquely identifying the public land mobile network), mobile country code (MCC, code uniquely identifying the country of domicile of the mobile subscriber) and predefined domain name. For example:

```
IMSI in use: 234150999999999; where:
MCC: 234;
MNC: 15;
MSIN: 0999999999; and
Private user identity is:
234150999999999@234.15.IMSI.3gppnetwork.org
```

Temporary public user identity

If there is no ISIM application to host the public user identity, a temporary public user identity will be derived, based on the IMSI. The temporary public user identity will take the form of a SIP URI, "sip:user@domain". The user and domain part are derived similarly to the method used for private user identity [3GPP TS 23.003]. Following our earlier example a corresponding temporary public user identity would be:

```
sip:234150999999999@234.15.IMSI.3gppnetwork.org
```

The IMS architecture imposes the following requirements for a temporary public user identity [3GPP TS 23.228]:

- It is strongly recommended that the temporary public user identity is set to "barred" for IMS non-registration procedures so that it cannot be used for IMS communication. The following additional requirements apply if the temporary public user identity is "barred":

  - The temporary public user identity will not be displayed to the user and will not be used for public usage (e.g., displayed on a business card).

  - The temporary public user identity will only be used during the registration to obtain implicitly registered public user identities (the concept of implicitly registered public user identities is explained in Section 3.14).

- Implicitly registered public user identities will be used for session handling, in other SIP messages and at subsequent registration processes.

- After the initial registration only the UE will use the implicitly registered public user identity(s).

- The temporary public user identity will only be available to CSCF and HSS nodes.

3.4.1.4 Relationship between private and public user identities

Here a basic example shows how different identities are linked to each other. Joe is working for a car sales company and is using a single terminal for both his work life and his personal life. To handle work-related matters he has two public user identities: sip:joe.smith@brandnewcar.com and tel:+358 50 1234567. When he is off-duty he uses two additional public user identities to manage his personal life: sip:joe.smith@ims.example.com and tel:+358503334444. By having two sets of public user identities he could have totally different treatment for incoming sessions: for example, he is able to direct all incoming work-related sessions to a messaging system after 5 p.m. and during weekends and holidays.

Joe's user and service-related data are maintained in two different service profiles. One service profile contains information about his work life identities and is downloaded to the S-CSCF from the HSS when needed: that is, when Joe registers a work life public user identity or when the S-CSCF needs to execute unregistered services for a work life public user identity. Similarly, another service profile contains information about his personal life identities and is downloaded to the S-CSCF from the HSS when needed. The concept of service profile is explained in Section 3.11.1.

Figure 3.3 shows how Joe's private user identity, public user identities and service profiles are linked together.

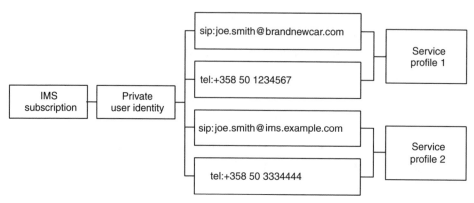

Figure 3.3 Relationship between user identities.

3.4.2 Identification of services (public service identities)

With the introduction of standardized presence, messaging, conferencing and group service capabilities it became evident that there must be identities to identify services and groups that are hosted by ASs. Identities for these purposes are also created on the fly: that is, they may be created by the user on an as-needed basis in the AS and are not registered prior to usage. Ordinary public user identities were simply not good enough; therefore, Release 6 introduced a new type of identity, the public service identity. Public service identities take the form of a SIP URI or are in tel URL format: for example, in messaging services there is a public service identity for the messaging list service (e.g., sip:messaginglist_joe@ims.example.com) to which the users send messages and then the messages are distributed to other members on the messaging list by the messaging list server. The same applies to conferencing services (i.e., audio/video and messaging sessions), where a URI for the conferencing service is created.

3.4.3 Identification of network entities

In addition to users, network nodes that handle SIP routing need to be identifiable using a valid SIP URI. These SIP URIs would be used when identifying these nodes in the header fields of SIP messages. However, this does not require that these URIs will be globally published in domain name system (DNS) [3GPP TS 23.228]. An operator could name its S-CSCF as follows:

Example of network entity naming	sip:finland.scscf1@ims.example.com

3.5 Identity modules

3.5.1 IP Multimedia Services Identity Module

IP Multimedia Services Identity Module (ISIM) is an application residing on the Universal Integrated Circuit Card (UICC), which is a physically secure device that can be inserted and removed from UE. There may be one or more applications in the UICC. The ISIM itself stores IMS-specific subscriber data mainly provisioned by an IMS operator. The stored data can be divided into six groups as shown in Figure 3.4. Most of the data are needed when a user performs an IMS registration [3GPP TS 31.103]:

- Security keys consist of integrity keys, ciphering keys and key set identifiers. Integrity keys are used to prove integrity protection of SIP signalling. Ciphering keys are used to provide confidential protection of SIP signalling. Confidential protection is not used in Release 5; however, in Release 6 it should be possible to use confidential protection. At the time of writing there is a need for key set identifiers (see Section 3.6 for further information).

- The private user identity simply contains the private user identity of the user. It is used in a registration request to identify the user's subscription (see Section 3.4.1.1 for further information).

- The public user identity contains one or more public user identities of the user. It is used in a registration request to identify an identity to be registered and is used

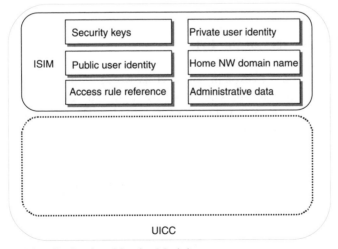

Figure 3.4 IP Multimedia Services Identity Module.

for request communication with other users (see Section 3.4.1.2 for further information).

- The home network domain name consists of the name of the entry point of the home network. It is used in a registration request to route the request to the user's home network.

- Administrative data include various data, which could be used, say, by IMS subscribers for IMS operations or by manufacturers to execute proprietary auto-tests.

- Access Rule Reference is used to store information about which personal identification number needs to be verified in order to get access to the application.

3.5.2 Universal Subscriber Identity Module

The Universal Subscriber Identity Module (USIM) is required for accessing the Packet Switched (PS) domain (GPRS) and unambiguously identifies a particular subscriber. Similarly to the ISIM, the USIM application resides on the UICC as a storage area for subscription and subscriber-related information. Additionally, it may contain applications that use the features defined in the USIM Application Toolkit.

The USIM contains such data as the following: security parameters for accessing the PS domain, IMSI, list of allowed access point names, MMS-related information [3GPP TS 31.102, TS 22.101, TS 21.111]. Section 3.4.1.3 describes how UE with the USIM could derive the necessary information to access the IMS.

3.6 Security services in the IMS

This section is intended to explain how security works in the IMS. It is intentionally thin in cryptography and thus will not discuss algorithms and key lengths in depth, nor will it perform any cryptanalysis on IMS security. There are much better books specifically for that purpose.[1]

Instead, what this chapter will do is give a high-level view of the security architecture and explain the components of that architecture, including the models and protocols used to provide the required security features. After reading this chapter the reader should be familiar with the main concepts in the IMS security

---

[1] See, for example, V. Niemi and K. Nyberg (2003) *UMTS Security*, John Wiley & Sons, Chichester, UK.

architecture and understand the underlying models, especially those related to trust and identity that shape IMS security as a whole.

3.6.1 IMS Security Model

The IMS security architecture consists of three building blocks, as illustrated in Figure 3.5. The first building block is the Network Domain Security (NDS) [3GPP TS 33.210], which provides IP security between different domains and nodes within a domain. Layered alongside NDS is IMS access security [3GPP TS 33.203]. The access security for SIP-based services is a self-sustaining component in itself, with the exception that the security parameters for it are derived from the Universal Mobile Telecommunications System (UMTS) Authentication and Key Agreement (AKA) Protocol [3GPP TS 33.102]. AKA is also used for bootstrapping purposes (namely, keys and certificates are derived from AKA credentials and subsequently used for securing applications that run on HTTP [RFC2616], among other things).

Intentionally left out of this architectural model are those security layers that potentially layer on top of the IMS access security or run below the NDS. For example, in the UMTS the radio access layer implements its own set of security features, including ciphering and message integrity. However, the IMS is designed in a way that does not depend on the existence of either access security or user-plane security.

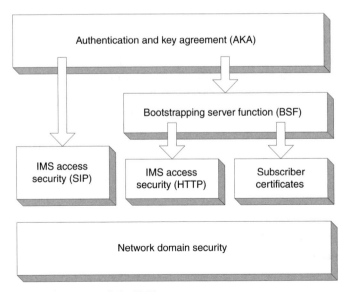

Figure 3.5 Security architecture of the IMS.

3.6.2 Authentication and Key Agreement

Security in the IMS is based on a long-term secret key, shared between the ISIM and the home network's Authentication Centre (AUC). The most important building block in IMS security is the ISIM module, which acts as storage for the shared secret (K) and accompanying AKA algorithms, and is usually embedded on a smartcard-based device called the Universal Integrated Circuit Card (UICC). Access to the shared secret is limited. The module takes AKA parameters as input and outputs the resulting AKA parameters and results. Thus it never exposes the actual shared secret to the outside world.

The device on which the ISIM resides is tamper-resistant, so even physical access to it is unlikely to result in exposing the secret key. To further protect the ISIM from unauthorized access, the user is usually subject to user domain security mechanisms. This in essence means that in order to run AKA on the ISIM, the user is prompted for a PIN code. The combination of ownership (i.e., access to a physical device (UICC/ISIM) and knowledge of the secret PIN code) makes the security architecture of the IMS robust. An attacker is required to have possession of both "something you own" and "something you know", which is difficult, as long as there is some level of care taken by the mobile user.

AKA accomplishes mutual authentication of both the ISIM and the AUC, and establishes a pair of cipher and integrity keys. The authentication procedure is set off by the network using an authentication request that contains a random challenge (RAND) and a network authentication token (AUTN). The ISIM verifies the AUTN and in doing so verifies the authenticity of the network itself. Each end also maintains a sequence number for each round of authentication procedures. If the ISIM detects an authentication request whose sequence number is out of range, then it aborts the authentication and reports back to the network with a synchronization failure message, including with it the correct sequence number. This is another top-level concept that provides for anti-replay protection.

To respond to the network's authentication request, the ISIM applies the secret key on the random challenge (RAND) to produce an authentication response (RES). The RES is verified by the network in order to authenticate the ISIM. At this point the UE and the network have successfully authenticated each other and as a by-product have also generated a pair of session keys: the cipher key (CK) and the integrity key (IK). These keys can then be used for securing subsequent communications between the two entities. Table 3.3 lists some of the central AKA parameters and their meaning.

3.6.3 Network domain security

3.6.3.1 Introduction

One of the main identified weaknesses of 2G systems is the lack of standardized security solutions for the core networks. Even though the radio access from the

Table 3.3 AKA parameters.

AKA Parameter	Length	Description
K	128 bits	Shared secret; authentication key shared between the network and the mobile terminal
RAND	128 bits	Random authentication challenge generated by the network
AUTN	128 bits	(Network) authentication token
SQN	48 bits	Sequence number tracking the sequence of the authentication procedures
AUTS	112 bits	Synchronization token generated by the ISIM on detecting a synchronization failure
RES	32 bits–128 bits[a]	Authentication response generated by the ISIM
CK	128 bits	Cipher key generated during authentication by both the network and the ISIM
IK	128 bits	Integrity key generated during authentication by both the network and the ISIM

[a] Note that the short key lengths are to accommodate backward compatibility with 2G authentication.

mobile terminal to the base station is usually protected by encryption, nodes in the rest of the system pass traffic in clear text. Sometimes these links even run over unprotected radio hops, so an attacker that has access to this medium can fairly easily eavesdrop on the communications.

Having learned from these shortcomings in 2G, 3G systems have set out to protect all IP traffic in the core network. NDS[2] accomplishes this by providing confidentiality, data integrity, authentication and anti-replay protection for the traffic, using a combination of cryptographic security mechanisms and protocol security mechanisms applied in IP security (IPsec).

3.6.3.2 Security domains

Security domains are central to the concept of NDS. A security domain is typically a network operated by a single administrative authority that maintains a uniform security policy within that domain. As a result, the level of security and the installed security services will in general be systematically the same within a security domain.

In many cases a security domain will correspond directly to an operator's core network. It is however possible to run several security domains each pertaining to a subset of the operator's entire core network. In the NDS/IP the interfaces between different security domains are denoted as Za, while interfaces between elements inside a security domain are denoted as Zb. While use of the Zb interface is in

---

[2] To specifically indicate that the protected traffic is IP-based, NDS is usually denoted by the abbreviation NDS/IP.

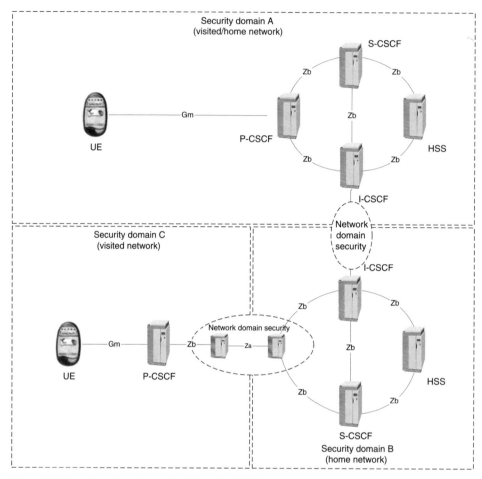

Figure 3.6 Security domains underlining the IMS.

general optional and up to the security domain's administrator, use of the Za interface is always mandatory between different security domains. Data authentication and integrity protection is mandatory for both interfaces, while use of encryption is optional for the Zb, as opposed to being recommended for the Za.

The IMS builds on the familiar concept of a home network and a visited network. Basically, two scenarios exist, depending on whether the IMS terminal is roaming or not. In the first scenario the UE's first point of contact to the IMS, called the P-CSCF, is located in the home network. In the second scenario the P-CSCF is located in the visited network, meaning that the UE is in fact roaming in such a way that its first point of contact to the IMS is not its home network. These two scenarios are illustrated in Figure 3.6.

Quite often, an IMS network corresponds to a single security domain, and therefore traffic between the operator's IMS networks is protected using the NDS/

IP. The same applies also in the above mentioned second scenario, where traffic between the visited and the home network is also protected using the NDS/IP.

In the IMS the NDS/IP only protects traffic between network elements in the IP layer; so, further security measures are required. These will be covered in subsequent chapters. Most importantly, the first hop[3] in terms of SIP traffic is not protected using the NDS/IP, but is using IMS access security measures [3GPP TS 33.203]. As will be explained in subsequent chapters, the above scenario in which the IMS elements are split across the home network and the visited network and therefore different security domains requires some special care in terms of authentication and key distribution.

3.6.3.3 Security gateways

Traffic entering and leaving a security domain passes through a security gateway (SEG). The SEG sits in the border of a security domain and tunnels traffic toward a defined set of other security domains. This is called a hub-and-spoke model; it provides for hop-by-hop security between security domains. The SEG is responsible for enforcing security policy when passing traffic between the security domains. This policy enforcement may also include packet filtering or firewall functionality, but such functionality is the responsibility of the domain administrator.

In the IMS all traffic within the IMS core network is routed via SEGs, especially when the traffic is inter-domain, meaning that it originates from a different security domain from the one where it is received. When protecting inter-domain IMS traffic, both confidentiality as well as data integrity and authentication are mandated in the NDS/IP.

3.6.3.4 Key management and distribution

Each SEG is responsible for setting up and maintaining IPsec security associations (SAs) [RFC2401] with its peer SEGs. These SAs are negotiated using the Internet Key Exchange (IKE) [RFC2409] protocol, where authentication is done using long-term keys stored in the SEGs. A total of two SAs per peer connection are maintained by the SEG: one for inbound traffic and one for outbound traffic. In addition, the SEG maintains a single Internet Security Association and Key Management Protocol (ISAKMP) SA [RFC2408], which is related to key management and used to build up the actual IPsec SAs between peer hosts. One of the key prerequisites for

---

[3] Referring here to the interface between the UE and the P-CSCF, denoted as Gm.

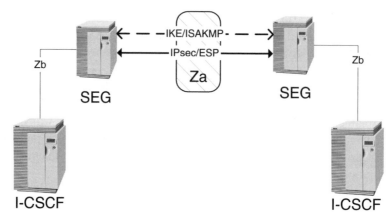

Figure 3.7 NDS/IP and SEGs.

the ISAKMP SA is that the peers are authenticated. In the NDS/IP, authentication is based on pre-shared secrets.[4] Figure 3.7 illustrates the model.

The security protocol used in the NDS/IP for encryption, data integrity protection and authentication is the IPsec Encapsulating Security Payload (ESP) [RFC2406] in tunnel mode. In tunnel-mode ESP the full IP datagram including the IP header is encapsulated in the ESP packet. For encryption, the Triple DES (3DES) [RFC1851] algorithm is mandatory, while for data integrity and authentication both MD5 [RFC1321] and SHA-1 [RFC2404] can be used. For the specifics of IPsec/IKE and ESP please refer to Chapter 18, where these protocols are discussed more extensively.

3.6.4 IMS Access Security for SIP-based services

3.6.4.1 Introduction

SIP is at the core of the IMS, as it is used for creating, managing and terminating various types of multimedia sessions. The key thing to accomplish in securing access to the IMS is to protect the SIP signalling in the IMS. As noted previously, in the IMS core network this is accomplished using the NDS/IP. But the first hop, meaning the interface for SIP communications between the UE and the IMS P-CSCF denoted as Gm, needs additional measures since it is outside the scope of the NDS/IP.

The security features and mechanisms for secure access to the IMS are specified in [3GPP TS 33.203]. This defines how the UE and network are authenticated as well as how they agree on used security mechanisms, algorithms and keys.

---

[4] Recently, work has been ongoing to add PKI support to NDS authentication. This work is denoted as NDS/AF [3GPP TS 33.310] for the Network Domain Security/Authentication Framework.

3.6.4.2 Trust model overview

The IMS establishes a trust domain, as described in [RFC3325], that encompasses the following IMS elements:

- P/I/S-CSCF.

- Breakout Gateway Control Function (BGCF).

- Media Gateway Control Function/Multimedia Resource Function Controller (MGCF/MRFC).

- All ASs that are not in third-party control.

The main component of trust is identity: in order to trust an entity accessing the IMS there needs to be an established relationship with that entity (i.e., its identity is known and verified). In the IMS this identity is passed between nodes in the trust domain in the form of an asserted identity. The UE can state a preference to this identity if multiple identities exist; but, it is ultimately at the border of the trust domain (namely, in the P-CSCF) that the asserted identity is assigned. Conversely, the P-CSCF plays a central role in authenticating the UE.

The level of trust is always related to the expected behaviour of an entity. For example, Alice may know Bob and trust him to take her children to school. She expects and knows that Bob will act responsibly, drive safely and so on. But she may not trust Bob enough to give him access to her bank account.

Another important property of the IMS trust model is that it is based on transitive trust. The existence of pairwise trust between a first and a second entity as well as a second and a third entity automatically instils trust between the first and the third entity: for example, Alice knows and trusts Bob, who in turn knows Celia and trusts her to take his children to school. Now, according to transitive trust, Alice can also trust Celia to take her children to school without ever actually having met Celia in person. It is enough that Alice trusts Bob and knows that Celia is also part of the trust domain of parenthood. The fact that Alice and Bob are both parents assures Alice that Bob has applied due diligence when judging whether Celia is fit to take children to school. In essence, the trust domain of parenthood forms a network of parents, all compliant with the predefined behaviour of a mother or a father.

In [RFC3325] terms both the expected behaviour of an entity in a trust domain and the assurance of compliance to the expected behaviour needs to be specified for a given trust domain T in what is called a "Spec(T)". The components that make up a Spec(T) are:

- Definition of the way in which users entering the trust domain are authenticated and definition of the used security mechanisms that secure the communications between the users and the trust domain. In the IMS this entails authentication using the AKA protocol and related specifications on Gm security in [3GPP TS 33.203].

- Definition of mechanisms used for securing the communications between nodes in a trust domain. In the IMS this bit is documented in the NDS/IP [3GPP TS 33.210].

- Definition of the procedures used in determining the set of entities that are part of the trust domain. In the IMS this set of entities is basically represented by the set of peer SEGs, of which a SEG in a security domain is aware.

- Assertion that nodes in a trust domain are both compliant with SIP and SIP-asserted identity specifications.

- Definition of privacy handling. This definition relies on SIP privacy mechanisms and the way they are used with asserted identities (Section 3.6.4.3 will discuss these issues more deeply).

3.6.4.3 User privacy handling

The concepts of trust domain and asserted identity enable passing a user's asserted identity around, potentially to entities that are not part of the trust domain. This creates obvious privacy issues, since the user may in fact require that her identity be kept private and internal to the trust domain.

In the IMS the user can request that her identity is not revealed to entities outside the trust domain. This is based on SIP privacy extensions [RFC3323]. A UE inserts its privacy preferences in a privacy header field, which is then inspected by the network. Possible values for this header are:

- User—indicates that user-level[5] privacy functions should be provided by the network. This value is usually set by intermediaries rather than user agents.

- Header—indicates that the user agent (UA) is requiring that header privacy be applied to the message. This means that all privacy-sensitive headers be obscured and that no other sensitive header be added.

- Session—indicates that the UA is requiring that privacy-sensitive data be obscured for the session (i.e., in the SDP payload of the message).

---

[5] By user-level privacy we mean privacy functions that the SIP UA itself is able to provide (e.g., using an anonymous identity in the From field of a request).

- Critical—indicates that the requested privacy mechanisms are critical. If any of those mechanisms is unavailable, the request should fail.

- ID—indicates that the user requires her asserted identity be kept inside the trust domain. In practice, setting this value means that the P-Asserted-Identity header field must be stripped from messages that leave the trust domain.

- None—indicates that the UA explicitly requires no privacy mechanisms to be applied to the request.

3.6.4.4 Authentication and security agreement

Authentication for IMS access is based on the AKA protocol. However, the AKA protocol cannot be run directly over IP; instead, it needs a vehicle to carry protocol messages between the UE and the home network. Obviously, as the entire objective of IMS access authentication is to authenticate for SIP access, SIP is a natural choice for such a vehicle. In practice, the way in which the AKA protocol is tunnelled inside SIP is specified in [RFC3310]. This defines the message format and procedures for using AKA as a digest authentication [RFC2617] password system for the SIP registration procedure. The digest challenge originating from the network will contain the RAND and AUTN AKA parameters, encoded in the server nonce value. The challenge contains a special algorithm directive that instructs the client to use the AKA protocol for that particular challenge. The RES is used as the password when calculating digest credentials, which means that the digest framework is utilized in a special way to tunnel the AKA protocol in IMS access security.

Concurrent with authentication of the user, the UE and the IMS also need to negotiate the security mechanisms that are going to be used in securing subsequent SIP traffic in the Gm interface. The protocol used for this security agreement is again SIP, as specified in [RFC3329]. The UE and the P-CSCF exchange their respective lists of supported security mechanisms and the highest commonly supported one is selected and used. At a minimum, the selected security mechanism needs to provide data integrity protection, as that is required to protect actual security mechanism negotiation. Once the security mechanism has been selected and its use started, the previously exchanged list is replayed back to the network in a secure fashion. This enables the network to verify that the security mechanism selection was correct and that the security agreement was not tampered with. An example of an attack that would be possible without this feature is a "bidding-down attack", where an attacker forces peers into selecting a known weak security mechanism. The important benefit from having secure negotiation of the used security mechanism is that new mechanisms can be later added and old ones removed. The mechanisms can coexist nicely as each UE always uses the strongest mechanism it has available to it.

3.6.4.5 Confidentiality and integrity protection

In IMS access security, both confidentiality as well as data integrity and authentication are mandatory. The protocol used to provide them is IPsec ESP [RFC2406], explained in further detail in Chapter 18.

AKA session keys are used as keys for the ESP SAs. The IK is used as the authentication key and the CK as the encryption key.

3.6.4.6 Key management and distribution

As described in previous chapters, the P-CSCF may also reside in the visited network. By virtue of the AKA protocol, the shared secret is only accessible in the home network, which means that, while authentication needs to take place in the home network, certain delegation of responsibility needs to be assigned to the P-CSCF, as IPsec SAs exist between the P-CSCF and the UE. In practice, while IMS authentication takes place in the home network, the session keys that are produced in AKA authentication and used in ESP are delivered to the P-CSCF piggybacked on top of SIP registration messages.

To renew the SAs the network has to re-authenticate the UE. This means that the UE has to re-register as well, which may be either network-initiated or due to the registration expiring. The net effect is the same: the AKA protocol is run and fresh keys are delivered to the P-CSCF.

3.6.5 IMS access security for HTTP-based services

3.6.5.1 Introduction

Parallel to SIP traffic, there is a need for the UE to manage data associated with certain IMS applications. The Ut interface, as explained in Section 2.3.14, hosts the protocols needed for that functionality. Securing the Ut interface involves confidentiality and data integrity protection of HTTP-based traffic [RFC2616]. As previously mentioned, the authentication and key agreement for the Ut interface is also based on AKA.

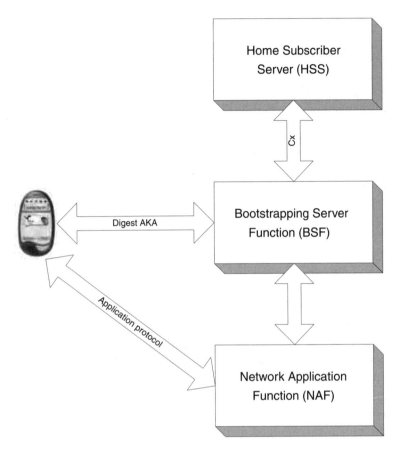

Figure 3.8 GBA.

3.6.5.2 The Generic Bootstrapping Architecture

As part of the Generic Authentication Architecture (GAA), the IMS defines the Generic Bootstrapping Architecture (GBA) [3GPP TS 33.220], illustrated in Figure 3.8. The Bootstrapping Server Function (BSF) and the UE perform mutual authentication based on AKA, allowing the UE to bootstrap session keys from the 3G infrastructure. Session keys are the result of AKA and enable further applications provided by a Network Application Function (NAF). One such example is a NAF that issues subscriber certificates[6] using an application protocol secured by the bootstrapped session keys.

---

[6] Such an entity is usually referred to as a Public Key Infrastructure (PKI) Portal.

3.6.5.3 Authentication and key management

Authentication in the Ut interface is performed by a specialized element, called the authentication proxy. In terms of the GBA the authentication proxy is another type of NAF. Traffic in the Ut interface goes through the authentication proxy and is secured using the bootstrapped session key.

3.6.5.4 Confidentiality and integrity protection

The Ut interface employs the Transport Layer Security (TLS) for both confidentiality and integrity protection [3GPP TS 33.222]. TLS is discussed more thoroughly in Chapter 14.

3.7 Discovering the IMS entry point

In order to communicate with the IMS, a UE has to know at least one IP address of the P-CSCF. The mechanism by which the UE retrieves these addresses is called P-CSCF discovery. Two mechanisms for P-CSCF discovery have been standardized in the Third Generation Partnership Project (3GPP): the Dynamic Host Configuration Protocol's (DHCP) domain name system (DNS) procedure and the GPRS procedure. Additionally, it is possible to configure either the P-CSCF name or the IP address of the P-CSCF in the UE.

In the GPRS procedure (Figure 3.9), the UE includes the P-CSCF address request flag in the PDP context activation request (or secondary PDP context activation request) and receives IP address(es) of the P-CSCF in the response. This information is transported in the protocol configuration options information element [3GPP TS 24.008]. The mechanism the Gateway GPRS Support Node (GGSN) used to get the IP address(es) of the P-CSCF(s) is not standardized. This mechanism does not work with pre-Release 5 GGSNs.

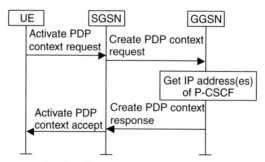

Figure 3.9 GPRS-specific mechanism for discovering P-CSCF.

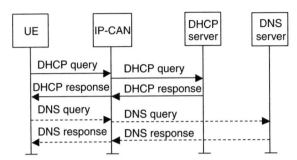

Figure 3.10 Generic mechanism for discovering P-CSCF.

In the DHCP DNS procedure (Figure 3.10), the UE sends a DHCP query to the IP connectivity access network (e.g., GPRS), which relays the request to a DHCP server. According to [RFC3319] and [RFC3315], the UE could request either a list of SIP server domain names of the P-CSCF(s) or a list of SIP server IPv6 addresses of the P-CSCF(s). When domain names are returned the UE needs to perform a DNS query (NAPTR/SRV) to find an IP address of the P-CSCF. The DHCP DNS mechanism is an access-independent way to discover the P-CSCF.

3.8 S-CSCF assignment

Section 3.7 explained how the UE discovers the IMS entry point, P-CSCF. The next entity on a session signalling path is the S-CSCF. There are three cases when S-CSCF is assigned:

- User registers in the network.

- Unregistered user receives a SIP request.

- Previously assigned S-CSCF is not responding.

3.8.1 S-CSCF assignment during registration

When a user is registering herself into a network the UE sends a REGISTER request to the discovered P-CSCF, which finds the user's home network entity, I-CSCF, as described in Section 3.2. Then the I-CSCF exchanges messages with the HSS (UAR and UAA) as described in Section 2.3.4. As a result the I-CSCF receives S-CSCF capabilities, as long as there is no previously assigned S-CSCF. Based on the received capabilities the I-CSCF selects a suitable S-CSCF.

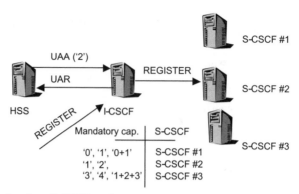

Figure 3.11 Example of an S-CSCF assignment.

Capability information is transferred between the HSS and the I-CSCF within the Server-Capabilities attribute value pair (AVP). The Server-Capabilities AVP contains [3GPP TS 29.228 and TS 29.229]:

- Mandatory-Capability AVP—the type of this AVP is unsigned and contains the mandatory capabilities of the S-CSCF. Each mandatory capability available in an individual operator's network will be allocated a unique value.

- Optional-Capability AVP—the type of this AVP is unsigned and contains the optional capabilities of the S-CSCF. Each optional capability available in an individual operator's network will be allocated a unique value.

- Server-Name AVP—this AVP contains a SIP URI used to identify a SIP server.

Based on the mandatory and optional capability AVPs, an operator is able to distribute users between S-CSCFs, depending on the different capabilities (required capabilities for user services, operator preference on a per-user basis, etc.) that each S-CSCF may have. It is the operator's responsibility to define (possibly based on the functionality offered by each S-CSCF installed in the network) the exact meaning of the mandatory and optional capabilities. As a first choice, the I-CSCF will select the S-CSCF that has all the mandatory and optional capabilities for the user. If that is not possible, then the I-CSCF applies a "best-fit" algorithm. None of the selection algorithms is standardized (i.e., solutions are implementation-dependent). Figure 3.11 shows one illustrative example.

Using the Server-Name AVP, an operator has the possibility to steer users to certain S-CSCFs; for example, having a dedicated S-CSCF for the same company/ group to implement a VPN service or just making S-CSCF assignment very simple.

3.8.2 S-CSCF assignment for an unregistered user

Section 3.3 and Figure 3.2 explained at a high level how a session is routed from UE *A* to UE *B*. It can be seen from the figure that the I-CSCF is a contact point within an operator's network. In Section 2.3.4 location retrieval procedures were explained (i.e., an incoming SIP request will trigger LIR/LIA commands to find out which S-CSCF is serving user *B*). If the HSS does not have knowledge of a previously assigned S-CSCF, then it returns S-CSCF capability information and the S-CSCF assignment procedure will take place in the I-CSCF as described in Section 3.8.1.

3.8.3 S-CSCF assignment in error cases

3GPP standards allow S-CSCF re-assignment during registration when the assigned S-CSCF is not responding: that is, when the I-CSCF realizes that it cannot reach the assigned S-CSCF it sends the UAR command to the HSS and explicitly sets the type of authorization information element to the value registration_and_capabilities. After receiving S-CSCF capabilities, the I-CSCF performs S-CSCF assignment as described in Section 3.8.1.

3.8.4 S-CSCF de-assignment

The S-CSCF is de-assigned when a user de-registers from the network or the network decides to de-register the user (e.g., because registration has timed out or the subscription has expired). It is the responsibility of the S-CSCF to clear the stored S-CSCF name from the HSS.

3.8.5 Maintaining S-CSCF assignment

When a user de-registers from the network or a registration timer expires in the S-CSCF an operator may decide to keep the same S-CSCF assigned for the unregistered user. It is the responsibility of the S-CSCF to inform the HSS that the user has been de-registered; however, the S-CSCF could indicate that it is willing to maintain the user profile. This optimizes the load of the Cx reference point because there is no need to transfer the user profile once the user registers again or receives sessions while she has a services-related unregistered state.

3.9 Mechanism for controlling bearer traffic

Separation of the control plane and the user plane was maybe one of the most important issues of IMS design. Full independence of the layers is not feasible

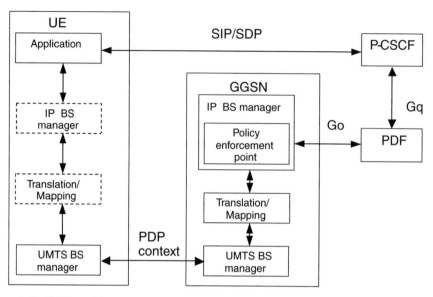

Figure 3.12 SBLP entities.

because, without interaction between the user plane and the control plane, operators are not able to control quality of service (QoS), source/destination of IMS media traffic and when the media starts and stops. Therefore, a mechanism to authorize and control the usage of the bearer traffic intended for IMS media traffic was created; it was based on the SDP parameters negotiated at the IMS session. This overall interaction between the GPRS and the IMS is called a service-based local policy (SBLP) control. Later on, charging correlation was specified as an additional capability. Figure 3.12 shows the functional entities involved in the SBLP. The figure shows a stand-alone policy decision function (PDF) and the Gq reference points that are being standardized in Release 6:

- IP bearer service (BS) manager—manages the IP BS using a standard IP mechanism. It resides in the GGSN and optionally in the UE.

- Translation/Mapping function—provides the inter-working between the mechanism and parameters used within the UMTS BS and those used within the IP BS. It resides in the GGSN and optionally in the UE.

- UMTS BS manager—handles resource reservation requests from the UE. It resides in the GGSN and in the UE.

- Policy enforcement point—is a logical entity that enforces policy decisions made by the PDF. It resides in the IP BS manager of the GGSN.

- Policy decision function—is a logical policy decision element that uses standard

IP mechanisms to implement SBLP in the IP media layer. In Release 5 it resides in the P-CSCF. In Release 6 it is a stand-alone entity. The PDF is effectively a policy decision point according to [RFC2753] that defines a framework for policy-based admission control.

3.9.1 SBLP functions

There are seven SBLP functions. These seven functions are described in the following subchapters:

- Bearer authorization.

- Approval of QoS commit.

- Removal of QoS commit.

- Indication of bearer release.

- Indication of bearer loss/recovery.

- Revoke authorization.

- Exchange of charging identifiers.

3.9.1.1 Bearer authorization

Session establishment and modification in the IMS involves an end-to-end message exchange using SIP and SDP. During the message exchange UEs negotiate a set of media characteristics (e.g., common codec(s)). If an operator applies the SBLP, then the P-CSCF will forward the relevant SDP information to the PDF together with an indication of the originator. The PDF notes and authorizes the IP flows of the chosen media components by mapping from SDP parameters to authorized IP QoS parameters for transfer to the GGSN via the Go interface.

When the UE is activating or modifying a PDP context for media it has to perform its own mapping from SDP parameters and application demands to some UMTS QoS parameters. PDP context activation or modification will also contain the received authorization token and flow identifiers as the binding information.

On receiving the PDP context activation or modification, the GGSN asks for authorization information from the PDF. The PDF compares the received binding information with the stored authorization information and returns an authorization decision. If the binding information is validated as correct, then the

PDF communicates the media authorization details in the decision to the GGSN. The media authorization details contain IP QoS parameters and packet classifiers related to the PDP context.

The GGSN maps the authorized IP QoS parameters to authorized UTMS QoS parameters and finally the GGSN compares the UMTS QoS parameters against the authorized UMTS QoS parameters of the PDP context. If the UMTS QoS parameters from the PDP context request lie within the limits authorized by the PDF, then the PDP context activation or modification will be accepted. Figure 3.13 shows the explained functionality and the PDF is shown as a part of the P-CSCF for simplicity.

From the above we can find two different phases: authorize QoS resources (steps 1–6) and resource reservation (steps 7–14). Next we take a deeper look at both steps and then the final step of bearer authorization, approval of QoS commit, is described.

Authorize QoS resources

Steps 2 and 5 in Figure 3.13 correspond to authorization of the QoS resources procedure. During the session set-up the PDF collects IP QoS authorization data. These data comprise:

- Flow identifier—used to identify the IP flows that are described within a media component associated with a SIP session. A flow identifier consists of two parts: (1) the ordinal number of the position of the "m =" lines in the SDP session description and (2) the ordinal number of the IP flow(s) within the "m =" line assigned (in order of increasing port numbers).

- Data rate—this information is derived from SDP bandwidth parameters. The data rate will include all the overheads coming from the IP layer and the layers above (e.g., UDP, RTP or RTCP). If multiple codecs per media are agreed to be used in a session, then the authorized data rate is set according to the codec requiring the highest bandwidth.

- QoS class—the QoS class information represents the highest class that can be used for the media component. It is derived from the SDP media description.

Let's say that Tobias (UE #1 in Figure 3.13) wants to talk to his sister Theresa (UE #2). In addition to an ordinary voice call, Tobias wants to activate bidirectional and unidirectional video streams. Therefore, his terminal builds a SIP INVITE containing an SDP that reflects Tobias's preferences and his UE capabilities. SDP contains supported codecs, bandwidth requirements (plus characteristics of each) and assigned local port numbers for each possible media flow. Here we concentrate only on those parameters that are necessary for the SBLP. Chapter 6 contains a

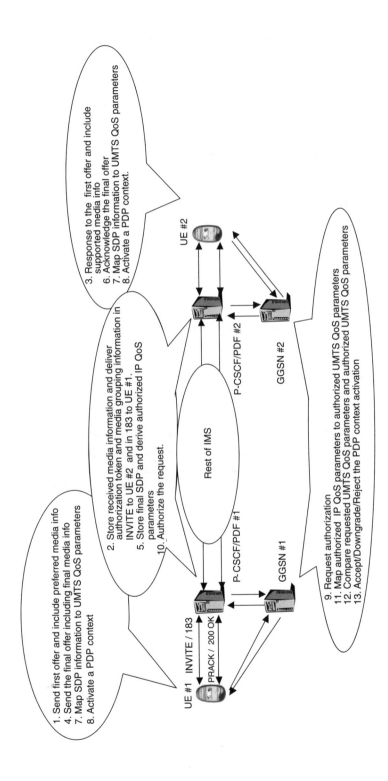

1. Send first offer and include preferred media info
4. Send the final offer including final media info
7. Map SDP information to UMTS QoS parameters
8. Activate a PDP context

2. Store received media information and deliver authorization token and media grouping information in INVITE to UE #2 and in 183 to UE #1.
5. Store final SDP and derive authorized IP QoS parameters
10. Authorize the request.

3. Response to the first offer and include supported media info
6. Acknowledge the final offer
7. Map SDP information to UMTS QoS parameters.
8. Activate a PDP context.

9. Request authorization
11. Map authorized IP QoS parameters to authorized UMTS QoS parameters
12. Compare requested UMTS QoS parameters and authorized UMTS QoS parameters
13. Accept/Downgrade/Reject the PDP context activation

UE #2

P-CSCF/PDF #2

GGSN #2

Rest of IMS

P-CSCF/PDF #1

GGSN #1

UE #1 INVITE / 183

PRACK / 200 OK

Figure 3.13 Bearer authorization using SBLP.

general description for the whole session set-up. SDP sent from UE #1 would look like this:

```
v=0
o=- 3262464865 3262464868 IN IP6 5555::1:2:3:
t=3262377600 3262809600
m=video 50230 RTP/AVP 31
c=IN IP6 5555::1:2:3:4
b=AS:35
b=RS:700
b=RR:700
m=video 50240 RTP/AVP 31
c=IN IP6 5555::1:2:3:4
b=AS:32
b=RS:640
b=RR:640
a=sendonly
m=audio 3456 RTP/AVP 97 96
c=IN IP6 5555::1:2:3:4
b=AS:25.4
b=RS:500
b=RR:500
```

When PDF #1 in Figure 3.13 receives this, it is able to formulate the authorization data for the uplink direction (from UE #1 to GGSN #1). When Theresa's UE responds, PDF #1 is able to formulate the authorization data for the downlink direction (from GGSN #1 to UE #1). Note that Theresa is not willing to receive unidirectional video, therefore the corresponding port number is set to zero:

```
v=0
o=- 3262464865 3262464868 IN IP6 5555::1:2:3:4
t=3262377600 3262809600
m=video 60230 RTP/AVP 31
c=IN IP6 5555::5:6:7:8
b=AS:35
b=RS:700
b=RR:700
m=video 0 RTP/AVP 31
c=IN IP6 5555::5:6:7:8
b=AS:32
b=RS:640
b=RR:640
```

```
a=recvonly
m=audio 3550 RTP/AVP 0
c=IN IP6 5555::5:6:7:8
b=AS:25.4
b=RS:500
b=RR:500
```

From this information, PDF #1 and PDF #2 are able to construct the necessary flow identifiers. Table 3.4 shows the flow identifiers in PDF #1.

Data rates PDF derives the data rate value for the media IP flow(s) from the "b = AS" SDP parameter. For possibly associated Real-time Transport Control Protocol (RTCP) IP flows, the PDF will use SDP "b = AS", "b = RR" and "b = RS" parameters, if present. When SDP "b = RR" or "b = RS" are missing the data rate for RTCP IP flows is derived from the available parameters as described in [3GPP TS 29.208]. Description of RTCP bandwidth:

- If b = RS and b = RR exist, then the RTCP bandwidth for UL and DL = (bRS + bRR)/1,000].

- If either b = RS or b = RR is missing, then the RTCP bandwidth for UL and DL = MAX[0.05 * bAS, bRS/1,000 or bRR/1,000].

- If both b = RS and b = RR are missing, then the RTCP bandwidth for UL and DL = 0.05 * bAS.

Table 3.5 shows maximum data rates per flow identifier as calculated in PDF #1.

QoS class The PDF maps media-type information into the highest QoS class that can be used for the media. The PDF will use an equal QoS class for both the uplink

Table 3.4 Flow identifier information in PDF #1.

Order of "*m*" line	Type of IP flow	Destination IP address	Port number of the IP flows	Flow identifier
1	RTP (video) UL	5555::1:2:3:4	50230	<1, 1>
1	RTP (video) DL	5555::5:6:7:8	60230	<1, 1>
1	RTCP UL	5555::1:2:3:4	50231	<1, 2>
1	RTCP DL	5555::5:6:7:8	60231	<1, 2>
3	RTP (audio) DL	5555::5:6:7:8	3550	<3, 1>
3	RTP (audio) UL	5555::1:2:3:4	3456	<3, 1>
3	RTCP DL	5555::5:6:7:8	3551	<3, 2>
3	RTCP UL	5555::1:2:3:4	3457	<3, 2>

Table 3.5 The maximum data rates per media type.

Media type (m-line in the SDP)	Maximum authorized QoS class
Bidirectional audio or video	A
Undirectional audio or video	B
Application	A
Data	E
Control	C
Others	F

Table 3.6 The maximum data rates and QoS class per flow identifier in PDF #1.

	Flow Identifier			
	< 1, 1 >	< 1, 2 >	< 3, 1 >	< 3, 2 >
Maximum data rate downlink (kbps)	35	0.7	25.4	0.5
Maximum data rate uplink (kbps)	35	0.7	25.4	0.5
Maximum QoS class	A	A	A	A

and the downlink directions when both directions are used [3GPP TS 29.207]. [3GPP TS 29.208] contains detailed derivation rules (a summary is presented in Table 3.5). Table 3.6 shows how the information in Table 3.5 is utilized in our example. The maximum authorized QoS class for a RTCP IP flow is the same as for the corresponding RTP media IP flow.

The authorized IP QoS was created and stored in the PDFs during steps 2 and 5 in Figure 3.13 (bearer authorization). The authorized IP QoS comprises the QoS class and data rate. At the same time the PDFs created the flow identifiers that will be used to create packet classifiers in the GGSNs later on. Table 3.6 summarizes things so far.

Authorization token   In Figure 3.13, step 2 states: "deliver authorization token to UE #1 and UE #2." But, what is an authorization token?:

- It is a unique identifier across all PDP contexts associated with an access point name.

- It is created in the PDF when the authorization data are created.

- It consists of the IMS session identifier and the PDF identifier.

- Its syntax conforms to [RFC3520].

- It is delivered to the UE by means of [RFC3313].

- The UE includes it in a PDP context activation/modification request.

- GGSN uses a PDF identifier within the authorization token to find the PDF that holds the authorized IP QoS information.

- The PDF uses the authorization token to find the right authorized data when receiving requests from the GGSN.

Media grouping SIP and the IMS allow multimedia sessions to be set up which may comprise a number of different components, such as audio and video. Any participating party may add or drop a media component from an ongoing session. As stated in Section 2.1.6, all components should be individually identifiable for charging purposes, and it must be possible to charge for each of these components separately in a session.

Unfortunately, Release 5 GGSN is able to produce only one GGSN call detail records (CDR) for a PDP context. Therefore, it is impossible to separate traffic for each media component within the same PDP context. As the current model for charging data generation and correlation does not allow multiplexing media flows in the same secondary PDP context, there must be a mechanism on the IMS level to force the UE to open separate PDP contexts for each media component. For this purpose a keep-it-separate indication was defined.

When the P-CSCF receives an initial INVITE request for a terminating session set-up or a 183 (Session Progress) response to an INVITE request for an originating session set-up, the P-CSCF may modify SDP according to [RFC3524] to indicate to the UE that a particular media stream(s) is grouped according to a local policy [3GPP TS 24.229].

[RFC3524] defines the single reservation flow (SRF) group type (a = group:SRF). SRF groups are used in the following way:

- If a network wants to carry particular medias in the same PDP context, then the P-CSCF sets the same SRF value for these media components.

- If a network wants to carry particular medias in different PDP contexts, then the P-CSCF sets different SRF values for each media component.

- If a network does not set the SRF indication, then the UE is free to group media streams as it likes.

The following further restrictions and guidelines are given in standards [3GPP TS 23.228] and [3GPP TS 24.229]:

- The P-CSCF will apply and maintain the same policy throughout the complete SIP session.

- If a media stream is added and grouping of media stream(s) was indicated in the initial INVITE or 183 (Session Progress) response, P-CSCF will modify SDP according to [RFC3524] to indicate to the UE that the added media stream(s) will be grouped into either a new group or into one of the existing groups.

- The P-CSCF will not apply [RFC3524] to SDP for additional media stream(s), if grouping of media stream(s) was not indicated in the initial INVITE request or 183 (Session Progress) response.

- The P-CSCF will not indicate re-grouping of media stream(s) within SDP.

- All associated IP flows (e.g., RTP/RTCP) used by the UE to support a single media component are assumed to be carried within the same PDP context.

- It is assumed that media components from different IMS sessions are not carried within the same PDP context.

There is ongoing work in Release 6 to introduce a capability to charge on an IP flow basis. This would allow more freedom to transport media components in the same PDP context.

Following our example, P-CSCF #1 forces a separate PDP context for all media types in the 183 (Session Progress) response as follows (183 toward UE #1):

```
v=0
o=- 3262464865 3262464868 IN IP6 5555::1:2:3:4
t=3262377600 3262809600
a=group:SRF 1
a=group:SRF 2
a=group:SRF 3
m=video 60230 RTP/AVP 31
a=mid: 1
c=IN IP6 5555::5:6:7:8
b=AS:35
b=RS:700
b=RR:700
m=video 0 RTP/AVP 31
a=mid: 2
c=IN IP6 5555::5:6:7:8
b=AS:32
b=RS:640
b=RR:640
a=recvonly
m=audio 3550 RTP/AVP 0
a=mid: 3
```

```
c=IN IP6 5555::5:6:7:8
b=AS:25.4
b=RS:500
b=RR:500
```

Forking issues   When the P-CSCF receives a forked response it will pass the information listed in Section 2.3.18 to the PDF. As the PDF receives the forking indication it also assigns the previously allocated authorization token to the forked response. Additionally, the PDF authorizes any additional media components and any increased QoS requirements for previously authorized media components, as requested by the forked response. Thus, the QoS authorized for a media component equals the highest QoS requested for that media component by any of the forked responses [3GPP TS 29.207]. This solution may be changed because it causes traffic over the Gq and Go reference point when the forked response is received by the P-CSCF and requires special handling in the PDF when the final answer is received. It may be desirable that the P-CSCF hides the forking from the PDF.

Resource reservation

UE functions   When the UE receives an authorization token within the end-to-end message exchange it knows that the SBLP is applied in the network. Therefore, it has to generate the requested QoS parameters and flow identifiers for a PDP context activation (modification) request. The requested QoS parameters include the information listed in Table 3.7.

From the SBLP point of view the first three rows in Table 3.7 presents values that are interesting. Interested readers can find detailed descriptions of other QoS parameters in [3GPP TS 23.107]. Here the traffic class, guaranteed bit rate and maximum bit rate are described:

- Traffic class—the four different traffic classes defined for UMTS are conversational, streaming, interactive and background. By including the traffic class,

Table 3.7   Requested QoS parameters.

Traffic class	Maximum bit rate for downlink
Guaranteed bit rate for downlink	Maximum bit rate for uplink
Guaranteed bit rate for uplink	Maximum SDU size
SDU format information	Residual BER
SDU error ratio	Traffic-handling priority
Delivery of erroneous SDUs	Allocation/Retention priority
Transfer delay	Delivery order
Source statistics descriptor	

UMTS can make assumptions about the traffic source and optimize the transport for that traffic type.

● Guaranteed bit rate (GBR)—describes the bit rate the UMTS bearer service will guarantee to the user or application.

● Maximum bit rate (MBR)—describes the upper limit a user or application can accept or provide. This allows different rates to be used for operation (e.g., between GBR and MBR).

Table 3.8 The maximum authorized traffic class per media type in the UE.

Media type (m-line in SDP)	UMTS traffic class
Bidirectional audio or video	Conversational
Unidirectional audio or video	Streaming
Application	Conversational
Data	Interactive
Control	Interactive
Others	Background

The traffic class values, GBR for downlink/uplink and MBR for downlink/uplink should not exceed the derived values of maximum authorized bandwidth and maximum authorized traffic class per flow identifier. The maximum authorized bandwidth in the UE is derived from SDP in the same way as was done in the PDF. The maximum authorized traffic class is derived according to Table 3.8. The exact derivation rules for both parameters are described in [3GPP TS 29.208].

[3GPP TS 26.236] gives recommendations on how other requested QoS parameters for conversational codec applications could be set. Correspondingly, [3GPP TS 26.234] gives recommendations on how other requested QoS parameters for streaming codec applications could be set.

Flow identifiers are derived in the UE same manner as in the PDF. Table 3.9 shows the maximum authorized UMTS QoS parameters per flow identifier as calculated by the UE.

Next, the UE needs to decide how many PDP contexts are needed. The key factors are the nature of media streams (i.e., required traffic class) and the received grouping indication from the P-CSCF. In our example there are two different types of bidirectional media: video and audio. Both media would require high QoS (low delay and preserved time relation); therefore, a single conversational traffic class PDP context would be suitable. However, the P-CSCF has indicated that a separate PDP context is required for each IMS media component. Therefore, UE #1 should activate two different PDP contexts. Otherwise, the PDP context activation would

Table 3.9 The values of the maximum authorized UMTS QoS parameters per flow identifier as calculated by UE #1 (Tobias) from the example.

	Flow identifier			
	< 1, 1 >	< 1, 2 >	< 3, 1 >	< 3, 2 >
Maximum data rate downlink (kbps)	35	0.7	25.4	0.5
Maximum data rate uplink (kbps)	35	0.7	25.4	0.5
Maximum QoS class	Conversational	Conversational	Conversational	Conversational

Table 3.10 The values of the maximum authorized UMTS QoS parameters per PDP context as calculated by UE #1 from the example.

	PDP context #	
	1	2
Maximum authorized bandwidth DL (kbps)	35.7	25.9
Maximum authorized bandwidth UL (kbps)	35.7	25.9
Maximum authorized traffic class	Conversational	Conversational

fail due to the SBLP decision enforced by the PDF. Finally, Table 3.10 presents the maximum authorized UMTS QoS parameters per PDP context as calculated by UE #1.

The UE has now completed step 7 in Figure 3.13. After deriving and choosing the suitable, requested QoS parameters, the UE activates the necessary PDP contexts. The authorization token and flow identifiers are inserted within the traffic flow template information element. A detailed description of how the authorization token and flow identifiers are carried in the traffic flow template information element is provided in [3GPP TS 24.008]. The requested QoS parameters are inserted within the QoS information element. A detailed description of how the requested QoS parameters are carried in the QoS information element is provided in [3GPP TS 24.008].

GGSN functions When a GGSN receives a secondary PDP context activation request to an access point name for which the Go reference point is enabled, GGSN:

• Identifies the correct PDF by extracting the PDF identify from the provided authorization token. If an authorization token is missing, then the GGSN may either reject the request or accept it within the limit imposed by a locally stored QoS policy.

- Requests authorization information from the PDF for the IP flows carried by a PDP context. This request is a Common Open Policy Service (COPS) request and contains the provided authorization token and the provided flow identifiers. The exact content of the request is described in Chapter 17.

- Enforces the decision after receiving an authorization decision. The authorization decision is given as a COPS authorization_decision message. The exact content of the decision is described in Chapter 17. The main components of the decision are:

  o Direction indication—uplink, downlink.

  o Authorized IP QoS—data rate, maximum authorized QoS class.

  o Packet classifiers (also called a gate description)—source IP address and port number(s), destination IP address and port number(s), protocol ID.

- Maps the authorized IP QoS to the authorized UTMS QoS.

- Compares the requested QoS parameters with the authorized UTMS QoS. If all the requested parameters lie within the limits, then the PDP context activation will be accepted. In other words, if the following criteria are fulfilled [3GPP TS 29.208]:

  o The requested GBR DL/UL (if the requested traffic class is conversational or streaming) or MBR DL/UL (if the requested traffic class is Interactive or background) is less than or equal to the maximum authorized data rate DL/UL.

  o And the requested traffic class is less than or equal to the maximum authorized traffic class.

  If the requested QoS exceeds the authorized UTMS QoS, then the requested UMTS QoS information is downgraded to the authorized UMTS QoS information.

- Constructs a gate description based on the received packet classifier. The gate description allows a gate function to be performed. The gate function enables or disables the forwarding of IP packets. If the gate is closed, then all packets of the related IP flows are dropped. If the gate is open, then the packets of the related IP flows are allowed to be forwarded. The opening of the gate may be part of the authorization decision event or may be a stand-alone decision as described in Section 3.9.1.2. The closing of the gate may be part of the revoke authorization decision.

- Stores the binding information.

- May cache the policy decision data of the PDF decisions.

During the secondary PDP context modification the GGSN may use previously cached information for a local policy decision in case the modification request does not exceed the previously authorized QoS. If the GGSN does not have cached information, then it performs above described functions. There is one exception: if the GGSN receives a secondary PDP context modification request to an access point name for which the Go interface is enabled and no binding information is received, then the GGSN rejects the secondary PDP context modification as long as binding information has been previously provided for the PDP context.

PDF functions   When a PDF receives a COPS request the PDF validates that:

- The authorization token is valid.

- The corresponding SIP session exists.

- The binding information contains valid flow identifier(s).

- The authorization token has not changed in an authorization modification request.

- The UE follows the grouping indication.

- If validation is successful, then the PDF will determine and communicate the authorized IP QoS, packet classifiers and the gate status to be applied to the GGSN. When valid binding information consists of more than one flow identifier, the information sent back to the GGSN will include the aggregated QoS for all the IP flows and suitable packet filter(s) for these IP flows.

In our example, UE #1 needs to activate two PDP contexts. When a secondary PDP context activation for the first PDP context (bidirectional video) arrives at GGSN #1, which extracts the authorization token and flow identifiers (1, 1 and 1, 2) from the traffic flow template and sends them to PDF #1. PDF #1 uses the authorization token to identify the corresponding IMS session. PDF #1 verifies that the session exists and returns the authorized IP QoS parameters and packet classifiers corresponding to the flow identifiers (1, 1 and 1, 2). GGSN #1 maps the authorized IP QoS to the authorized UMTS QoS; it compares the values and realizes that everything is OK. Finally, GGSN #1 accepts the request and installs the gate, based on the received packet classifiers. The same is applied to other, related PDP contexts.

Additionally, the PDF is able to send a new stand-alone decision to the GGSN when it receives modified SDP information from the P-CSCF. This may be needed, for example, in the case of forking.

3.9.1.2 Approval of the QoS commit function

During the resource reservation procedure a PDF sends packet classifiers to a GGSN. Based on the packet classifiers, the GGSN formulates a gate to policy-control incoming and outgoing traffic. It is the PDF's decision when to open the gate. When the gate is open, the GGSN allows traffic to pass through the GGSN. Opening the gate could be sent as a response to an initial authorization request from the GGSN or the decision can be sent as a stand-alone decision. With a stand-alone decision an operator can ensure that user-plane resources are not used before the IMS session is finally accepted (i.e., when a SIP 200 OK message is received).

3.9.1.3 Removal of the QoS commit function

This function closes a gate in the GGSN when a PDF does not allow traffic to traverse through the GGSN. This function is used, for example, when a media component of a session is put on hold due to media re-negotiation.

3.9.1.4 Indication of bearer release function

When the GGSN receives a delete PDP context request and the PDP context has been previously authorized via the Go reference point, the GGSN informs the PDF of the bearer release related to the SIP session by sending a COPS delete request-state message. The PDF removes the authorization for the corresponding media component(s).

When the PDF receives a report that a bearer has been released, it could request the P-CSCF to release the session(s) and revoke all the related media authorization with the procedure described in Section 3.9.1.6.

3.9.1.5 Indication of bearer loss/recovery

When the MBR value equals 0 kbit/s in an update PDP context request, the GGSN needs to send a COPS report message to the PDF. Similarly, when the MBR is modified from 0 kbit/s, the GGSN sends a COPS report message to the PDF after receiving an update from the serving GPRS Support Node (SGSN).

Using this mechanism the IMS is able to learn that the UE has lost/recovered its radio bearer(s) when a streaming or conversational traffic class is in use in the GPRS system. [3GPP TS 23.060] states that the SGSN needs to send an update PDP context request to the GGSN when the radio network controller (RNC) informs the SGSN about Iu release or radio access bearer release.

When the PDF receives a report that the MBR equals 0 kbit/s, it could request the P-CSCF to release the session(s) and revoke all the related media authorization with the procedure described in Section 3.9.1.6.

3.9.1.6 Revoke function

This function is used to force the release of previously authorized bearer resources in a GPRS network. With this mechanism the PDF is able, for example, to ensure that the UE releases a PDP context when a SIP session is ended or that the UE modifies the PDP context when a media component bound to a PDP context is removed from the session. If the UE fails to do so within a predefined time set by an operator, then PDF revokes the resources.

3.9.1.7 Charging identifiers exchange function

The Go reference point is the link between the IMS and the GPRS networks. For charging correlation to be carried out as described in Section 3.10.2, the IMS layer needs to know the corresponding GPRS layer charging identifier and vice versa. These charging identifiers are exchanged during the bearer authorization phase. An IMS charging identifier is delivered to the GGSN within the authorization_ decision message, while a GPRS charging identifier is delivered to the PDF as part of the authorization report.

3.10 Charging

This section explains the charging architecture for offline and online charging and describes how GPRS charging information is correlated with IMS charging informa-tion. The description here is based on Release 5 charging principles.

3.10.1 Charging architecture

The IMS architecture supports both online and offline charging capabilities. Online charging is a charging process where IMS entities, such as an application server (AS), interact with the online charging system. The online charging system in turn interacts in real time with the user's account and controls or monitors the charges related to service usage: for example, the AS queries the online charging system prior to allowing session establishment or it receives information about how long a user can participate in a conference. Offline charging is a charging process where charging information is mainly collected after the session and the charging system does not affect in real time the service being used. In this model a user typically receives a bill on a monthly basis, which shows the chargeable items during a particular period. Due to the different nature of charging models different architecture solutions are required.

3.10.1.1 Offline charging architecture

The central point in the offline charging architecture is the Charging Collection Function (CCF). The CCF receives accounting information from IMS entities via the Rf reference point. It further processes the received data and then constructs and formats the actual CDR. The CDR is passed to the billing system, which takes care of providing the final CDR, taking into account information received from other sources as well (e.g., Charging Gateway Function, or CGF). Figure 3.14 depicts the offline charging architecture in a case where both the calling party and the called party are using IMS roaming. When the user is not roaming there will be only one CCF involved.

Charging Collection Function

The usage of the CCF enables an operator to have a single reference point toward the billing system, as the CCF transfers charging information from IMS entities

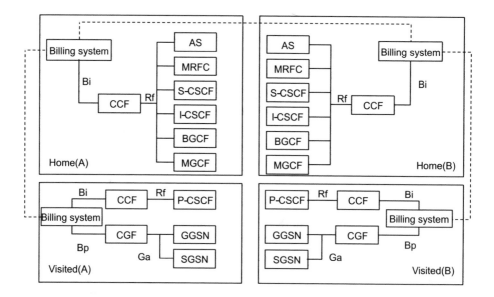

Figure 3.14 IMS offline charging architecture.

(AS, MRFC, S-CSCF, I-CSCF, P-CSCF, BGCF, MGCF) to the network operator's chosen billing system(s). The main functions of the CCF are:

- To collect accounting information from the IMS entities and generate accounting information.

- To correlate, consolidate, filter unnecessary fields and add operator-specific information to the received account information.

- To create CDRs after pre-processing.

- To transfer CDRs to the billing system.

- To buffer the CDR when the billing system is busy [3GPP TS 32.200].

The CCF can be implemented as a centralized, separate network element or as an integrated functionality resident in the IMS entities. Having a stand-alone CCF reduces the load of the actual IMS entity because it does not need to buffer and render the actual CDRs.

Charging Gateway Function

The CGF within the PS domain provides a mechanism to transfer charging information from SGSN and GGSN nodes to the network operator's chosen billing systems. The CGF's main functionalities for the PS domain are in principle equivalent to the CCFs that are used in the IMS domain [3GPP TS 32.200]. One difference is that the CGF receives valid CDRs from the SGSN and the GGSN.

Billing system

The CCF and CGF send CDRs to the billing system that creates the actual bill (e.g., sent to a subscriber on a monthly basis). The bill could contain, for example, the number of sessions, destinations, duration and type of sessions (audio, video).

Rf reference point

The IMS session traverses through various IMS entities and all entities that perform SIP session control are able to generate offline charging information. The charging information is sent from the IMS entities to the CCF using Diameter accounting requests (ACRs) via the Rf interface. SIP signalling relates either to IMS sessions or IMS events. IMS session-related ACRs are called start, interim and stop and are sent at the start, during and at the end of a session, as the name implies. Non-session-related ACRs are called event ACRs. Event ACRs cause the CCF to generate corresponding CDRs, while session ACRs cause the CCF to open, update and

Table 3.11 Offline charging messages reference table.

Command-Name	Purpose	Abbreviation	Source	Destination
Accounting-Request	ACR is used to report/ stop accounting information to the CCF	ACR	S-CSCF, I-CSCF, P-CSCF, MRFC, MGCF, BGCF, AS	CCF
Accounting-Answer	ACA is used to acknowledge the ACR and report the result	ACA	CCF	S-CSCF, I-CSCF, P-CSCF, MRFC, MGCF, BGCF, AS

close corresponding CDRs. The CCF also has timers for closing partial session CDRs. All the elements apart from the I-CSCF send session ACRs and all the elements apart from the MRFC send event ACRs.

It is the operator's choice which SIP method or ISDN User Part (ISUP) message triggers the ACR. However, two mandatory items have been defined:

- Whenever SIP 200 OK, acknowledging an initial SIP INVITE, is received or MGCF receives an ISUP answer, ACR start will be sent to the CCF.

- Whenever SIP BYE is received or MGCF receives an ISUP release, ACR stop will be sent to the CCF.

Table 3.11 shows the use of these messages for offline charging. The Accounting-Request command contains suitable DIAMETER protocol AVPs and 3GPP DIAMETER accounting AVPs. The use of AVPs is specified per IMS entity and ACR type: for example, ACRs generated by the S-CSCF could contain information about the contacted AS and ACRs generated by the P-CSCF and could contain authorized QoS information. The Accounting-Answer command contains suitable DIAMETER-based protocol AVPs (see [3GPP TS 32.225] for the detailed coding of different commands).

Bi reference point

The CCF uses the Bi reference point to transfer the created CDRs to the billing system. Because there is a lot of variation among existing billing systems, the 3GPP has not specified any particular protocol for the Bi reference point. However, the 3GPP has set a minimum requirement that all implementations support a file-based bulk interface for the transfer of CDRs from the CCF to the billing system: the recommendation is FTP over TCP/IP [3GPP TS 32.225]. The previous paragraph stated that the use of AVPs is specified per IMS entity; hence, the CCF will send different CDRs to the billing system through the Bi reference point. The following CDR types exist:

- S-CSCF—CDR generated based on information from the S-CSCF.

- I-CSCF—CDR generated based on information from the I-CSCF.

- P-CSCF—CDR generated based on information from the P-CSCF.

- BGCF—CDR generated based on information from BGCF.

- MGCF—CDR generated based on information from MGCF.

- MRFC—CDR generated based on information from the MRFC.

- AS—CDR generated based on information from the AS.

See [3GPP TS 32.225] for the detailed coding of different CDRs.

3.10.1.2 Online charging architecture

The S-CSCF, AS and MRFC are the IMS entities that are able to perform online charging. The AS and MRFC use the Ro reference point, while the S-CSCF uses the IMS Service Control (ISC) reference point for communicating with the Online Charging System (OCS). Figure 3.15 shows the online charging architecture.

Event Charging Function

When the UE requests something from either the AS or the MRFC that requires charging authorization, the AS or the MRFC contacts the Event Charging Function

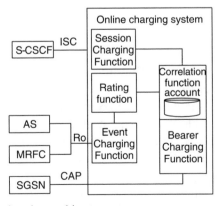

Figure 3.15 IMS online charging architecture.

(ECF) through the Ro reference point before delivering the service to the user: for example, the user could send a SUBSCRIBE request to a news server asking for the latest betting odds or asking for a voice conference to be set up. The ECF supports two different authorization models: immediate event charging and event charging with unit reservation.

In the immediate event charging model the ECF uses the rating function to find the appropriate tariff for an event. After resolving the tariff and the price, the ECF deducts a suitable amount of money from the user's account and grants the ACRs from the AS or the MRFC. When using this model the AS or the MRFC should know that it could deliver the requested service to the user itself. For example, the AS could send an ACR and inform the ECF of the service (say, a game of chess) and the number of items (say, 2) to be delivered. Then the ECF uses the rating function to resolve the tariff (€0.3) and to calculate the price based on the number of delivered units (€0.6). Finally, €0.6 is deduced from the user's account and the ECF informs the AS that 2 units have been granted within the Accounting-Answer (ACA).

In the event charging with unit reservation model the ECF uses the rating function to determine the price of the desired service according to service-specific information, if the cost was not given in the ACR. Then the ECF reserves a suitable amount of money from the user's account and returns the corresponding amount of resources to the AS or the MRFC. The amount of resources could be time or allowed data volume. When resources granted to the user have been consumed or the service has been successfully delivered or terminated, the AS or the MRFC informs the ECF of the amount of resources consumed. Finally, the ECF deducts the used amount from the user's account [3GPP TS 32.200, TS 32.225], but may require further interaction with the rating function. This model is suitable when the AS or the MRFC is not able to determine beforehand whether the service could be delivered or when the required amount of resources are not known prior to the use of a specific service (e.g., duration of the conference).

Session Charging Function

The Session Charging Function (SCF) is intended to perform charging according to session resource usage, based on received requests from the S-CSCF via the ISC reference point. The SCF should be able to control session establishment by allowing or denying a session establishment request after checking the user's account. In addition, the SCF should be able to terminate an existing session when, say, the user's account is empty. The SCF supports the event charging with unit reservation model for event charging.

The current design imposes severe problems. It would mean, for instance, that the SCF should support the SIP protocol stack, act as an AS, maintain the call-state model and perform budget control for IMS sessions. Having all these functions as part of an online charging system would overload the system and would lead to an incoherent online charging architecture. In practice, there are two options to resolve this problem: either extending the ISC reference point or choosing another suitable reference point. It is expected that the reference point toward the SCF will be changed in Release 6 to the Ro reference point. It may lead to some kind of gateway or interworking function being introduced between the S-CSCF and the SCF.

Bearer Charging Function

The SGSN uses the CAMEL Application Part (CAP)-based reference point for requesting permission for bearer usage from the Bearer Charging Function (BCF). The BCF controls bearer usage (e.g., in terms of time or traffic volume). The BCF interacts with the rating function and the user's account. In Release 6 the BCF functions need to be extended to cover Wireless Local Area Network (WLAN) and IP-flow based charging requests from the GGSN.

Rating function

The rating function performs unit, price and tariff determination. In a unit determination process the rating function calculates the number of session-related non-monetary units (e.g., service units, data volume, time), based on the requested service. Tariff determination means calculation of network utilization charges for the use of a particular service: for example, an ordinary IMS session tariff could be €0.1 per minute. Price determination is used to calculate the price of a given number of non-monetary units. The price is used for account balance updates (debit/credit).

It is possible to execute the rating function before and/or after service consumption.

Correlation function

As seen in Figure 3.15, there are multiple sources that are able to produce charging data regarding a single IMS session. If an operator wants to correlate information coming from different sources (ECF, SCF and BCF), it needs to ensure that unique charging identifiers are assigned to each chargeable event. The correlation function is the entity that links different CDRs, based on the charging identifiers. The following sections show what identities are used and how this information is distributed in the network.

Table 3.12 Online charging messages reference table.

Command-Name	Purpose	Abbreviation	Source	Destination
Accounting-Request	ACR is used to report/stop accounting information to the CCF	ACR	MRFC, AS	ECF
Accounting-Answer	ACA is used to acknowledge the ACR and report the result	ACA	ECF	MRFC, AS

Ro reference point

The AS and the MRFC use ACR and ACA messages of the base DIAMETER protocol for sending online accounting information through the Ro reference point to the ECF, as for online charging. As the messages and protocol are the same for offline and online charging, the AS and the MRFC should be able to distinguish whether to apply online or offline charging. This decision could be based on information provided by an operator or information received in SIP signalling (CCF and/or ECF address). The architecture allows the use of both online and offline charging reference points simultaneously [3GPP TS 32.225].

When the AS or the MRFC applies the immediate event charging model, an ACR-type event is used to report the accounting information to the ECF. On the contrary, when the event charging with unit reservation model is applied, the ACR types start, stop and interim are used. The ACR types start, interim and stop are used for accounting data related to successful SIP sessions. Event accounting data are used for session-unrelated accounting data, such as simple registration or interrogation, and for accounting data related to unsuccessful SIP session establishment attempts [3GPP TS 32.225]. Table 3.12 summarizes the online charging messages over the Ro reference point.

Compared with the messages used in offline charging, online charging messages include additional Diameter credit control AVPs. Unfortunately, the 3GPP Release 5 online charging solution refers to an outdated version of the Diameter Credit-Control Application Internet draft [Draft-hakala-diameter-credit-control] that was used for accounting and thus had accounting commands. The 3GPP community was not able to reach consensus to align the work with the latest version of the Diameter Credit-Control Application during 2003. Version 01 of the Diameter Credit-Control Application draft also takes into account requirements coming from the IETF side, and it is very likely that it will reach RFC status in the first half of 2004 [Draft-ietf-aaa-diameter-cc]. The Diameter Credit-Control application is an authorization application and it no longer uses accounting. However, it keeps the basic credit control mechanism as well as the model used to send Credit-Control-

Requests to OCS unchanged. Therefore, there should be no obstacles to correcting this issue in 3GPP.

3.10.2 Charging information correlation

Due to the layering design, IMS entities are not aware of user-plane traffic volumes related to IMS sessions and IP connectivity network entities (e.g., SGSN and GGSN) are not aware of the status of control-plane signalling (i.e., status of IMS sessions). From the operator's perspective, it is desirable to have a possibility to correlate charging information created at the user plane and the control plane. Exchanging charging identifiers—the IMS charging identifier (ICID) and the GPRS charging identifier (GCID)—through the Go reference point enables charging correlation between the IMS and the GPRS networks.

During a session establishment phase the UE activates the necessary secondary PDP context(s). During the PDP context authorization process the GGSN and the PDF exchange charging identifiers as follows:

1. The PDF passes the ICID to the GGSN in the authorization decision.

2. The GGSN passes the GCID to the PDF in the report about the authorization decision.

The PDF also passes the GCID to the P-CSCF, which forwards the GCID to the IMS entities in its own network where it is included on IMS CDRs. The GGSN includes the ICID on its G-CDR (i.e., a Gateway GPRS Support Node Charging Data Record), but does not pass the ICID to the SGSN. When a single IMS session requires several secondary PDP contexts, one or more GCIDs are mapped to one ICID. In addition, the GGSN is responsible for updating GCID information at the IMS level when secondary PDP context or media flows are removed or added during the session. As a last link, the SGSN creates an S-CDR (i.e., a Serving GPRS Support Node Charging Data Record) that includes GCID and GGSN addresses. This is a unique identifier for each PDP context. Figure 3.16 shows an example of an IMS session that contains two media components which are transported in separate PDP contexts.

As seen from the example above, the 3GPP IMS architecture defines the ICID and GCID for charging data correlation and a mechanism for exchanging the identifiers between the IMS and the PS domain.

However, the mechanism lacks the ability to measure a single media flow at the packet core layer and to correlate that with IMS charging data when media flows are multiplexed to the same secondary PDP context. At the time of writing, improvement work is ongoing, and it is estimated that the work will be ready within Release 6. The work is called "IP flow-based charging".

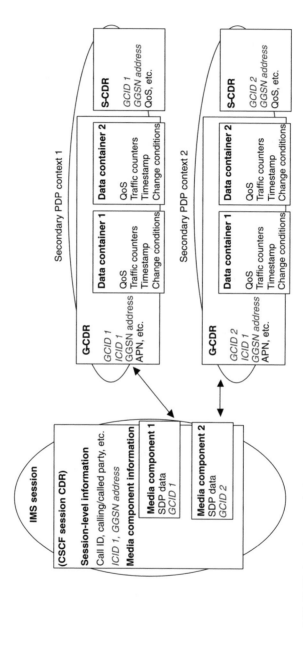

Figure 3.16 IMS charging correlation.

3.10.3 Charging information distribution

Section 3.10.2 explained how charging information is correlated. This section shows how charging information is distributed between different IMS entities.

The first IMS entity within the SIP signalling path generates an ICID. This ICID is passed along the SIP signalling path to all entities involved, except the UE: that is, the P-CSCF in the terminating network will remove the ICID. The ICID is used for correlating charging data between IMS components. The ICID applies for the duration of the event with which it is associated: for example, an ICID assigned for session establishment is valid until session termination, etc. We can see from Figure 3.17 that IMS and GPRS charging identifiers are exhanged when the bearer is authorized. In addition, Figure 3.17 indicates when accounting requests are sent to the CCF. The address of the CCF is distributed during registration or, alternatively, it is configured in the IMS entities.

3.11 User profile

When a user obtains an IMS subscription from an operator, the operator needs to assign a user profile. The user profile contains at least one private user identity and single service profile. Figure 3.18 depicts the general structure of a user profile [3GPP TS 29.228]. The private user identity was described in Section 3.4.1.1, but it should be understood that a user profile may contain more than one private user identity, if a user is using a shared public user identity as described in Section 3.15. Figure 3.18 shows that a single IMS subscription may contain multiple service profiles; this allows different treatment for different public user identities as explained in Section 3.4.1.4.

3.11.1 Service profile

A service profile is a collection of user-specific information that is permanently stored in the HSS. It is transferred from the HSS to an assigned S-CSCF in two user data-handling operations, Server-Assignment-Answer (SAA) and Push-Profile-Request (PPR), as described in Sections 2.3.4.1 and 2.3.4.2. The service profile is carried in one Diameter AVP, where it is included as an XML (Extensible Markup Language) document. The service profile is further divided into three parts:

- Public Identification.

- Core Network Service Authorization.

- Initial Filter Criteria.

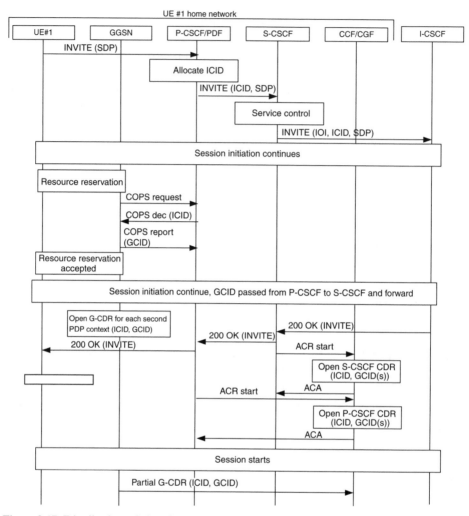

Figure 3.17 Distribution of charging information.

3.11.1.1 Public Identification

Public Identification comprises those user public identities that are associated with a service profile. Identities can be either SIP URIs or tel URIs. Each public user identity contains an associated barring indication. If the barring indication is set, then the S-CSCF will prevent that public identity (e.g., a temporary public user identity) from being used in any IMS communication other than registrations and re-registrations.

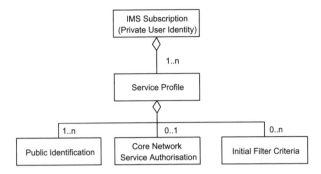

Figure 3.18 Structure of IMS user profile.

3.11.1.2 Media policy information

Media policy information is carried in the Core Network Service Authorization. It contains an integer that identifies a subscribed media profile in the S-CSCF (e.g., allowed SDP parameters). This information allows operators to define different subscriber profiles in their IMS networks. They may define different customer classes, such as gold, silver and bronze. Gold could mean that a user is able make video calls and all ordinary calls. Silver could mean that a user is able to use wideband AMR (adaptive multi-rate) as a speech codec, but she is not allowed to make video calls and so on. Transferring just the integer value between the HSS and the S-CSCF saves the storage space in the HSS and optimizes the usage of the Cx reference point.

The S-CSCF needs to have a static database that contains the mapping between the integer value and the subscribed media profile. The meaning of the integer value is not standardized (i.e., it is operator-specific). Figure 3.19 gives an illustrative example.

3.11.1.3 Service-triggering information

Service-triggering information is presented in form of Initial Filter Criteria. Initial Filter Criteria describe when an incoming SIP message is further routed to a specific

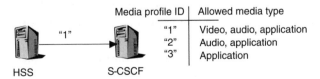

Figure 3.19 Media authorization in the S-CSCF.

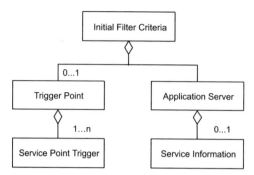

Figure 3.20 Structure of Initial Filter Criteria.

application server. Figure 3.20 shows that Initial Filter Criteria are composed of either zero or one instance of a Trigger Point and one instance of an Application Server [3GPP TS 29.228]. Each Initial Filter Criterion within the service profile has a unique priority value (integer) that is utilized in the S-CSCF. When multiple initial filter criteria are assigned the S-CSCF assesses them in numerical order: that is, an initial filter criterion with a higher priority number will be assessed after one with a smaller priority number.

Trigger Point

The Trigger Point describes conditions that should be checked to discover whether the indicated Application Server should be contacted. The absence of a Trigger Point will indicate unconditional triggering to an AS. Each Trigger Point contains one to multiple instances of the Service Point Trigger. Service Point Triggers may be linked by means of logical expressions (AND, OR, NOT). Section 3.12 will give a more detailed explanation of how trigger points are used.

Application Server

The Application Server defines the application server (AS) that is contacted if the trigger points are met. The Application Server may contain information about the default handling of the session if contact with the AS fails. Default handling will either terminate the session or let the session continue based on the information in the Initial Filter Criteria. In addition, the Application Server contains zero or one instance of the Service Information. Service Information enables provisioning of information that is to be transferred transparently via the S-CSCF to an AS when the conditions of Initial Filter Criteria are satisfied during registration.

3.12 Service provision

3.12.1 Introduction

The IMS is not a service in itself; on the contrary it is a SIP-based architecture for enabling an advanced IP service and application on top of the PS network. IMS provides the necessary means for invoking services; this functionality is called "service provision". IMS service provisioning contains three fundamental steps:

1. Define possible service or service sets.

2. Create user-specific service data in the format of Initial Filter Criteria when a user orders/modifies a subscription.

3. Pass an incoming initial request to an application server.

Item (1) is not addressed in this book because it is up to operators and service providers to define what kind of services they are willing to offer their subscribers. The other two steps are described next.

3.12.2 Creation of the filter criteria

Whenever a user obtains an IMS subscription and her subscription contains some value-added services or an operator is willing to utilize ASs as part of its IMS infrastructure, they need to create service-specific data. These service-specific data are part of the user's user profile. More precisely, service-specific data are represented as Initial Filter Criteria. Hereafter, we only concentrate on Initial Filter Criteria. Section 3.11 describes how Initial Filter Criteria fit into a user profile. When constructing Initial Filter Criteria an operator needs to consider these questions:

● What is a Trigger Point?

● What is the correct AS when the Trigger Point is met?

● What is the priority of an initial filter criterion?

● What should be done if the application server is not responding?

The Trigger Point is used to decide whether an application server is contacted. It contains one to multiple instances of a Service Point Trigger [3GPP TS 29.228]. The Service Point Trigger comprises the items shown in Figure 3.21:

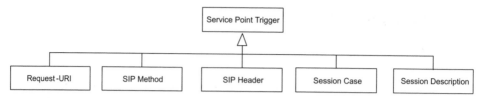

Figure 3.21 Structure of service point trigger.

- Request-URI—identifies a resource that the request is addressed to (e.g., sportnews@ims.example.com).

- SIP Method—indicates the type of request (e.g., INVITE or MESSAGE).

- SIP header—contains information related to the request. A Service Point Trigger could be based on the presence or absence of any SIP header or the content of any SIP header. The value of the content is a string that is interpreted as a regular expression. A regular expression could be as simple as a proper noun (e.g., John) in the FROM header that indicates the initiator of the request.

- Session Case—has three possible values, Originating, Terminating or Terminating_Unregistered, that indicate whether the filter should be used by the S-CSCF that is handling the originating, terminating or terminating for an unregistered end user services. An originating case refers to when the S-CSCF is serving the calling user. A terminating case refers to when the S-CSCF is serving the called user.

- Session Description—defines a Service Point Trigger for the content of any SDP field within the body of a SIP method. Regular expressions can be used to match the trigger.

Based on the above an operator could build, for example, Initial Filter Criteria to handle unregistered users: an IMS user who has not registered any of her public user identities. The following initial filter criterion routes an incoming session to a voice-mail server (sip:vmail@ims.example.com) when the user is not registered. To make this happen the operator has to set a SIP Method to match INVITE and a Session Case to match the value of Terminating_Unregistered. If the voicemail server cannot be contacted, then the default handling should be that the session is terminated. Initial Filter Criteria are coded in XML, as shown below (see [3GPP TS 29.228] for the exact coding rules of Initial Filter Criteria):

```
Method="INVITE" AND SessionCase="2"
<?xml version="1.0" encoding="UTF-8"?>
<testDatatype xmlns:xsi="http://www.w3.org/2001/XMLSchema-instance"
xsi:noNamespaceSchemaLocation="D:\ CxDataType.xsd">
  <IMSSubscription>
```

```
<PrivateID>privatexzyjoe@ims.example.com </Identity>
<ServiceProfile>
  <PublicIdentity>
    <Identity>sip: joe.doe@ims.example.com </Identity>
  </PublicIdentity>
  <PublicIdentity>
    <Identity>tel:+358503334444</Identity>
  </PublicIdentity>
  <InitialFilterCriteria>
    <Priority>0</Priority>
    <TriggerPoint>
      <ConditionNegated>0</ConditionTypeCNF>
      <SPT>
        <ConditionNegated>0</ConditionNegated>
        <Group>0</Group>
        <Method>INVITE</Method>
      </SPT>

      <SPT>
        <ConditionNegated>0</ConditionNegated>
        <Group>0</Group>
        <SessionCase>2</SessionCase>
      </SPT>
    </TriggerPoint>
    <ApplicationServer>
      <ServerName>sip:vmail@ims.example.com</ServerName>
      <DefaultHandling>1</DefaultHandling>
      </ApplicationServer>
  </InitialFilterCriteria>
  </ServiceProfile>
<IMSSubscription>
</testDatatype>
```

3.12.3 Selection of AS

Initial Filter Criteria are downloaded to the S-CSCF on user registration or on a terminating initial request for an unregistered user. After downloading the user profile from the HSS, the S-CSCF assesses the filter criteria for the initial request alone, according to the following steps [3GPP TS 24.229]:

1. Check whether the public user identity is barred; if not, then proceed.

2. Check whether this request is an originating request or a terminating request.

3. Select the Initial Filter Criteria for a session case (originating, terminating or terminating for an unregistered end user).

4. Check whether this request matches the initial filter criterion that has the highest priority for that user by comparing the service profile with the public user identity that was used to place this request:

    ○   If this request matches the initial filter criterion, then S-CSCF will forward this request to that AS, check to see whether it matches the next following filter criterion of lower priority and apply the filter criteria on the SIP method received from the previously contacted AS.

    ○   If this request does not match the highest priority initial filter criterion, then check to see whether it matches the following filter criterion's priorities until one does match.

    ○   If no more (or none) of the Initial Filter Criteria apply, then the S-CSCF will forward this request based on the route decision.

There exists one clear difference in how the S-CSCF handles originating and terminating Initial Filter Criteria. When the S-CSCF realizes that an AS has changed the Request-URI in the case terminating Initial Filter Criteria, it stops checking and routes the request based on the changed value of the Request-URI. In an originating case the S-CSCF will continue to evaluate Initial Filter Criteria until all Initial Filter Criteria have been evaluated.

If the contacted AS does not respond, then the S-CSCF follows the default-handling procedure associated with Initial Filter Criteria: that is, either terminate the session or let the session continue based on the information in the filter criteria. If the Initial Filter Criteria do not contain instructions to the S-CSCF regarding the failure to contact the AS, then the S-CSCF will let the call continue, as the default behaviour [3GPP TS 24.229].

According to our Initial Filter Criteria example, incoming INVITE requests will be routed to a voicemail server, vmail@ims.example.com, when Joe is not registered in the network. In exceptional cases, when the voicemail server is not responding, the S-CSCF is instructed to release a session attempt.

3.12.4 AS behaviour

Section 3.12.3 described how the request is routed to an AS. After receiving the request the AS initiates the actual service. To carry the service out the AS may act in three different modes:

●   Terminating UA—the AS acts as the UE. This mode could be used for providing a voicemail service.

●   Redirect server—the AS informs the originator about the user's new location or about alternative services that might be able to satisfy the session. This mode could be used for redirecting the originator to a particular Web page.

- SIP proxy—the AS processes the request and then proxies the request back to the S-CSCF. While processing, the AS may add, remove or modify the header contents contained in the SIP request according to the proxy rules specified in [RFC3261].

- Third-party call control/back-to-back UA—the AS generates a new SIP request for a different SIP dialog, which it sends to the S-CSCF.

These modes are described in more detail in Section 8.3. In addition to these modes, an AS can act as an originating UA. When the application is acting as an originating UA it is able to send requests to the users: for example, a conferencing server may send SIP INVITE requests to a pre-defined number of people at 9 a.m. for setting up a conference call. Another example could be a news server sending a SIP MESSAGE to a soccer fan to let him know that his favourite team has scored a goal.

3.13 Connectivity between traditional circuit-switched users and IMS users

For the time being, most users are utilizing traditional circuit-switched (CS) UE: that is, fixed line telephones and all kinds of cellular terminals. Therefore, it is desirable for the IMS to interwork with legacy CS networks to support basic voice calls between IMS users and CS network users. This requires interworking both at the user plane and the control plane because the used protocols are different in both planes. Control-plane interworking is tasked to MGCF. It performs mapping from SIP signalling to Bearer Independent Call Control (BICC) or ISUP used in CS legacy networks, and vice versa. IMS-MGW in turn translates protocols at the user plane. It terminates the bearer channels from the CS (PSTN/ISDN/GSM) networks as well as media streams from IP or ATM-based PS networks and provides the translation between these terminations. Additional functions, such as codec interworking, echo cancellation and continuity check, can be also provided. The terminations are controlled by MGCF. Network configurations for handling both IMS and CS-originated calls are explained next.

3.13.1 IMS-originated session toward a user in the CS core network

When an IMS user initiates a session she does not need to bother about whether the called user is an IMS user or a CS user. She simply makes a call and the IMS takes care of finding the called party. The session request from the calling user will always arrive at the S-CSCF serving the calling user, based on a route learned during IMS registration. When the S-CSCF receives a session request using a tel URL type of user identity (tel:+358 50 1234567), it has to perform an ENUM query for converting tel URL to SIP URI, as IMS routing principles do not allow routing with tel URLs. If the S-CSCF is able to convert the identity to SIP URI format it will route the session further to the target IMS network, and when this conversion fails the S-CSCF will try to reach the user in the CS network. To break out to the CS network, the S-CSCF routes the session request further to BGCF in the same network. The selected BGCF has two options: either selecting the breakout point in the same network or selecting another network to break out to the CS network. In the former case BGCF selects MGCF in the same network in order to convert SIP signalling to ISUP/BICC signalling and control the IMS-MGW. In the latter case BGCF selects another BGCF in a different IMS network to select MGCF in its network for handling breakout. MGCF acts as an end point for SIP signalling; so, it negotiates media parameters together with the IMS UE and, similarly, negotiates media parameters together with the CS entity (e.g., with an MSC server). Figure 3.22 visualizes the interworking concept when an IMS-originated session is terminated in the CS network. The arrows in the figure show how the first signalling message traverses from the S-CSCF to the CS network.

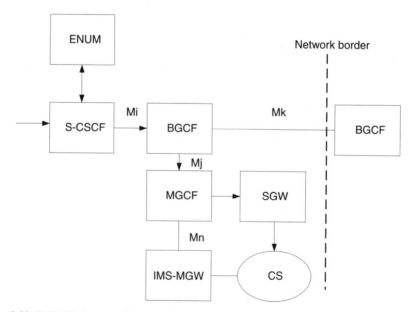

Figure 3.22 IMS–CS interworking configuration when an IMS user calls a CS user.

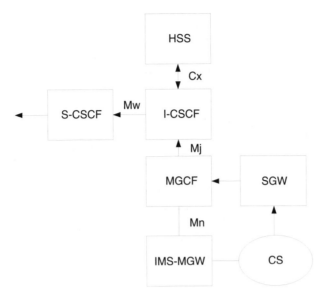

Figure 3.23 IMS–CS interworking configuration when a CS user calls an IMS user.

3.13.2 CS-originated session toward a user in IMS

When a CS user dials an E.164 number that belongs to an IMS user, it will be handled in the CS network like any other E.164 number; however, after routing analysis it will be sent to MGCF in the IMS user's home network. After receiving the ISUP/BICC signalling message, the MGCF interacts with the IMS-MGW to create a user-plane connection, converts ISUP/BICC signalling to SIP signalling and sends a SIP INVITE to the I-CSCF, which finds the S-CSCF for the called user with the help of the HSS (as described in Section 2.3.4.1). Then the S-CSCF takes the necessary action to pass the SIP INVITE to the UE. Thereafter, MGCF continues communication with the UE and the CS network to set the call up. Figure 3.23 shows how the functions interwork when a CS-originated call is terminated by the IMS network. The arrows in the figure show how the first signalling message traverses from the CS to the IMS user.

3.14 Mechanism to register multiple user identities at once

SIP allows one public user identity to be registered at a time; so, if a user has more than one public user identity, then she has to register every public user identity individually. This may be frustrating and time-consuming from the end user perspective. Obviously, registering four public user identities would consume four times

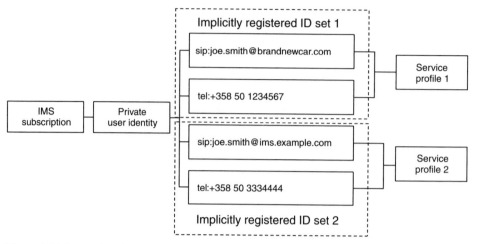

Figure 3.24 Example of implicit registration sets.

as much radio resource in the case of the UMTS than registering one public user identity. It was for these reasons that the 3GPP developed a mechanism to register more than one public user identity at a time. This concept is called "implicit registration".

An implicit registration set is a group of public user identities that are registered via a single registration request. When one of the public user identities within the set is registered, all public user identities associated with the implicit registration set are registered at the same time. Similarly, when one of the public user identities within the set is de-registered, all public user identities that have been implicitly registered are de-registered at the same time. Public user identities belonging to an implicit registration set may point to different service profiles. Some of these public user identities may point to the same service profile [3GPP TS 23.228].

To get implicitly registered public user identities the UE must send a SUBSCRIBE request for a registration event package to the S-CSCF. When the S-CSCF receives the SUBSCRIBE request it will return the implicitly registered public user identity with a NOTIFY request. For example, a user has four public user identities that are grouped in two implicit registration sets (Figure 3.24). The first set contains joe.smith@brandnewcar.com and tel:+358 50 1234567. The second set contains joe.smith@ims.example.com and tel:+358 50 3334444. When Joe sends a REGISTER request containing joe.smith@brandnewcar.com as an identity to be registered, the allocated S-CSCF performs a normal registration procedure and, after successful authorization, the S-CSCF downloads the service profiles that are associated with the public user identities belonging to the implicit registration set (service profile 1). To obtain the implicitly registered public user identities, Joe's UE must send a SUBSCRIBE request to the S-CSCF. When the S-CSCF receives the

SUBSCRIBE request it will return the implicitly registered public user identity, tel:358 50 1234567, within NOTIFY.

3.15 Sharing a single user identity between multiple terminals

Traditionally, in the CS every single user has her own Mobile Subscriber International ISDN Number (MSISDN) number that is used to reach the user. It is not possible for a single user to use multiple terminals with the same MSISDN number simultaneously. Having two mobile stations with identical MSISDN numbers would cause significant conflicts in the network. Nowadays, users may have more than one item of UE with totally different capabilities: big/small screen, camera/no camera, full keyboard and so forth. Different UEs may serve different purposes (e.g., one for gaming, another for ordinary voice and video sessions). From the user's point of view, the user should be reachable via the same identity regardless of the number of items of UE that she is using simultaneously. The IMS makes this feature possible.

Release 6 IMS allows users to register the same public user identity from a number of items of UE. In addition, a user is able to indicate her preferences regarding a single UE at the registration phase. Different registrations can be differentiated by means of the private user identity and the used IP address. Figure 3.25 shows an example in which a user has two items of UE: one for video sessions and another for chat and gaming applications. When someone is calling the user, Joe, it is his S-CSCF that makes the decision as to which UE is going to be contacted in the first place. This decision can be done based on the preferences given at the registration phase: for example, if the incoming session contains a video component, then the S-CSCF could select UE #2, which is Joe's primary preference for video sessions.

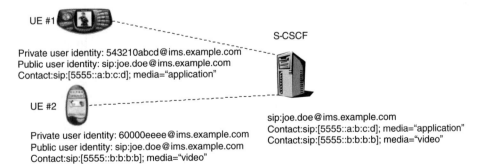

UE #1

S-CSCF

Private user identity: 543210abcd@ims.example.com
Public user identity: sip:joe.doe@ims.example.com
Contact:sip:[5555::a:b:c:d]; media="application"

UE #2

sip:joe.doe@ims.example.com
Contact:sip:[5555::a:b:c:d]; media="application"
Contact:sip:[5555::b:b:b:b]; media="video"

Private user identity: 60000eeee@ims.example.com
Public user identity: sip:joe.doe@ims.example.com
Contact:sip:[5555::b:b:b:b]; media="video"

Figure 3.25 Multiple terminals.

In addition to preference-based routing, the S-CSCF may perform forking. There are two types of forking:

- Sequential forking.

- Parallel forking.

Sequential forking means that different items of UE are contacted one by one: for example, the S-CSCF first sends the request to UE #2 and, if Joe fails to respond, within a certain time limit the S-CSCF then tries to reach Joe through UE #1.

Parallel forking means that different items of UE are contacted at the same time: for example, when two items of UE are ringing, Joe can decide which UE to use for the incoming session; however, in the end the session can only be connected to a single item of UE.

3.16 SIP compression

The IMS supports multimedia services using the SIP call control mechanism. SIP is a client server, text-based signalling protocol used to create and control multimedia sessions with two or more participants. The messages also contain a large number of headers and header parameters, including extensions and security-related information. Setting up a SIP session is a tedious process involving codec and extension negotiations as well as quality of service (QoS) interworking notifications. In general, this provides a flexible framework that allows sessions with differing requirements to be set up. However, the drawback is the large amount of bytes and the many messages exchanged over the radio interface. The increased message size means that:

- Call set-up procedures using SIP will take much more time to be completed compared with those using existing cellular-specific signalling, which means that the end user will experience a delay in call establishment that will be unexpected and likely unacceptable.

- Intra-call signalling will in some way adversely affect voice quality/system performance.

Therefore, support for real-time multimedia applications requires particular attention when SIP call control is used. To speed up session establishment, the 3GPP has mandated the support of SIP compression by both the UE and the P-CSCF [3GPP TS 23.221]. Although the support of compression is mandatory, the 3GPP was not happy to mandate its usage because in the future WLAN terminals may not need to use SIP compression at all. At the time of writing, it is required that the UE and the P-CSCF implement compression functionalities as defined in [RFC3320], [RFC3485], [RFC3486] and [3GPP TS 24.229]. The first mentioned RFC gives an overall solution to how SIP messages between two entities can be compressed. The second RFC defines a SIP/SDP-specific static dictionary that a signalling compression solution may use in order to achieve higher efficiency. The third RFC explains how the UE could signal that compression is desired for one or more SIP messages (see Section 8.13.15 and Chapter 19 for more detail).

Part II

Detailed Procedures

4

Introduction

This part gives a detailed example of Session Initiation Protocol (SIP) and Session Description Protocol (SDP)-related procedures in the Internet Protocol (IP) Multimedia Subsystem (IMS). Signalling flows and elements are described and explained based on IMS registration and a subsequent IMS session between two users.

The reader will see how IMS-signalling works and how previously described concepts and architecture are realized at the protocol level. Nevertheless, this part does not handle error or abnormal procedures in detail.

To give a better understanding of the procedures applied, the part is split into several chapters that concentrate on different concepts, such as routing, authentication or media negotiation. Because of this, different call flows will not be followed step by step. Each chapter will describe those parts of individual SIP and SDP messages that are necessary for their understanding.

An overview section is included in each chapter to give an introduction to the basic operation. At the end of each chapter the related standards and specifications are listed, to allow the interested reader to obtain more detail by reading the base specifications.

4.1 The example scenario

This section gives a detailed example of a normal IMS session between two users and all the required prerequisites. It is based on the assumption that both users are attached to the General Packet Radio Service (GPRS), which is used as the example access technology throughout the section.

Tobias, who is a student in France and currently visiting Finland, is calling his sister Theresa, who is working in Hungary and currently on a business trip to Austria (see Table 4.1 and Figure 4.1).

Tobias's home operator is located in France. As he is roaming in Finland, the Finnish operator provides the Proxy Call Session Control Function (P-CSCF), as the

The IMS. Miikka Poikselkä, Georg Mayer, Hisham Khartabil and Aki Niemi
Copyright 2004 by John Wiley & Sons, Ltd. ISBN 0-470-87113-X

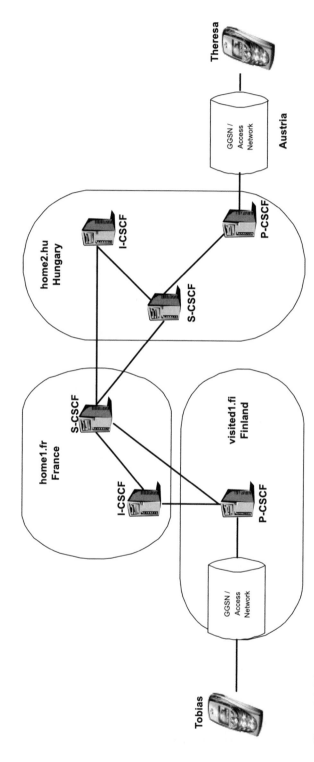

Figure 4.1 The example scenario.

Table 4.1 Location of CSCFs and GPRS access for the example scenario.

User	Home operator S-CSCF location	P-CSCF location	GPRS access
Tobias	France	Finland	Finland
Theresa	Hungary	Hungary	Austria

home operator and the Finnish operator have signed an IMS roaming agreement. Consequently, the Gateway GPRS Support Node (GGSN), that Tobias is using is also located in Finland.

Theresa's home operator in Budapest has no IMS roaming agreement with the operator in Austria. Therefore, her terminal gets attached to the P-CSCF in her Hungarian home network, where the GGSN is also located. Theresa's access to the IMS is based on the GPRS-level roaming agreement between the operators of her home network and the visited network.

It is assumed that Theresa has already registered her SIP URI (uniform resource identifier), sip:theresa@home2.hu, as Tobias is just switching on his mobile phone. He wants to call his sister to show her one of the beautiful wooden buildings in Oulu and, therefore, points his camera, which is connected to the phone, toward the building. In parallel to this, his phone will also send a second video stream, showing his face to Theresa. The built-in camera of his phone records this second stream. However, Tobias first has to register his public user identity, sip:tobias@home1.fr, before he can call his sister.

4.2 Base standards

The following specifications define the basic procedures and architecture as used in the following chapters:

- 3GPP TS 23.228   IP Multimedia Subsystem (IMS).

- 3GPP TS 24.229   IP Multimedia Call Control Protocol based on SIP and SDP.

- RFC3261   SIP: Session Initiation Protocol

[3GPP TS 24.228] (Signaling flows for the IP multimedia call control based on SIP and SDP) provides example call flows for procedures within the IMS.

5

An example IMS registration

5.1 Overview

Session Initiation Protocol (SIP) registration is performed in order to bind the Internet Protocol (IP) address that is currently used by the user and the user's public user identity, which is a SIP URI (uniform resource identifier).

If Tobias wants to call Theresa, he will send a SIP INVITE request to her address "sip:theresa@home2.hu"; he does not need to be aware of which terminal Theresa is using. The INVITE then gets routed to Theresa's registrar, which is located in home2.hu. This registrar became aware of Theresa's current terminal address during her registration. Therefore, it will replace the address sip:theresa@ home2.hu with the registered contact, which is an IP address. Afterward, the request can be routed to Theresa's terminal.

Therefore, even for non-IMS cases, Theresa needs to be registered at a SIP registrar so that her current terminal address can be discovered. The IP Multimedia Subsystem (IMS) couples more functionality to SIP registration procedures, which makes it necessary that Tobias registers as well, before he can call his sister.

The following procedures are performed during Tobias's IMS registration (see Figure 5.1):

- The dedicated signalling Packet Data Protocol (PDP) context is established between Tobias's user equipment (UE) and the Gateway GPRS Support Node (GGSN) in the case of General Packet Radio Service (GPRS)—Section 5.2.

- The UE discovers the address of the Proxy Call Session Control Function (P-CSCF), which it uses as a SIP outbound proxy during registration and for all other SIP signalling while it is registered—Section 5.3.

- The UE sends a REGISTER message to Tobias's home network to perform SIP registration for Tobias's public user identity—Section 5.5.2.

The IMS. Miikka Poikselkä, Georg Mayer, Hisham Khartabil and Aki Niemi
Copyright 2004 by John Wiley & Sons, Ltd. ISBN 0-470-87113-X

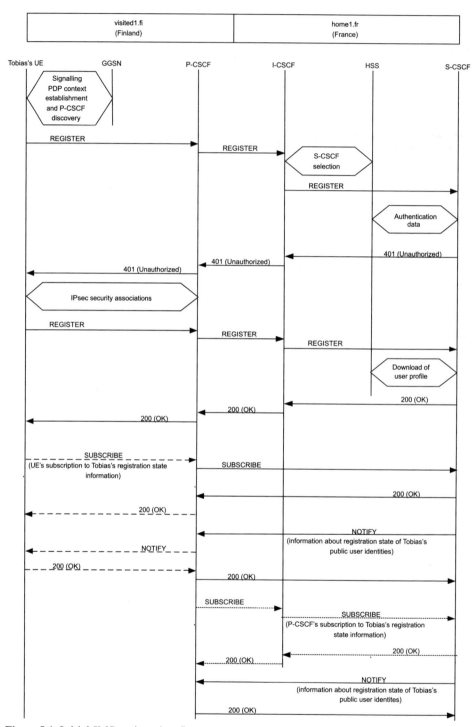

Figure 5.1 Initial IMS registration flow.

- The Interrogating-CSCF (I-CSCF) selects the Serving-CSCF (S-CSCF) that serves the user while it is registered—Section 5.5.5.

- The S-CSCF downloads the authentication data of the user from the Home Subscriber Server (HSS)—Section 5.6.4.

- The UE and the P-CSCF agree on a security mechanism—Section 5.8.

- The UE and the network (S-CSCF) authenticate each other—Section 5.6.

- IP security (IPsec) associations between the UE and the P-CSCF are established—Section 5.7.

- SIP compression starts between the UE and the P-CSCF—Section 5.9.

- The UE learns the route to the S-CSCF—Section 5.5.8.

- The S-CSCF learns the route to the UE—Section 5.5.9.

- The S-CSCF downloads the user profile of the user from the HSS—Section 5.5.6.

- The S-CSCF registers the default public user identity of the user—Section 5.5.6.

- The S-CSCF may, based on the user profile, implicitly register further public user identities of the user—Section 5.12.

- The UE becomes aware of all public user identities that are assigned to Tobias and his current registration state—Section 5.12.

- The P-CSCF becomes aware of all public user identities that are assigned to Tobias and his current registration state—Section 5.12.

Due to all these required basic actions, Tobias would not be able to send the INVITE to his sister had he not registered earlier.

5.2 Signalling PDP context establishment

Before Tobias's UE can start the IMS registration procedures, it needs to establish an IP connection with the network. In the case of GPRS such an IP connection is provided by either a dedicated signalling PDP context or a general purpose PDP context. The concepts and procedures for PDP context establishment and usage are described in Section 3.7 and Chapter 13.

In this example it is assumed that Tobias's UE establishes a dedicated signalling PDP context with the GGSN in Finland. After the UE has established the signalling PDP context, it will be able to send SIP signalling over the air interface.

5.3 P-CSCF discovery

The P-CSCF is the single entry point for all SIP messages that are sent from Tobias's UE to the IMS. Therefore, the P-CSCF address needs to be known by the UE before the first SIP message is sent. As this address is not pre-configured in our example, it needs to be discovered by the UE.

In the case of the GPRS the UE can request the addresses of a P-CSCF during the establishment of the general or signalling PDP context. The GGSN then will return the IPv6 prefix of an P-CSCF in response to the activate PDP context request.

Alternatively, the UE can choose to use DHCPv6 (Dynamic Host Configuration Protocol for IPv6) in order to discover the P-CSCF. If the P-CSCF address is returned from DHCP as a fully qualified domain name (FQDN) rather than an IP address, then the P-CSCF address will be resolved via the domain name system (DNS) as the address of any other SIP server. The related procedures are described in Chapter 12.

5.4 Transport protocols

The IMS puts no further restrictions on the transport protocol for SIP used between the UE and the P-CSCF. In this example it is assumed that the User Datagram Protocol (UDP) is the default transport protocol. UDP will be used for the transport of SIP messages that are sent between Tobias's UE and the P-CSCF as long as these messages do not exceed 1,300 bytes. When they exceed this limit, the Transmission Control Protocol (TCP) must be used.

Due to the fact that SIP also allows large content in the SIP message body (e.g., pictures can be attached to the body of a MESSAGE request), it is likely that both UDP and TCP will be used in parallel while a user is registered.

5.5 SIP registration and registration routing aspects

5.5.1 Overview

This section concentrates on the SIP aspects of Tobias's registration (see Table 5.1 and Figure 5.2).

Tobias's UE will first of all construct a REGISTER request, which it sends to the home domain of Tobias's operator. The relevant information is obtained from the IP Multimedia Services Identity Module (ISIM) application on Tobias's Uni-

Table 5.1 Routing-related headers.

Header	Function	Set up
Via	Routing of responses	By every traversed SIP entity, which puts its address to the Via header during the routing of the request
Route	Routing of requests	Initial requests: by the request-originating UE, which puts the P-CSCF (outbound proxy) address and entries of the Service-Route header
		Initial requests: by CSCFs, which find the next hop from the public user identity in the request URI (by querying DNS and HSS) or the received Path header
		Subsequent requests: by the request-originating UE, which put entries to the Route header as collected in the Record-Route header during initial request routing
Record-Route	Records the Route header entries for subsequent requests within a dialog	By CSCFs, which put their addresses into the Record-Route header if they need to receive subsequent requests within a dialog
Service-Route	Indicates the Route header entries for initial requests from the UE to the user's S-CSCF (originating case)	By the S-CSCF, which sends this header back to the UE in the 200 (OK) response for the REGISTER request
Path	Collects the Route header entries for initial requests from the S-CSCF to the user's P-CSCF (terminating case)	By the P-CSCF, which adds itself to the Path header in the REGISTER request and sends it to the S-CSCF

versal Subscriber Identity Module (USIM). The request will traverse P-CSCF and the I-CSCF, which—if not previously assigned—will select an S-CSCF for Tobias.

The S-CSCF will create, based on the information given in the REGISTER request, the binding between Tobias's public user identity and the IP address of Tobias's UE. This makes it possible for requests from other users to be routed from the S-CSCF to Tobias's UE. The S-CSCF will update the registration information in the HSS, download Tobias's user profile and will, based on the received initial filter criteria from the HSS, inform any application servers (ASs) that are interested in Tobias's registration state.

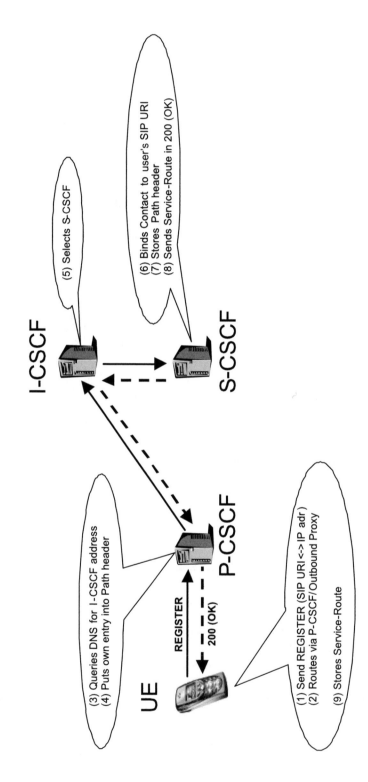

Figure 5.2 Routing during registration.

During the registration procedures the UE will learn the direct route to the S-CSCF from the Service-Route header. After that the I-CSCF will no longer need to be contacted when Tobias's UE sends out an initial request.

The S-CSCF will become aware of the address of Tobias's P-CSCF from the Path header. This is necessary as all initial requests that are destined for Tobias (e.g., an INVITE request) need to first traverse the P-CSCF before they can be sent to the UE.

5.5.2 Constructing the REGISTER request

After establishing the signalling PDP context and discovering the P-CSCF address, Tobias's UE can finally start to construct the initial REGISTER request:

```
REGISTER sip:home1.fr SIP/2.0
Via: SIP/2.0/UDP [5555::a:b:c:d];branch=0uetb
Route: sip:[5555::a:f:f:e];lr
Max-Forwards: 70
From: <sip:tobias@home1.fr>;tag=pohja
To: <sip:tobias@home1.fr>
Contact: <sip:[5555::1:2:3:4]>;expires=600000
Call-ID: apb03a0s09dkjdfglkj49111
CSeq: 25 REGISTER
Content-Length: 0
```

How the used public and private user identities as well as the registrar address are obtained from the ISIM is described in Section 5.12.2.

The above message is not a complete IMS REGISTER request: there are some headers and parameters missing from it. It only includes the information required to explain the procedures in this section, as is the case with all the following messages.

The final destination of the request is the registrar, which is identified in the request URI as sip:home1.fr: the domain name of the home network of Tobias read from the ISIM.

In the To header we find the public user identity that is going to be registered (read from the sip:tobias@home1.fr ISIM). SIP registration takes place to tell the registrar that the public user identity sip:tobias@home1.fr will be reachable under the IP address that is indicated in the Contact header. This IP address includes the IPv6 prefix, which the UE got assigned during the establishment of the dedicated signalling PDP context (see Section 5.2).

Also within the Contact header, the UE indicates that this binding of the IP address to the SIP URI is intended to last 600,000 seconds (nearly a week). In IMS the UE is forced to register for such a long time. Nevertheless, the network can adjust this time:

- During registration procedures by setting the expires value in the Contact header of the 200 (OK) response to the REGISTER request to a smaller value.

- After the user has registered, by making use of registration-state event notifications (e.g., Section 5.13.2 for network-initiated re-authentication).

The UE puts its IP address into the Via header of the request as well. This ensures that all responses to this request will be routed back to the UE. A branch parameter that uniquely identifies the transaction is also put into the Via header. Every entity on the route will add its own Via header.

The P-CSCF, which was resolved in the previous step, is put into the Route header. The P-CSCF is the next hop to receive the REGISTER message, as it is the topmost—and only—entry of the Route header. The ;lr parameter indicates that the P-CSCF is a loose router (see Section 8.12.2).

The From header identifies the user who is performing the registration. We find in the From header the same public user identity as in the To header, as Tobias is performing a so-called first-party registration (i.e., he is registering himself)

Note that the From header includes a tag, while the To header does not. The recipient of the request (i.e., the registrar), will set the To tag when sending the response to the UE.

A Call-ID header is included which, together with the value of the CSeq header, identifies the REGISTER transaction.

Finally, there is the indication that the REGISTER request is empty of text, as the Content-Length header is set to 0.

The example shown on the previous page gives the header names in their long form. In order to avoid unnecessary signalling over the air interface, Tobias's UE would use the compact form, which would make the REGISTER request look like:

```
REGISTER sip:home1.fr SIP/2.0
v: SIP/2.0/UDP [5555::1:2:3:4];branch=0uetb
Route: sip:[5555::a:b:c:d];lr
Max-Forwards: 70
f: <sip:tobias@home1.fr>;tag=pohja
t: <sip:tobias@home1.fr>
m: <sip:[5555::1:2:3:4]>;expires=600000
i: apb03a0s09dkjdfglkj49111
CSeq: 25 REGISTER
l: 0
```

To make reading of SIP messages more convenient, only the long form of the header names will be used in this example.

5.5.3 From the UE to the P-CSCF

Now Tobias's UE can send out the REGISTER request to the next hop, which is the topmost entry of the Route header (i.e., the P-CSCF). It sends the request via the UDP protocol, as its length does not exceed the strict limit of 1,300 bytes. As no port is indicated in the Route header, the request gets sent to the default SIP port (i.e., 5060).

5.5.4 From the P-CSCF to the I-CSCF

When receiving the initial REGISTER request the P-CSCF becomes aware for the first time that Tobias's UE is using it as a SIP outbound proxy. As Tobias is not authenticated at this moment, it can only act as a SIP outbound proxy and, therefore, tries to route the REGISTER request to the next hop.

The P-CSCF removes its own entry from the Route header. After doing so the Route header will be empty. The only routing-related information left now is the registrar address in the request URI, which points to Tobias's home network. In order to discover the address of a SIP proxy in Tobias's home network the P-CSCF needs to resolve the domain name (as given in the request URI) via DNS. By using DNS NAPTR, SRV and AAAA queries, the P-CSCF will resolve the address of an I-CSCF in Tobias's home network (see Chapter 12). Nevertheless, the P-CSCF will not put the address of the I-CSCF into the Route header, as it cannot be sure whether the I-CSCF will act as a loose router or not. Therefore, the P-CSCF will put the address of the I-CSCF into the UDP packet that transports SIP requests.

As the P-CSCF will send the UDP packet directly to the resolved I-CSCF address anyway, it is not really necessary for the P-CSCF to add a Route header that points to the I-CSCF. In our example it is assumed that it does, nevertheless.

Before sending the REGISTER message the P-CSCF also adds itself to the Via header, in order to receive the response to the request. It also adds a branch parameter to the Via header:

```
REGISTER sip:home1.fr SIP/2.0
Via: SIP/2.0/UDP sip:pcscf1.visited1.fi;branch=0pctb
Via: SIP/2.0/UDP [5555::a:b:c:d];branch=0uetb
Max-Forwards: 69
From: <sip:tobias@home1.fr>;tag=pohja
To: <sip:tobias@home1.fr>
```

```
Contact: <sip:[5555::1:2:3:4]>;expires=600000
Call-ID: apb03a0s09dkjdfglkj49111
CSeq: 25 REGISTER
Content-Length: 0
```

5.5.5 From the I-CSCF to the S-CSCF

The I-CSCF is the entry point to Tobias's home network and will receive every REGISTER request that is originated by Tobias's UE. It will query the HSS for the S-CSCF that is assigned to serve the user who is registering. If no S-CSCF has been selected up to now, it is the task of the I-CSCF to select one. These procedures are described in Section 3.8.

After putting its own entry in the topmost Via header, the I-CSCF sends the REGISTER request to the S-CSCF address that it either got from the HSS or that it selected:

```
REGISTER sip:home1.fr SIP/2.0
Via: SIP/2.0/UDP sip:icscf1.home1.fr;branch=0ictb
Via: SIP/2.0/UDP sip:pcscf1.visited1.fi;branch=0pctb
Via: SIP/2.0/UDP [5555::a:b:c:d];branch=0uetb
Route: sip:scscf1.home1.fr;lr
Max-Forwards: 68
From: <sip:tobias@home1.fr>;tag=pohja
To: <sip:tobias@home1.fr>
Contact: <sip:[5555::1:2:3:4]>;expires=600000
Call-ID: apb03a0s09dkjdfglkj49111
CSeq: 25 REGISTER
Content-Length: 0
```

5.5.6 Registration at the S-CSCF

After receiving the initial REGISTER request, the S-CSCF will request Tobias to authenticate himself, as described in Section 5.6. This will result in another REGISTER request from Tobias. This second REGISTER request will include the same registration-related information and will also be routed exactly in the same way as the initial REGISTER request. Nevertheless, for the second REGISTER a new Call-ID will be created. Consequently, new CSeq numbers, branch parameters and a new From tag will be included in it. The second REGISTER received by the S-CSCF will look like:

```
REGISTER sip:home1.fr SIP/2.0
Via: SIP/2.0/UDP sip:icscf1.home1.fr;branch=3ictb
Via: SIP/2.0/UDP sip:pcscf1.visited1.fi;branch=2pctb
Via: SIP/2.0/UDP [5555::a:b:c:d];branch=1uetb
Route: sip:scscf1.home1.fr;lr
Max-Forwards: 67
From: <sip:tobias@home1.fr>;tag=ulkomaa
To: <sip:tobias@home1.fr>
Contact: <sip:[5555::1:2:3:4]>;expires=600000
Call-ID: apb03a0s09dkjdfglkj49222
CSeq: 47 REGISTER
Content-Length: 0
```

Assuming that the authentication procedures are successful, the S-CSCF will then register Tobias; this means S-CSCF will create a binding for the public user identity that was indicated in the To header of the REGISTER request (sip:tobias@home1.fr) and the contact address (sip:[5555::a:b:c:d]). This binding will exist for exactly 600,000 seconds, which is the value that the UE entered into the "expires" parameter of the Contact header, unless the S-CSCF decides to reduce this time due to local policy.

The S-CSCF will also update the information in the HSS to indicate that Tobias has now been registered. The HSS will download Tobias's user profile to the S-CSCF via the Cx interface (see Section 3.12).

5.5.7 The 200 (OK) response

Afterwards, the S-CSCF will send back a 200 (OK) response to the UE, to indicate that the registration procedure has succeeded:

```
SIP/2.0 200 OK
Via: SIP/2.0/UDP icscf1.home1.fr;branch=3ictb
Via: SIP/2.0/UDP pcscf1.visited1.fi;branch=2pctb
Via: SIP/2.0/UDP [5555::1:2:3:4]:1357;branch=1uetb
From: <sip:tobias@home1.fr>;tag=ulkomaa
To: <sip:tobias@home1.fr>;tag=kotimaa
Contact: <sip:[5555::a:b:c:d]>;expires=600000
Call-ID: apb03a0s09dkjdfglkj49222
CSeq: 47 REGISTER
Content-Length: 0
```

The S-CSCF has added a tag to the To header.

The response is routed back to the UE over all the CSCFs that received the REGISTER request; it manages to do this because CSCFs put their own address in the top most Via header list when they receive REGISTER requests. Now, when receiving the 200 (OK) response, they just remove their own entry from the Via list and send the request forward to the address indicated in the topmost Via header.

The UE, when receiving this response, will know that the registration was successful.

5.5.8 The Service-Route header

We have seen that neither the UE nor the P-CSCF were aware of the address of the S-CSCF during the registration procedures; consequently the I-CSCF had to be contacted to discover the S-CSCF address from the HSS.

In order to avoid the I-CSCF as an extra hop for every initial message sent from the UE, the S-CSCF will return its address in the Service-Route header in the 200 (OK) response for the REGISTER request:

```
SIP/2.0 200 OK
Service-Route: sip:orig@scscf1.home1.fr;lr
```

The UE, when receiving the 200 (OK) response, will store the entries in the Service-Route header. Whenever the UE sends out any initial request other than a REGISTER message, it will:

- include the addresses that were received in the Service-Route header within a Route header of the initial request; and

- include the P-CSCF address as the topmost Route entry in the initial request.

Examples of how initial requests are routed are given in Section 5.12.5 for a SUBSCRIBE request and in Section 6.3.2 for an INVITE request.

The S-CSCF in this example puts a user part ("orig") in its Service-Route entry as it needs to distinguish between two types of requests:

- requests originated from the served user (i.e., Tobias); and

- requests destined for Tobias's UE.

Whenever the S-CSCF receives an initial request (e.g., an INVITE request) it needs to determine whether this request is originated from or destined to the served user. The user part entry in the Route header makes it easy for the S-CSCF to find out whether a received request was originated from the served user, as Tobias's UE will include the S-CSCF's Service-Route entry as a Route entry within all requests that it originates.

5.5.9 The Path header

The S-CSCF will receive all initial requests that are destined to Tobias, as it acts as his registrar. Normal SIP procedures allow the registrar to send requests directly to the UE. In the case of the IMS this is not possible, because the P-CSCF needs to be contacted first; this is because the P-CSCF has established IPsec security associations (SAs) with the UE that guarantee that all messages will be sent and received integrity-protected (see Section 5.7). Furthermore, the P-CSCF has an important role in media authorization (see Section 6.7.2) as it is the only network element in the IMS that has a direct connection to the GGSN.

Therefore, the S-CSCF needs to ensure that every request that is sent to the UE first traverses the P-CSCF. To make this possible, the P-CSCF includes its own address in every REGISTER request within a Path header:

```
REGISTER sip:home1.fr SIP/2.0
Path: sip:pcscf1.visited1.fi;lr
```

After successful registration of the user, the S-CSCF saves this P-CSCF address. Whenever a request for Tobias is received, the S-CSCF will include a Route header with the address that was received in the Path header.

An example of routing an initial INVITE request toward the served user is given in Section 6.3.3.5.

5.5.10 Third-party registration to application servers

After successful registration the S-CSCF will check the downloaded filter criteria of the user (see Section 3.12). We assume that there is a presence server that provides its services to Tobias; this presence server needs to know that Tobias has now been registered and is therefore available. To inform the presence server about this, filter criteria have been set which trigger all the REGISTER requests that originate from Tobias's public user identity (Table 5.2).

Due to these filter criteria, the S-CSCF will generate a third-party REGISTER request (Figure 5.3) and send it to the presence server whenever Tobias performs a successful registration:

Table 5.2 Filter criteria in Tobias's S-CSCF.

Element of filter criteria	Filter criteria
SPT: session case	Originating
SPT: public user identity	sip:tobias@home1.fr
SPT: SIP method	REGISTER
Application server	sip:presence.hom1.fr;lr

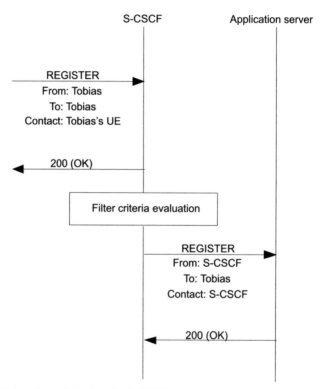

Figure 5.3 Third-party registration by S-CSCF.

```
REGISTER sip:presence.home1.fr SIP/2.0
Via: SIP/2.0/UDP scscf1.home1.fr;branch=99sctb
Max-Forwards: 70
From: <sip:scscf1.home1.fr>;tag=6fa
To: <sip:tobias@home1.fr>
Contact: <sip:scscf1.home1.fr>;expires=600000
Call-ID: las22kdoa45siewrf
CSeq: 87 REGISTER
Content-Length: 0
```

This REGISTER request is destined to the presence server at presence.home1.fr, as indicated in the request URI. As no Route header is included, the request will be sent directly to that address.

The To header includes the public user identity of Tobias, as this is the URI that was registered.

The S-CSCF indicates its own address in the From header, as it is registering Tobias's public user identity on behalf of Tobias (i.e., as a third party).

Furthermore, the S-CSCF indicates its own address within the Contact header. This ensures that the presence server never routes directly to Tobias's UE, but will always contact the S-CSCF first.

The presence server will send back a 200 (OK) response for this REGISTER request to the S-CSCF, but will not start acting as a registrar for Tobias. It will take the REGISTER request as an indication that Tobias has been successfully registered at the S-CSCF that is Tobias's registrar. If the presence server needs more information about Tobias's registration state (e.g., all other public user identities that have been implicitly registered for Tobias), it can subscribe to the registration-state information of Tobias in the same way as the UE and the P-CSCF do (see Section 5.12).

5.5.11 Related standards

Specifications relevant to Section 5.5 are:

• RFC3327   Session Initiation Protocol (SIP) Extension Header Field for Registering Non-Adjacent Contacts.

• RFC3608   Session Initiation Protocol (SIP) Extension Header Field for Service Route Discovery During Registration.

5.6 Authentication

5.6.1 Overview

As shown in Section 3.6, the IMS is based on several security relations. Two of them—authentication between user and network and the SAs between the UE and the P-CSCF—have an influence on SIP signalling (Figure 5.4). Authentication and SA establishment procedures in the IMS are directly coupled to SIP registration procedures. IMS authentication is based on a shared secret and a sequence number (SQN), which is only available in the HSS and the ISIM application that

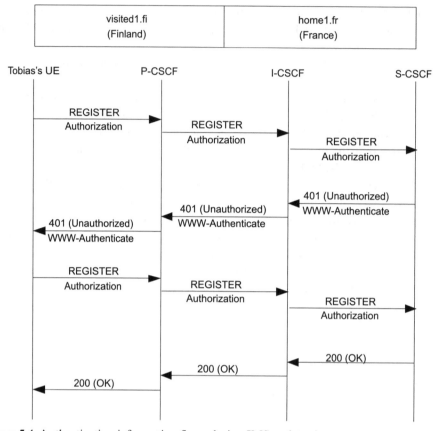

Figure 5.4 Authentication information flows during IMS registration.

is located in Tobias' phone. As the HSS never directly communicates with the UE, the S-CSCF performs the authentication procedures and all security related parameters that are needed by the S-CSCF. The so-called authentication vector (AV) is downloaded by the S-CSCF from the HSS during registration.

In order to authenticate Tobias sends his private user identity (in our example this is tobias_private@home1.fr) in the initial REGISTER request. This private user identity is stored within the ISIM application and is only used for authentication and registration procedures.

When receiving this REGISTER request, S-CSCF downloads the AV from the HSS. The AV does not include the shared secret and the SQN itself, but (among other parameters):

- a random challenge (RAND);

- the expected result (XRES);

- the network authentication token (AUTN);

- the integrity key (IK); and

- the ciphering key (CK).

These parameters enable the S-CSCF to perform authentication without knowing the shared secret or the SQN.

In order to authenticate, the S-CSCF rejects the initial REGISTER request from the user with a 401 (Unauthorized) response, which includes (among other parameters) the RAND, the AUTN, the IK and the CK.

The P-CSCF, when receiving the 401 (Unauthorized) response, removes the IK and the CK from the response before sending it to the UE. The IK is the base for the SAs that get established between the P-CSCF and the UE immediately afterwards (see Section 5.7).

After receiving the response, the UE hands the received parameters over to the ISIM application, which:

- Verifies the AUTN based on the shared secret and the SQN. When AUTN verification is successful the network is authenticated (i.e., the UE can be sure that the authentication data were received from the home operator's network).

- Calculates the result (RES) based on the shared secret and the received RAND.

- Calculates the IK, which is then shared between the P-CSCF and the UE and will serve as the base for the SAs.

Afterwards, the UE sends the authentication challenge response (RES) in the second REGISTER request back to the S-CSCF, which compares it with the XRES that was received in the AV from the HSS. If the verification is successful, the S-CSCF will treat the user as authenticated and will perform the SIP registration procedures (see Section 5.5.6).

Whenever the UE sends out another REGISTER request (i.e., due to either re or de-registration), it will always include the same authentication parameters as included in the second REGISTER request, until the S-CSCF re-authenticates the UE.

5.6.2 HTTP digest and 3GPP AKA

The Hyper Text Transfer Protocol (HTTP) digest is specified in [RFC2617], and how it is used with SIP is described in [RFC3261]. The IMS on the contrary is part of the Third Generation Partnership Project/Universal Mobile Telecommunications System (3GPP/UMTS) architecture, which uses the 3GPP Authentication and Key Agreement (AKA) mechanism for authentication.

In order to achieve 3GPP AKA-based authentication within the IMS, [RFC3310] defines how 3GPP AKA parameters (as described above) can be mapped onto HTTP digest authentication. Therefore, the signalling elements (SIP headers and parameters) used to transport 3GPP AKA information are identical to those used for the HTTP digest. Nevertheless, their meanings (i.e., their interpretation at the UE, the P-CSCF and the S-CSCF) are different.

In order to distinguish the 3GPP AKA authentication mechanism from the other HTTP digest mechanisms (e.g., MD5), it was given a new algorithm value: "AKAv1-MD5".

5.6.3 Authentication information in the initial REGISTER request

Within the initial REGISTER request Tobias's UE utilizes the HTTP Digest Authorization header to transport Tobias's private user identity. In order to fulfil HTTP digest requirements, the UE includes the following fields in the Authorization header:

- The authentication scheme—set to the value "Digest", as the 3GPP AKA is mapped onto the HTTP Digest mechanism.

- The username field—set to Tobias's private user identity, which will be used by the S-CSCF and the HSS to identify the user and to find the corresponding AV.

- The realm and URI fields—set to the home domain of Tobias.

- The response and nonce fields—which are left empty. These fields are mandated by the HTTP digest, but not used in the initial REGISTER request.

The REGISTER now looks like:

```
REGISTER sip:home1.fr SIP/2.0
Authorization: Digest username="tobias_private@home1.fr",
               realm="home1.fr",
               nonce="",
               uri="sip:home1.fr",
               response=""
```

As the UE and the P-CSCF did not establish any kind of mutual security mechanism at the SIP signalling level, the P-CSCF cannot guarantee that the REGISTER request really does originate from Tobias: for example, a malicious user could have constructed the request and sent it to the P-CSCF, without the P-CSCF knowing. Therefore the P-CSCF adds the integrity-protected field with the value

"no" to the Authorization header, before sending the request toward Tobias's home
network:

```
REGISTER sip:home1.fr SIP/2.0

Authorization: Digest username="tobias_private@home1.fr",
               realm="home1.fr",
               nonce="",
               uri="sip:home1.fr",
               response="",
               integrity-protected="no"
```

5.6.4 S-CSCF challenges the UE

The S-CSCF, after receiving the REGISTER request, identifies the user by the
private user identity found in the username field and downloads the AV from the
HSS. Based on the data in the AV, it returns the WWW-Authenticate header in
the 401 (Unauthorized) response and populates its fields as follows:

- In the nonce field it has the RAND and AUTN parameters, both 32 bytes long
  and Base64-encoded (the nonce field may include additional server-specific data.

- In the algorithm field it has the value "AKAv1-MD5", which identifies the
  3GPP AKA mechanism.

- And in the ik and ck extension fields it has the integrity and ciphering keys.
  Note that these two fields are not part of the original definition of the WWW-
  Authenticate header, which is defined in [RFC3261]. These fields are defined in
  [3GPP TS 24.229].

The WWW-Authenticate fields look like:

```
SIP/2.0 401 Unauthorized
WWW-Authenticate: Digest realm="home1.fr",
                  nonce=A34Cm+Fva37UYWpGNB34JP,
                  algorithm=AKAv1-MD5,
                  ik="0123456789abcdeedcba9876543210",
                  ck="9876543210abcdeedcba0123456789"
```

After receiving the 401 (Unauthorized) response, the P-CSCF must remove and
store the ik and ck fields from the WWW-Authenticate header, before sending the
response toward the UE:

```
SIP/2.0 401 Unauthorized
WWW-Authenticate: Digest realm="home1.fr",
                nonce=A34Cm+Fva37UYWpGNB34JP, algorithm=AKAv1-MD5
```

5.6.5 UE's response to the challenge

From the received AUTN parameter the ISIM application in Tobias's UE now discovers that it was really Tobias's home operator network that sent the 401 (Unauthorized) response. It can also derive from the AUTN that the SQN (sequence number) is still in sync between the HSS and the ISIM.

The received parameters as well as the shared secret allow the ISIM to generate the values for the response and hand them over to the UE. The UE adds the Authorization header to the second REGISTER request, including (among others) the following fields:

- The username field—which includes Tobias's private user identity.

- The nonce field—which is returned with the same value as it was received in the WWW-Authenticate header of the 401 (Unauthorized) response.

- The response field—which includes the authentication challenge RES that was derived by the ISIM from the received RAND and the shared secret.

The ISIM will also calculate the IK, which is also known by the P-CSCF. Based on this key (and other information—see Section 5.7) the UE and the P-CSCF establish IPsec SAs, over which the UE sends the second REGISTER request:

```
REGISTER sip:home1.fr SIP/2.0
Authorization: Digest username="user1_private@home1.fr",
                realm="home1.fr",
                nonce=A34Cm+Fva37UYWpGNB34JP, algorithm=AKAv1-MD5,
                uri="sip:home1.fr",
                response="6629fae49393a05397450978507c4ef1"
```

5.6.6 Integrity protection and successful authentication

The P-CSCF is now in a position to discover whether the received REGISTER request was modified on its way from the UE to the P-CSCF, as it can now check its integrity. If this check is successful, the P-CSCF adds the "integrity-protected" field with the value "yes" to the Authorization header and sends the REGISTER request toward Tobias's home network:

```
REGISTER sip:home1.fr SIP/2.0
Authorization: Digest username="user1_private@home1.fr",
                realm="home1.fr",
                nonce=A34Cm+Fva37UYWpGNB34JP, algorithm=AKAv1-MD5,
                uri="sip:home1.fr",
                response="6629fae49393a05397450978507c4ef1",
                integrity-protected="yes"
```

The S-CSCF now compares the received RES and the XRES that was included in the AV. If these two parameters are identical, then the S-CSCF has successfully authenticated the user. Only after that, will it proceed with normal SIP registration procedures.

5.6.7 Related standards

Specifications relevant to Section 5.6 are:

- 3GPP TS 33.102   Security architecture.

- 3GPP TS 33.203   Access security for IP-based services.

- RFC2401   Security Architecture for the Internet Protocol.

- RFC2403   The Use of HMAC-MD5-96 within ESP and AH.

- RFC2404   The Use of HMAC-SHA-1-96 within ESP and AH.

- RFC2617   HTTP Authentication: Basic and Digest Access Authentication.

- RFC3310   Hypertext Transfer Protocol (HTTP) Digest Authentication Using Authentication and Key Agreement (AKA).

5.7 Access security—IPsec SAs

5.7.1 Overview

Section 3.6.4 describes how access security works in principle. Security via the Gm interface is achieved by means of IPsec SAs, which require specific handling at the

SIP signalling level. This section describes how the UE and P-CSCF negotiate the security mechanism, how IPsec-related parameters are exchanged and how SAs are established and handled.

As the establishment of IPsec SAs is based on authentication of the user, new SAs are established during every re-authentication process. Consequently, new pairs of IPsec SAs have to be established between the UE and the P-CSCF.

5.7.2 Establishing an SA during initial registration

The initial REGISTER request as well as the 401 (Unauthorized) response are sent between the UE and the P-CSCF without any kind of protection. These two messages transport information that allows the UE and the P-CSCF to negotiate the security mechanism and to agree on the parameters and ports that will be used for the SAs.

During the registration process two pairs of IPsec SAs are established between the UE and the P-CSCF. Unless otherwise stated, such a set of two pairs of security associations is referred to as a "set of SAs", while a single or specific IPsec security association from these four is referred to as an "SA".

The four IPsec SAs are not static connections (e.g., TCP connections). They can be regarded as logical associations between the UE and the P-CSCF that allow the secure exchange of SIP messages.

A set of SAs facilitates four ports:

- The protected client port at the UE (uc1).

- The protected server port at the UE (us1).

- The protected client port at the P-CSCF (pc1).

- And the protected server port at the P-CSCF (ps1).

These ports are negotiated between the UE and the P-CSCF during initial registration (Figure 5.5) by using the Security-Client, Security-Server and Security-Verify headers of the SIP Security Mechanism Agreement (see Section 5.8).

The set of SAs needs to be established with a shared key. Unfortunately, the P-CSCF knows nothing about the security parameters that are shared between Tobias's ISIM application and the HSS in the home network. Therefore, the S-CSCF sends the IK and the CK to the P-CSCF within the WWW-Authenticate header in the 401 (Unauthorized) response. The P-CSCF must remove these two keys from the header and store them locally before sending the 401 (Unauthorized) response toward the UE. The IK is then used by the P-CSCF as the shared key for the set of SAs. The UE on the other side of the Gm interface calculates the IK from

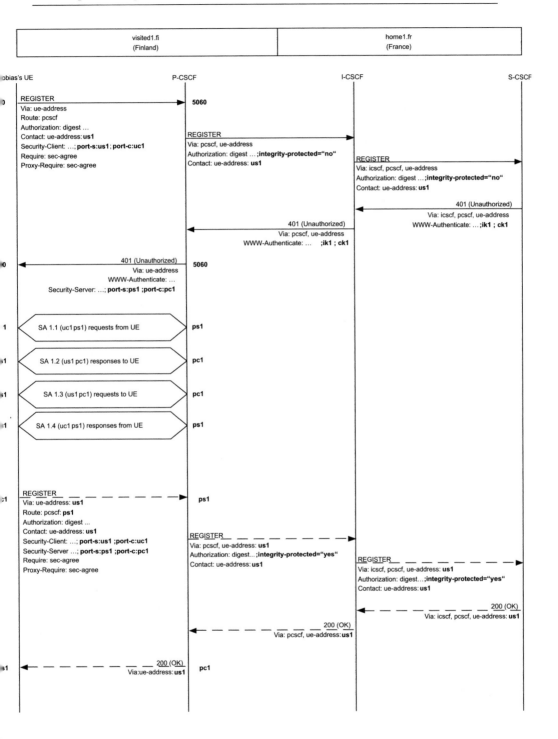

Figure 5.5 SA establishment during initial registration.

the received challenge in the 401 (Unauthorized) response and also uses it as the shared key (see Section 5.6.6).

By means of the IK, the P-CSCF and the UE can then establish the set of SAs between the four ports that were exchanged beforehand in the initial REGISTER request and its reponse:

- Between uc1 and ps1 for sending SIP requests from the UE to the P-CSCF.

- Between us1 and pc1 for sending SIP responses from the P-CSCF to the UE.

- Between us1 and pc1 for sending SIP requests from the P-CSCF to the UE.

- And between uc1 and ps1 for sending SIP responses from the UE to the P-CSCF.

After their establishment the set of SAs gets assigned a temporary lifetime. Although the UE will send all subsequent requests and responses via this temporary set of SAs, the set of SAs cannot be taken into use until the authentication procedure between the UE and the S-CSCF has been finished. This is done in order to ensure that the security mechanism between the UE and the P-CSCF is based on successful authentication of the user.

When sending the 200 (OK) response to the UE, the P-CSCF will update the lifetime of the set of SAs with the lifetime of the registration (as indicated in the expires value of the Contact header) plus 30 seconds. The UE will do the same after receiving the 200 (OK) response.

In the case of initial registration (as described here), both sides (i.e., P-CSCF and UE) will immediately afterwards take this set of SAs into use. This means that the P-CSCF will send all SIP messages that are directed toward the UE via the established set of SAs. The UE will in the same way send all SIP messages via the established set of SAs.

5.7.3 Handling of multiple sets of SAs in case of re-authentication

We have now seen how the first set of SAs is established during initial registration. As the establishment of a set of SAs is based on the authentication data that are sent from the S-CSCF in the 401 (Unauthorized) response, every re-authentication will generate a new set of SAs between the UE and the P-CSCF. Re-authentication procedures are described in Section 5.13.

After successful re-authentication the UE and the P-CSCF will maintain two sets of SAs (Figure 5.6):

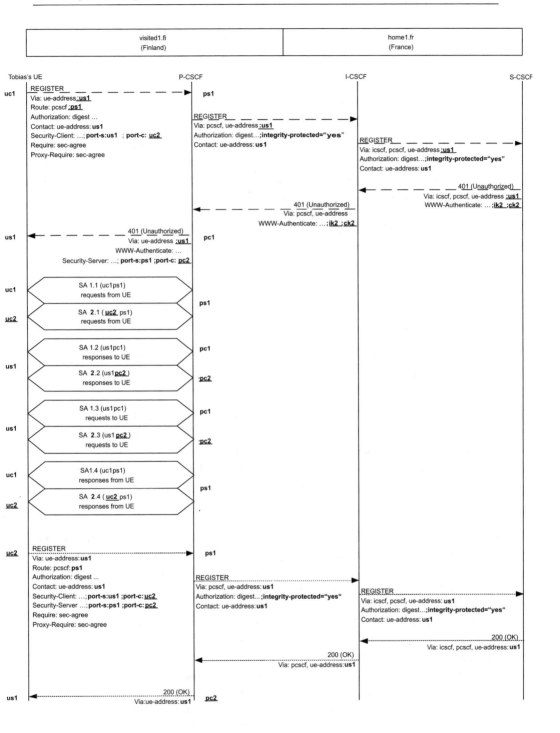

Figure 5.6 Two sets of SAs during re-authentication.

- the set of SAs that was already established and in use before the re-registration took place, which is now called the old set of SAs; and

- a new set of SAs that was established based on re-authentication, which is now called the new set of SAs.

The major complication in this situation is that the P-CSCF cannot be sure whether the 200 (OK) response for the second REGISTER request has been received by Tobias's UE, as SIP defines no acknowledgement mechanism for received responses for any other request than an INVITE. If the UE has not received the 200 (OK) response for the second REGISTER, then it will not take into use the new set of SAs. Therefore, it has to wait until the UE sends a new request on the new set of SAs before it can take the new set of SAs into use. This means that, as long as the P-CSCF does not receive a request from the UE on the new set of SAs, it will:

- send incoming requests to the UE over the old set of SAs (i.e., from its protected client port pc1 to the UE's protected server port us1); and

- keep both sets of SAs active until one or both of them either expires or a new request from the UE is received.

In our example we assume that the UE has received the 200 (OK) for the second REGISTER request and, therefore, is aware that the authentication procedure was successful and the new set of SAs can be used. Unfortunately, the P-CSCF does not know this and will send incoming requests to the UE over the old set of SAs; therefore, the UE also needs to maintain both sets of SAs.

When the UE needs to send out a new request, it will send it by means of the new set of SAs, which will confirm to the P-CSCF that the new set of SAs can be taken into full use (Figure 5.7). Furthermore, at this moment the old set of SAs will not be immediately dropped, as the UE might have received or sent a request over it, which the remote side has not yet responded to. Therefore, the old set of SAs is kept for another 64*T1 seconds (usually 128 seconds in an IMS environment), before it is dropped.

Note also that the UE cannot take the new set of SAs into use by sending a response (e.g., a 200 (OK) response) for a request (e.g., a MESSAGE request) that was received over the old set of SAs. The UE is forced either by the Via header of the P-CSCF or due to a TCP connection to send the response to the same port and over the same set of SAs as the request was received.

Whenever a set of temporary SAs is established the UE will drop all other SAs, other than the one over which it sent the last REGISTER request. Consequently, the UE never needs to handle more than two sets of SAs at the same time.

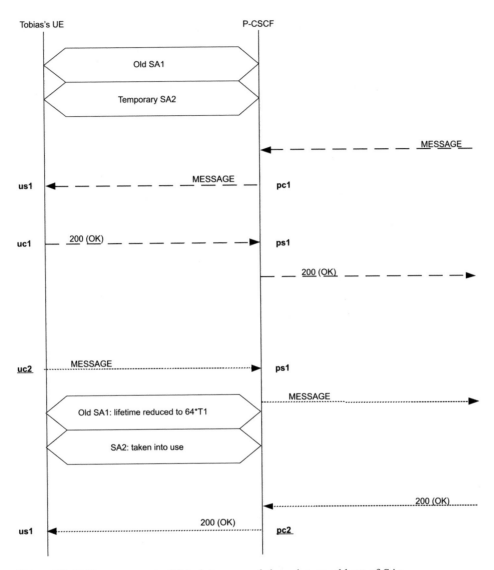

Figure 5.7 Taking a new set of SAs into use and dropping an old set of SAs.

5.7.4 SA lifetime

During an ongoing authentication procedure the lifetime of a temporary set of SAs is restricted to 4 minutes. This guarantees that the authentication procedure can be finished. After successful authentication the lifetime of the new set of SAs is set to:

- Either the expiration time of the concluded registration plus 30 seconds. The expiration time of the registration is indicated in the expires parameter that is returned in the Contact header of the 200 (OK) response to the REGISTER.

- Or, if another set of SAs does already exist, to the lifetime of that already-existing set of SAs as long as its lifetime is longer than the expiration time of the just-concluded registration plus 30 seconds.

Whenever a re-registration takes place and is successful the P-CSCF and the UE have to update the lifetime of all existing SAs with the expiration time of the concluded re-registration plus 30 seconds, if that value is bigger than the already-assigned lifetime of the SAs.

Consequently, the SAs between the UE and the P-CSCF will be kept 30 seconds longer than Tobias is registered to the IMS network.

When the P-CSCF becomes aware that Tobias is no longer registered (e.g., by receiving a NOTIFY with Tobias's registration-state information which indicates network-initiated de-registration—see Section 5.14.3), the P-CSCF will drop all SAs toward the UE after 64*T1 seconds.

5.7.5 Port setting and routing

Special attention has to be paid when it comes to the usage of SA ports, as they heavily influence the routing between the P-CSCF and the UE. As shown in Figure 5.6, Tobias's UE:

- Will send all requests from its protected client port (2468).

- Expects all responses to be received on its protected server port (1357).

- Expects all requests to be received at its protected server port (1357).

- Will send all responses to received requests from its protected client port (2468).

The P-CSCF, on the other hand:

- Will send all requests toward the UE from its protected client port (8642).

- Expects to receive all responses from the UE at its protected server port (7531).

- Expects to receive all requests from the UE at its protected server port (7531). and

- And will send all responses toward the UE from its protected client port (8642).

To ensure that all requests are sent via IPsec SAs:

- The UE will set its protected server port as part of its address:

  o In the Contact header of every request (including all REGISTER requests).

  o In the Via header of every request, besides the initial REGISTER.

- The UE will set the protected server port of the P-CSCF as part of the outbound proxy (i.e., P-CSCF) address in the Route header of every initial request that it sends.

- The P-CSCF will set its protected server port as part of its address:

  o In the Record-Route header of every initial request that is sent toward the UE.

  o In the Record-Route header of every response that carries the P-CSCF's Record-Route entry toward the UE (for detailed setting of port numbers in the Record-Route header see Section 6.3).

5.7.5.1 Port setting during registration

For example, Tobias's UE initially registers with the following information:

```
REGISTER sip:home1.fr SIP/2.0
Via: sip:[5555:1:2:3:4];branch=0uetb
Route: <sip:[5555::a:b:c:d];lr>
Security-Client: digest, IPsec-3gpp; alg=hmac-sha-1-96
         ;spi-c=23456789 ;spi-s=12345678
         ;port-c=2468; port-s=1357
Contact: sip:[5555::1:2:3:4]:1357
```

This means that the UE:

- Is going to establish IPsec SA with:

  o Port 2468 as the protected client port (port-c parameter of the Security-Client header).

  o Port 1357 as the protected server port (port-s parameter of the Security-Client header).

- Expects all incoming requests to be routed to its protected server port (port value in the Contact header).

- Will send this initial REGISTER request to the unprotected port 5060 of the P-CSCF, as no port value is given in the Route header.

- Will await all responses to this initial REGISTER request on the unprotected port 5060, as no port value is given in the Via header.

The 401 (Unauthorized) response that is received afterwards by the UE will look like this:

```
SIP/2.0 401 Unauthorized
Via: sip:[5555:1:2:3:4];branch=0uetb
Security-Server: tls ;q=0.2, IPsec-3gpp; q=0.1
          ;alg=hmac-sha-1-96
          ;spi-c=98765432 ;spi-s=87654321
          ;port-c=8642 ;port-s=7531
```

This means that the P-CSCF is going to establish IPsec SA with:

- Port 8642 as the protected client port (port-c parameter of the Security-Server header).

- And port 7531 as the protected server port (port-s parameter of the Security-Server header).

After this exchange the UE and the P-CSCF will set up the temporary set of SAs and the UE will then send the second REGISTER request already protected, which then will look like:

```
REGISTER sip:home1.fr SIP/2.0
Via: sip:[5555:1:2:3:4]:1357;branch=1uetb
Route: <sip:[5555::a:b:c:d]:7531;lr>
Contact: sip:[5555::1:2:3:4]:1357
```

Note that the Security-Client and Security-Verify headers are also included in this request (see Section 5.8), but as they no longer have any influence on SA establishment and the routing, they are not shown here. This means that the UE:

- Expects all incoming initial requests to be routed to its protected server port (port value in the Contact header).

- Sends this REGISTER request already over the temporary IPsec SA (i.e., to the protected server port of the P-CSCF—port value in the Route header).

- And expects all responses to this REGISTER request to be sent via the temporary IPsec SA (i.e., on its protected server port 1357—port value in the Via header).

5.7.5.2 Port setting during re-authentication

When exchanging the security parameter indexes and protected port numbers for the new set of SAs according to the SIP Security Mechanism Agreement, the P-CSCF and the UE only change their protected client ports:

- the UE receives requests and responses for both sets of SAs via its protected server port (us1);

- the P-CSCF receives requests and responses for both sets of SAs via its protected server port (ps1);

- the UE uses a new protected client port (uc2) for sending requests and responses toward the P-CSCF over the new set of SAs; and

- the P-CSCF also uses a new protected client port (pc2) for sending requests and responses toward the UE over the new set of SAs.

This is due to the fact that two sets of SAs must not use the same port parameters. Furthermore, if the protected server ports change, this would cause major problems and would mean that:

- the UE would need to perform re-registration, as its registered contact includes the protected server port;

- the UE would need to send re-INVITE on all established sessions, as its contact information that was sent to the remote side includes the protected server port;

- the P-CSCF would receive from the UE all subsequent requests to every already-established dialog (including all subscriptions of the UE) on the P-CSCF's old, protected server port, as there is no possibility in SIP to change the route information for an already-established dialog.

This list is not complete, but it shows that changing the protected server port would cause a lot of problems for SIP routing. Therefore, it is essential that this value is not changed as long as the user stays registered.

5.7.5.3 Port settings for other SIP requests than REGISTER

The setting of the protected ports in non-REGISTER requests is described in more detail in Section 6.3.

5.7.5.4 Usage of ports with UDP and TCP

The previous sections showed how requests and responses are routed via one or more sets of SAs. In the chosen example only UDP was used as a transport protocol. For TCP, however, there is a slight difference in these procedures.

When a request is sent out via UDP (Figure 5.8) the Via header indicates the IP address and port number to which all related responses should be routed. When the TCP is used to send the request (Figure 5.9) the information in the Via header is overridden and the response is routed back to the same address and port that the request was received from. This draws attention to the nature of TCP as a connection-oriented transport protocol. By applying this rule it is ensured that no additional TCP connection needs to be opened to send the response to a request that was received via TCP. This causes the routing of SIP messages between the P-CSCF and the UE to behave differently. The UE will set its protected server port (us1) in the Via header of every request that it sends out, regardless of whether UDP or TCP is used. All requests will originate from the UE's protected client port (uc1).

In the case of UDP the responses to such a request will be sent to the UE's protected server port (us1), as indicated in the Via header.

In the case of TCP the responses to such a request will be sent to the UE's protected client port (uc1), as the request originated from there.

The same is true in the other direction (i.e., for requests sent from the P-CSCF toward the UE and their responses).

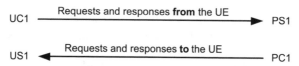

Figure 5.8 Request and response routing between the UE and the P-CSCF over UDP.

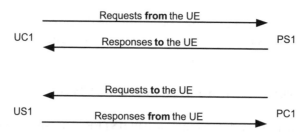

Figure 5.9 Request and response routing between the UE and the P-CSCF over TCP.

5.7.6 Related standards

Specifications relevant to Section 5.7 are:

- 3GPP TS 33.102   Security architecture.

- 3GPP TS 33.203   Access security for IP-based services.

- 3GPP TS 33.210   Network Domain Security (NDS); IP network layer security.

- RFC2401   Security Architecture for the Internet Protocol.

- RFC2403   The Use of HMAC-MD5-96 within ESP and AH.

- RFC2404   The Use of HMAC-SHA-1-96 within ESP and AH.

- RFC2451   The ESP CBC-Mode Cipher Algorithms.

5.8 SIP Security Mechanism Agreement

5.8.1 Why the SIP Security Mechanism Agreement is needed

The IMS in 3GPP Releases 5 and 6 makes use of IPsec as the security mechanism between the P-CSCF and the UE. IPsec is only one of several possible security mechanisms. The IMS was designed to allow alternative security mechanisms over the Gm interface as well. Allowing such an openness usually creates backward compatibility problems because, for example, a Release 6-compliant UE would not be able to understand any alternative security mechanism, while it could be attached to a P-CSCF of a higher release that would already support alternatives to IPsec.

Therefore, the SIP Security Mechanism Agreement (Sip-Sec-Agree) was introduced to allow the UE and the P-CSCF to negotiate a common security mechanism for use between them. For current releases the only security mechanism is IPsec; however, it might be that some entities already support alternative mechanisms on a proprietary basis.

5.8.2 Overview

To make the example not too simple and boring, we assume that the UE supports IPsec and the HTTP digest and the P-CSCF supports IPsec and the Transport Layer Security (TLS), with a preference toward TLS. It is not necessary that the reader of this chapter has any knowledge of any of these mechanisms.

As we have seen, the initial REGISTER request is sent without any protection from the UE to the P-CSCF. To guarantee that a common security mechanism can be established, Tobias's UE advertises the mechanisms it already supports in this initial REGISTER request within the Security-Client header, which includes a list of supported mechanisms.

The P-CSCF sends back in the 401 (Unauthorized) response a Security-Server header, which includes the list of supported mechanisms from the P-CSCF's side. Furthermore, the P-CSCF adds a preference (q-value) to each of the mechanisms.

Based on this information, both sides now know which common mechanisms are supported by the UE and the P-CSCF. If there is more than one common mechanism, the mechanism which was given the highest preference by the P-CSCF will be selected and applied. To guarantee that this mechanism can be established immediately, the P-CSCF will send further information in the 401 (Unauthorized) response to enable the UE to set up the mechanism: for example, in a non-IMS environment it could send a Proxy-Authenticate header when the HTTP digest is the chosen mechanism.

As we saw in Section 5.7, the UE and the P-CSCF will then establish the security mechanism, which is in our case based on IPsec SAs. Afterwards, all messages between the two entities will be sent protected over these SAs.

Nevertheless, the initial REGISTER request and its response are still not protected. There is the slight chance that a malicious user has tampered with the messages or that an error has occurred over the vulnerable air interface.

As shown in earlier chapters, the second REGISTER request from the UE repeats all the information necessary for authentication and registration, both of which are performed with the S-CSCF. In order to guarantee that SIP Security Mechanism Agreement-related information also have not been changed, the UE:

- repeats the Security-Client header that it sent in the initial REGISTER in the second REGISTER request as well; and

- copies the content of the Security-Server header that was received in the 401 (Unauthorized) response from the P-CSCF into a Security-Verify header and sends it along with the second REGISTER request as well.

As long as the established Security-Association is used, the UE will always repeat the same Security-Verify header in every request that it sends to the P-CSCF.

During the exchange between the Security-Client (from the UE) and the Security-Server (from the P-CSCF) headers, the two sides also agree on some parameters for IPsec SAs: that is, they indicate to each other the protected client and server ports (port-c and port-s) as well as the security parameter indexes (SPIs: spi-c and spi-s).

5.8.3 SIP Security Mechanism Agreement-related headers in the initial REGISTER request

In order to activate the agreement on the security mechanism, the UE includes the following information in the initial REGISTER request:

```
REGISTER sip:home1.fr SIP/2.0
Require: sec-agree
Proxy-Require: sec-agree
Security-Client: digest, IPsec-3gpp ;alg=hmac-sha-1-96
  ;spi-c=23456789 ;spi-s=12345678
  ;port-c=2468 ;port-s=1357
```

The Proxy-Require header includes the option tag "sec-agree"; this indicates that the next hop proxy, in this case the P-CSCF, must support the procedures for the SIP Security Mechanism Agreement in order to process the request further. If the next hop proxy does not support SIP Security Mechanism Agreement procedures, it would—based on the handling of the Proxy-Require header, which is defined in the main SIP [RFC3261]—send back a 420 (Bad Extension) response, including an Unsupported header with the option tag "sec-agree". As the P-CSCF in our example is fully IMS Releases 5 and 6-compliant, it of course supports SIP SA procedures and will not send this response to the UE.

Furthermore, the Require header includes the sec-agree option tag. This is mandated to be included by [RFC3329], which defines the SIP Security Mechanism Agreement. The Require header is used in the same way as the Proxy-Require header, but by the remote UE (not the proxy). It is there just in case a request (in this case the REGISTER request) is sent directly from the sending UE to the final destination (the S-CSCF), which would not look at the Proxy-Require header at all; this would mean that no negotiation of the security mechanism would take place. The Require header forces the receiving side to perform the sec-agree procedures.

As the P-CSCF is able to perform SIP Security Mechanism Agreement procedures, it removes the sec-agree option tags from the Require and Proxy-Require headers before sending the request toward Tobias's home operator network.

Tobias's UE sends the list of supported security mechanisms to the P-CSCF within the Security-Client header. The P-CSCF will discover, based on the information given in this header, that Tobias's UE supports two security mechanisms: one is the HTTP digest ("digest") and the other is IPsec as used in the 3GPP ("IPsec-3gpp"). These two mechanisms are separated by commas in the header. The list of parameters (separated by ";") for the latter includes:

- the algorithm (alg parameter)—used for IPsec encryption and protection—in this case it is the HMAC SHA 1-96 algorithm, which is defined in [RFC2404];

- the protected client port (port-c) and the protected server port (port-s)—used from the UE's side for IPsec SAs; and

- the SPI—used for the IPsec SA that relates to the protected client port (spi-c) as well as the SPI used for the IPsec SA that relates to the protected server port (spi-s).

The P-CSCF will also remove the Security-Client header before sending the REGISTER request further.

Note that for the IMS only the IPsec-3gpp security mechanism is relevant. The example given here uses digest and TLS as possible additional security mechanisms in SIP Security Mechanism Agreement-related headers. This is only done to explain the procedures behind the negotiation process.

5.8.4 The Security-Server header in the 401 (Unauthorized) response

When receiving a 401 (Unauthorized) response from the S-CSCF for a REGISTER request, the P-CSCF includes a list of supported security mechanisms in a Security-Server header in the response:

```
SIP/2.0 401 Unauthorized
Security-Server: tls ;q=0.2, IPsec-3gpp; q=0.1
  ;alg=hmac-sha-1-96
  ;spi-c=98765432 ;spi-s=87654321
  ;port-c=8642 ;port-s=7531
```

In this example the P-CSCF supports two security mechanisms: 3GPP-specific usage of IPsec and TLS. It even gives a higher preference to TLS: should the UE also support TLS, this would be chosen to protect the messages between the UE and the P-CSCF.

Furthermore, the P-CSCF sends IPsec-related information about SPIs and protected client and server ports in the same way as the UE.

At the point of sending out the 401 (Unauthorized) response to the UE, the P-CSCF is already aware that the IPsec will be used as the security mechanism, as it knows that this is the only mechanism that is supported by both the UE and itself.

5.8.5 SIP Security Mechanism Agreement headers in the second REGISTER

After receiving the 401 (Unauthorized) response the UE is able to set up IPsec SAs. When this has been done, it can use already-established SAs to send the second REGISTER request over it. In this REGISTER request it now includes the following related information:

```
REGISTER sip:home1.fr SIP/2.0
Require: sec-agree
Proxy-Require: sec-agree
Security-Verify: tls ;q=0.2, IPsec-3gpp ;q=0.1
  ;alg=hmac-sha-1-96
  ;spi-c=98765432 ;spi-s=87654321
  ;port-c=8642 ;port-s=7531
Security-Client: digest, IPsec-3gpp ;alg=hmac-sha-1-96
  ;spi-c=23456789 ;spi-s=12345678
  ;port-c=2468 ;port-s=1357
```

Once again the Require and Proxy-Require headers with the option tag "sec-agree" are there. They serve the same purpose as in the initial REGISTER (see Section 5.8.3) and will be repeated in every REGISTER request that is sent from the UE. The P-CSCF will always remove them before sending the request on, in the same way as it did for the initial REGISTER request. If either the Proxy-Require or the Require header (or both) are found empty after the sec-agree option tag has been removed, the P-CSCF will also remove this or these empty headers.

The Security-Verify header includes a copy of the received Security-Server header. The Security-Client header is simply re-sent as in the initial REGISTER request.

The P-CSCF will compare the two Security-Client headers that were received in the initial and this second REGISTER request and see whether they match. It will also compare the content of the Security-Server header that it sent in the 401 (Unauthorized) response and with the content of the Security-Verify header that it received in this second REGISTER request.

Before sending the REGISTER request any further, the P-CSCF will remove the Security-Client and Security-Server headers from it.

5.8.6 SIP Security Mechanism Agreement and re-registration

The S-CSCF can decide to re-authenticate the UE during every re-registration procedure, and by doing so it will force the UE and the P-CSCF to establish a new set of IPsec SAs, as these IPsec SAs are based on the IK, which changes during each re-authentication procedure (see Section 5.13.2). Establishing a new set of IPsec SAs also means that a new set of SPIs and new, protected client and server ports are negotiated.

When sending the new REGISTER request for re-registration the UE cannot be sure whether the S-CSCF will request re-authentication. Therefore, it will add in every new REGISTER request a new Security-Client header with new values for the SPIs and the protected client and server ports:

```
REGISTER sip:home1.fr SIP/2.0
Require: sec-agree
Proxy-Require: sec-agree
Security-Verify: tls ;q=0.2, IPsec-3gpp ;q=0.1
  ;alg=hmac-sha-1-96
  ;spi-c=98765432 ;spi-s=87654321
  ;port-c=8642 ;port-s=7531
Security-Client: digest, IPsec-3gpp ;alg=hmac-sha-1-96
  ;spi-c=23456790 ;spi-s=12345679
  ;port-c=2470 ;port-s=1357
```

Note that the values for the SPIs and the protected client port number have changed in the Security-Client header, in order to allow the set-up of a new set of SAs, should the S-CSCF re-authenticate the UE. The protected server port of the UE has not changed and will be kept throughout the user's registration (see Section 5.7.3).

The content of the Security-Verify header is sent unchanged, because it is a copy of the latest received Security-Server header.

Both the P-CSCF and the UE will know, at the moment of receiving the response to this REGISTER request from the S-CSCF, whether new IPsec SAs have to be established: that is, whether a 401 (Unauthorized) response is received or whether a 200 (OK) response is received.

When a 401 (Unauthorized) response is received from the S-CSCF, the P-CSCF will add a new Security-Server header to the response, providing new values for the protected ports and new SPIs.

```
SIP/2.0 401 Unauthorized
Security-Server: tls ;q=0.2, IPsec-3gpp ;q=0.1
  ;alg=hmac-sha-1-96
  ; spi-c=98765434 ;spi-s=87654322
  ;port-c=8644 ;port-s=7531
```

Furthermore, the P-CSCF will not change the value of the protected server port (7531). Consequently, the UE and the P-CSCF will now establish a new set of temporary SAs (see Section 5.7.3). The REGISTER request, which includes the response to the re-authentication challenge (see Section 5.6), will be sent over this new, temporary set of SAs and will include the following headers:

```
REGISTER sip:home1.fr SIP/2.0
Require: sec-agree
Proxy-Require: sec-agree
Security-Verify: tls ;q=0.2, IPsec-3gpp ;q=0.1
  ;alg=hmac-sha-1-96
  ;spi-c=98765434 ;spi-s=87654322
  ;port-c=8644 ;port-s=7531
Security-Client: digest, IPsec-3gpp ;alg=hmac-sha-1-96
  ;spi-c=23456790 ;spi-s=12345679
  ;port-c=2470 ;port-s=1359
```

Once again, as during the initial registration procedure (Figure 5.10), the second REGISTER request repeats the Security-Client header that was sent in the latest REGISTER request (with the new values) and copies into the Security-Verify header the values of the Security-Server header that was received in the last 401 (Unauthorized) response: in other words, this second REGISTER request within the re-registration procedure no longer carries any information related to any previously established set of SAs.

5.8.7 Related standards

Specifications relevant to Section 5.8 are:

- 3GPP TS 33.203   Access security for IP-based services.

- RFC2246   The TLS Protocol Version 1.0.

- RFC2617   HTTP Authentication: Basic and Digest Access Authentication.

- RFC3329   Security Mechanism Agreement for the Session Initiation Protocol (SIP).

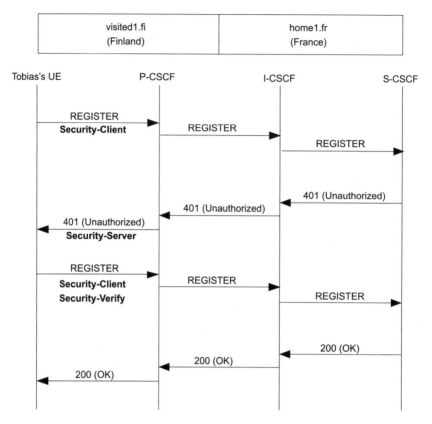

Figure 5.10 SIP Security Mechanism Agreement during initial registration.

5.9 Compression negotiation

5.9.1 Overview

The ability to compress SIP messages over the air interface is essential for the IMS. How signalling compression (SigComp) works is described in Section 3.16. This section shows how the UE and the P-CSCF indicate that they support SigComp and are both willing to use it.

P-CSCF and IMS UE must support SIP signalling compression (SigComp), but they are not mandated to use it. Therefore, they need a mechanism to express whether they are willing to apply signalling compression.

[RFC3486] defines a new URI parameter "comp", which can be set to "comp=SigComp" by either the UE or a SIP proxy (in the case of the IMS this applies only to the P-CSCF) in order to express its willingness to route certain SIP messages compressed.

Tobias's UE will express its willingness to use signalling compression with the P-CSCF that is already in the initial REGISTER request. The P-CSCF will give a similar indication in the 401 (Unauthorized) response. As these two SIP messages are sent without any protection, the P-CSCF and the UE will not create states (compartments) for signalling compression at this point in time; this is to ensure that a malicious user, who wants, say, to start a denial of service (DOS) attack against the P-CSCF, cannot overload the P-CSCF by forcing it to reserve memory for a huge number of unnecessary SigComp compartments.

State creation will only be done after IPsec SA (see Section 5.7) between the UE and the P-CSCF has been established.

5.9.2 Indicating willingness to use SigComp

The "comp" parameter can be set:

- By the UE in the Contact header of the REGISTER request—this means that the UE is willing to receive every initial request that is destined for it compressed, as initial requests that are destined to the UE are routed based on the registered contact address.

- By the UE in the Contact header of any other initial request or the first response to an initial request—this means that the UE is willing to receive every subsequent request within this dialog compressed, as subsequent requests are routed based on the address in the Contact header of the initial request (from the originating side) or the first response to an initial request (from the terminating side).

- By the UE in the Via header of any request—this means that the UE is willing to receive all responses to this request compressed, as responses are routed based on the Via header in the related request.

- By the P-CSCF in its own entry to the Record-Route header that is sent toward the UE—this means that the P-CSCF is willing to receive subsequent requests within this dialog compressed, as subsequent requests are routed toward SIP proxies based on the entries in the Route header (which is generated from the Record-Route header).

- And by the P-CSCF in the Via header of any request—this means that the P-CSCF is willing to receive all responses to this request compressed, as responses are routed based on the Via header in the related request.

5.9.3 comp=SigComp parameter during registration

The initial REGISTER request by the UE will include the following compression-related information:

```
REGISTER sip:home1.fr SIP/2.0
Via: SIP/2.0/UDP sip:[5555::1:2:3:4];comp=SigComp ;branch=0uetb
Route: sip:[5555::a:b:c:d];lr
Contact: <sip:[5555::1:2:3:4]:1357;comp=SigComp>;expires=600000
```

The comp=SigComp parameter is included in the Via header and indicates that the UE is willing to receive all responses to this request compressed. Consequently, the P-CSCF may send the 401 (Unauthorized) response already compressed, but it will not create a state (i.e., a compartment) because of this.

The comp=SigComp parameter can also be found in the Contact header. This parameter will be included in every initial request that is received by the UE, as the S-CSCF will replace the request URI (which points to sip:tobias@home1.fr) of every initial request with the registered contact address (i.e., sip:[5555::1:2:3:4]:1357; comp=SigComp).

The 401 (Unauthorized) response from the P-CSCF will not include any further information on the P-CSCF's ability to perform signalling compression. The P-CSCF address that was discovered before the initial registration (see Section 5.3) cannot be discovered with the comp=SigComp parameter. As SIP messages should only be sent compressed when the comp=SigComp parameter is set in the address of the next hop, the UE would therefore not send any initial request to the P-CSCF compressed.

Subsequent requests (such as ACK, PRACK, UPDATE or BYE) could be sent compressed, as the routing from the UE to the P-CSCF would be based on the Record-Route entry of the P-CSCF (see Section 6.3.2), in which the P-CSCF can include the comp=SigComp parameter. The same is true for responses sent from the UE to the P-CSCF, as they are routed based on the Via header entry of the P-CSCF, which is also set by the P-CSCF itself.

Although it is a requisite for the comp parameter to indicate whether compression is used, 3GPP TS 24.229 version 5.6.0 does not make a clear requirement on compression of the initial message. One possibility would be that the UE just sends every initial request compressed, as the P-CSCF must support the SigComp no matter what.

Another possibility would be that the UE queries the P-CSCF with an OPTIONS request after successful registration. The P-CSCF then would return its address, including the comp=SigComp parameter, in the Contact header of the 200 (OK) response to the OPTIONS request. As already mentioned, this issue needs further clarifications; it is also a good example of the ongoing activities in 3GPP standardization.

For this example we assume that the UE just adds the comp=SigComp parameter to the P-CSCF address that was discovered previously. Therefore, it can send out the second REGISTER request already compressed:

```
REGISTER sip:home1.fr SIP/2.0
Via: SIP/2.0/UDP sip:[5555::a:b:c:d];comp=SigComp;branch=1uetb
Route: sip:[5555::a:f:f:e]:7531;lr;comp=SigComp
Contact: <sip:[5555::a:b:c:d]:1357;comp=SigComp>;expires=600000
```

This REGISTER request is routed on the basis of the topmost Route header, which includes the P-CSCF address and the comp=SigComp parameter. As the parameter is already there, the UE can send the request compressed.

The 200 (OK) response to this REGISTER request will be sent from the P-CSCF to the UE on the basis of the Via header, and, as the UE also includes the comp=SigComp parameter, the P-CSCF will send it compressed.

5.9.4 comp=SigComp parameter in other requests

The handling of the comp=SigComp parameter in requests other than REGISTER is described in Section 6.4.

5.9.5 Related standards

The comp parameter is defined in [RFC3486]: Compressing the Session Initiation Protocol (SIP).

5.10 Access and location information

5.10.1 P-Access-Network-Info

When the P-Access-Network-Info header is sent in an INVITE request that is sent for an emergency call, the P-CSCF and S-CSCF can determine from the Cell-ID which emergency centre is closest to the user and should be contacted. When writing this chapter the details for IMS emergency calls were still under discussion in 3GPP standardization. In the future there may be more applications that use the information contained in this header.

The P-Access Network-Info header is a 3GPP-specific header and indicates to the IMS network over which access technology the UE is attached to the IMS. In our example the access technology is GPRS. It also includes the cell global ID (CGI), which indicates the location of the user.

Tobias's UE will include the P-Access-Network-Info header in every request (along with ACK and CANCEL requests) and every response (along with responses to the CANCEL request) that it sends out, but only if that request or response is sent integrity-protected (i.e., via an SA, see Section 5.7).

The first time this header is sent out is therefore within the second REGISTER request, which is sent after the 401 (Unauthorized) response has been received by the UE. The header looks like:

```
REGISTER sip:home1.fr SIP/2.0
P-Access-Network-Info: 3GPP-UTRAN-TDD
 ;utran-cell-id-3gpp=234151D0FCE11
```

Tobias's S-CSCF will remove the P-Access-Network-Info header from every request or response that it sends toward another entity. The only exception from this rule is the ASs that are in the same trust domain as the S-CSCF (see Section 5.5.10).

5.10.2 P-Visited-Network-ID

The P-Visited-Network-ID header indicates to Tobias's home network the identification of the network within which Tobias is currently roaming. The header is included by the P-CSCF to which Tobias's UE is attached. The information in this header will be used by the S-CSCF to check the roaming agreement with that visited network.

In this example it is assumed (see Section 4.1) that Tobias is roaming in Finland and is attached to the fictitious Finish operator Musta Kissa. As the P-CSCF is also provided by this operator, it will include a P-Visited-Network-ID header in every REGISTER request that it sends toward Tobias's home network. Within this header will be a string, from which the S-CSCF will recognize the visited network:

```
REGISTER sip:home1.fr SIP/2.0
P-Visited-Network-ID: "Kaunis Musta Kissa"
```

5.10.3 Related standards

3GPP-specific SIP headers are defined in [RFC3455]: Private Header (P-Header) Extensions to the Session Initiation Protocol (SIP) for the 3rd-Generation Partnership Project (3GPP).

5.11 Charging-related information during registration

Charging in the IMS does involve much more than signalling between SIP entities. The charging concept and the relevant entities in the network are described in Section 3.10. The current section only explains the handling and content of SIP headers that are related to charging during registration. A more sophisticated way of charging IMS sessions is described in Section 6.7.7.

When the P-CSCF receives the initial REGISTER request, it creates the IMS charging ID (ICID), which is valid for all IMS-related signalling as long as the user stays registered. The ICID value is transported from the P-CSCF to the S-CSCF in the P-Charging-Vector header:

```
REGISTER sip:home1.fr SIP/2.0
P-Charging-Vector: icid-value=
   "AyretyU0dm+6O2IrT5tAFrbHLso=023551024"
```

The S-CSCF, when receiving this header, will store the ICID and will perform the charging procedures as described in Section 6.7.7.

The P-Charging-Vector header is defined in [RFC3455]. Extensions to this header and its usage within the IMS are described in [3GPP TS 24.229].

5.12 User identities

5.12.1 Overview

Tobias needs to register within his home network in order to be able to originate a call toward his sister. In the example so far he has used the SIP URI sip:tobias@ home1.fr for registration. This is the user identity Tobias uses when he uses IMS services that are not work-related.

Nevertheless, Tobias has a whole set of user identities that are registered at his operator in France, which are shown in Table 5.3.

Table 5.3 Tobias's public user identities.

Registration set	SIP URI	tel URL
1	sip:tobias@hom1.fr	+44-123-456-789
2	sip:tobi@hom1.fr	+44-123-456-111
3	sip:gameMaster@home1.fr	

During the initial registration procedure, Tobias can explicitly register only one of those URIs, which in the our example is sip:tobias@home1.fr. Nevertheless, the IMS allows implicit and explicit registration of further public user identities:

- Some of the above-listed identities might automatically (implicitly) be registered by the network during the initial registration phase.

- Others might stay unregistered until Tobias explicitly requests them to be registered.

When receiving the 200 (OK) response for the second REGISTER request, both Tobias's terminal and the P-CSCF discover Tobias's default public user identity, which is received as the first URI in the P-Associated-URI header.

To find out more about the registration status of the other public user identities that are assigned to Tobias, the UE automatically subscribes to the registration-state event information that is provided by the S-CSCF in the home network. It is mandatory that the UE performs this subscription immediately after the initial registration has succeeded, because:

- The UE needs to get the registration status of the associated URIs.

- The subscription enables the network (S-CSCF) to force the UE to perform re-authentication (see Section 5.13.2).

- The subscription enables the network (S-CSCF) to de-register the user (see Section 5.14.3).

In parallel, the P-CSCF also performs a subscription to the user's registration-state information, mainly to be informed about network-initiated de-registration (see Section 5.14.3).

5.12.2 Public and private user identities for registration

The identities that go into first REGISTER request are read from the ISIM, one of the applications contained on the Universal Integrated Circuit Card (UICC) within the UE. Data read from the ISIM include:

- The private user identity of the user.

- The public user identity of the user which is used for registration.

- And the address of the SIP registrar of the user.

The private user identity is only used for authentication, which is described in Section 5.6. The public user identity is the SIP URI that Tobias is going to initially register. There may be more public user identities available for Tobias, some of them may even be stored on the ISIM; however, at the beginning only one is explicitly registered.

If the UE is not equipped with an ISIM, it will derive the identities and the address of the registrar from the USIM application that also resides in the UICC. The USIM includes all user related data that are needed for circuit switched (CS) and packet-switched (PS) domain registration and authentication. This is described in more detail in Section 5.12.3.

Armed with these parameters the UE can fill in the following fields of the initial REGISTER request:

```
REGISTER sip:home1.fr SIP/2.0
From: <sip:tobias@home1.fr>;tag=pohja
To: <sip:tobias@home1.fr>
Authorization: Digest username="tobias_private@home1.fr",
               realm="home1.fr",
               nonce="",
               uri="sip:home1.fr",
               response=""
```

The public user identity, as read from the ISIM, is put into the To and From headers. The value of the username field of the Authorization header takes the value of the private user identity and the address of the registrar is put into the request URI of the request as well as in the realm and uri fields of the Authorization header.

5.12.3 Identity derivation without ISIM

When Tobias registers, his UE takes the SIP URI "sip:tobias@home1.fr" from the ISIM application that is running on the UICC that he got from his operator and put into his UE. The ISIM always holds at least one valid public user identity.

However, IMS services can also be provided to users who own UICC cards on which no ISIM application—and therefore no valid public user identity—is present. Therefore, the UE needs to create a temporary public user identity from the data available from the USIM application (see Section 3.5) and use this temporary identity for registration.

As the temporary public user identity is constructed from security-related data on the USIM, it must not be exposed to any entity outside the IMS. Therefore, it is treated as a "barred identity": that is, it is strongly recommended that the network reject any usage of this identity outside user's registration.

In this case also the private user identity will be derived from USIM data. It will take the format of the IMSI (International Mobile Subscriber Identity) as the user part, followed by a host part, which includes the MCC (mobile country code) and the MNC (mobile network code) that are included in the IMSI: for example, the private user identity of Tobias could look like: 222330999999999@33.222. IMSI.3gppnetwork.org.

The domain name of Tobias's home network would also be derived from the USIM and would look like the domain part of the private user identity (i.e., 33.222.IMSI.3gppnetwork.org).

5.12.4 Default public user identity/P-Associated-URI header

If Tobias had used a temporary public user identity for his initial registration, he would now have the problem that he would be registered but could not perform any other action (e.g., call his sister or subscribe to a service), as he is registered with an identity that he must not use further (barred identity). His terminal needs to know an identity that has been implicitly registered.

Whenever a user has successfully been authenticated and registered, the S-CSCF therefore sends in the 200 (OK) response for the REGISTER to request the P-Associated-URI header, which lists all the SIP URIs and tel URLs (i.e., public user identities), which are associated but not necessarily registered for the user. Only the first URI listed in this header is always a valid, registered public user identity and can be used by the UE and the P-CSCF for further actions.

The P-Associated-URI in the 200 (OK) response to Tobias's REGISTER request looks like:

```
SIP/2.0 200 OK
P-Associated-URI: <sip:tobias@home1.fr>, <sip:tobi@home1.fr>,
                  <sip:gameMaster@home1.fr>,
                  <sip:+44-123-456-789@home1.fr;user=phone>,
                  <sip:+44-123-456-111@home1.fr;user=phone>
```

From this information Tobias knows that at least the public user identity "sip:tobias @home1.fr" is registered. He also becomes aware that there are two more SIP URIs and two more tel URIs that he can use, but he does not know whether they are currently registered or not.

As the P-Associated-URI is only defined to transport SIP URIs, it includes the tel URLs that are associated with Tobias (tel:+44-123-456-789 and tel:+44-123-456-111) in the format of SIP URIs.

5.12.5 UE's subscription to registration-state information

After the initial registration and authentication has succeeded, Tobias's terminal sends out a SUBSCRIBE request with the following information:

```
SUBSCRIBE sip:tobias@home1.fr SIP/2.0
Via: SIP/2.0/UDP [5555::1:2:3:4]:1357;comp=sigcomp;branch=4uetb
Route: <sip:[5555::a:b:c:d]:7531;lr>
Route: <sip:orig@scscf1.home1.fr;lr>
From: "Tobias" <sip:tobias@home1.fr>;tag=sipuli
To: "Tobias" <sip:tobias@home1.fr>
P-Preferred-Identity: "Tobias" <sip:tobias@home1.fr>
Event: reg
Expires: 600000
Accept: application/reginfo+xml
Contact: <sip:[5555::1:2:3:4]:1357>
Content-Length: 0
```

Again, not all the information that is included in the SUBSCRIBE request is shown here—the above headers are only those that are necessary to understand the nature of the registration-state event subscription and the routing of the request.

The subscription is intended for an event named "reg", which is the registration-state event package; it is identified in the Event header of the request.

The request URI identifies the user whose registration-state information is requested and, therefore, has to be set to registered public user identity of Tobias and given in the To header.

In order to identify itself, Tobias's UE sets the To and P-Preferred-Identity headers to a SIP URI that it knows is currently registered. This is:

- Either the default public user identity that was received in the P-Associated-URI header (see Section 5.12.4).

- Or the public user identity that was explicitly registered during initial registration as long as that was not a temporary public user identity (see Section 5.12.3). If no temporary public user identity was used, it is possible that this explicitly registered public user identity is identical with the default public user identity.

The relationship between the P-Preferred-Identity header, the P-Asserted-Identity header and the To header is described in Section 6.2. The To header does not include a tag, as SUBSCRIBE is an initial request and, therefore, the tag will be assigned by the remote side (i.e., in this case the S-CSCF).

The Expires header is set to the same value as the expiration time of the initial registration (i.e., 600,000 seconds which is about 7 days).

The Accept header indicates that only information of the type "reginfo + xml" can be processed by the UE for this subscription, which is the XML (Extensible Markup Language) format for registration-state information.

The Contact header is set to the same contact information as used during registration: that is, the IP-Address of the UE which was assigned by the access network (see Section 5.2) and the protected server port that is used by the IPsec SA (see Section 5.7).

Finally, the Route headers are worth looking at: they include the route set that was received in the Service-Route header within the 200 (OK) response for the REGISTER request (see Section 5.5.8) and on top of it the address of the P-CSCF, which acts as an outbound proxy. This forces the SUBSCRIBE request to be routed first to the P-CSCF and then onward directly to the S-CSCF that was assigned during registration.

The P-CSCF, when receiving this SUBSCRIBE request from the UE, will check whether the information set in the P-Preferred-Identity header is a valid public user identity of Tobias. If this is the case, then it replaces the P-Preferred-Identity header with the P-Asserted-Identity header:

```
SUBSCRIBE sip:tobias@home1.fr SIP/2.0
P-Asserted-Identity: "Tobias" <sip:tobias@home1.fr>
```

The S-CSCF, when receiving this SUBSCRIBE request, will check whether the user identified by the P-Asserted-Identity header is registered at the S-CSCF. Afterwards, it checks whether it can provide the requested registration-state information of Tobias to the subscribing user (Figure 5.11). As Tobias is subscribing to his own registration-state information in this case, this is allowed. Therefore, the S-CSCF will immediately:

- Return a 200 (OK) response for the SUBSCRIBE request, indicating that the subscription was successful.

- Generate an XML document of type reginfo, including the current registration-state information for the URIs that are associated with Tobias.

- And send the generated XML document in a NOTIFY message toward the subscriber (in this case Tobias's UE).

As the 200 (OK) response and the NOTIFY request are sent approximately at the same time, the NOTIFY request may be received at the terminal before the 200 (OK) response. In this exceptional case, the UE must be able to create the related subscription dialog based on the NOTIFY request: that is, it must not discard the

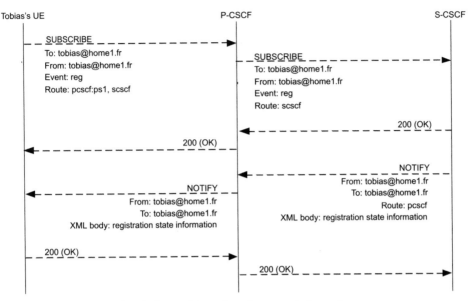

Figure 5.11 Tobias's subscription to his registration-state information.

information received in the NOTIFY request just because it did not receive a prior
200 (OK) response to the SUBSCRIBE request.

5.12.6 P-CSCF's subscription to registration-state information

The P-CSCF also needs to subscribe to Tobias's registration-state information and,
therefore, creates a SUBSCRIBE request, which looks similar to the one that the
terminal generates:

```
SUBSCRIBE sip:tobias@home1.fr SIP/2.0
Via: SIP/2.0/UDP pcscf1.visited1.fi
From: <sip:pcscf1.visited1.fi>;tag=retiisi
To: "Tobias" <sip:tobias@home1.fr>
P-Asserted-Identity: <sip:pcscf1.visited1.fi>
Event: reg
Expires: 600000
Accept: application/reginfo+xml
Contact: <sip:pcscf1.visited1.fi>
Content-Length: 0
```

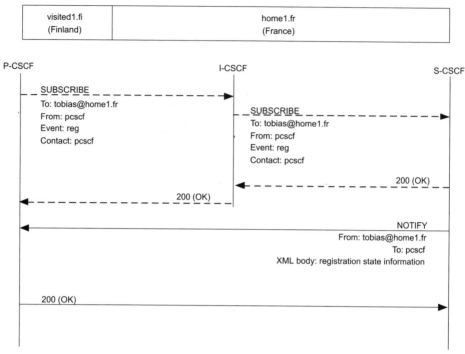

Figure 5.12 P-CSCF subscription to Tobias's registration-state information.

The main difference here is that it is the P-CSCF that subscribes to the registration-state information of Tobias (Figure 5.12); therefore, it has to identify itself in the From header and the P-Asserted-Identity header. As the P-CSCF is a trusted entity (see Section 3.6.4.2) it immediately puts a P-Asserted-Identity header into the request.

As the P-CSCF did not save any routing information during the initial registration phase for its own routing purposes, it has no knowledge about the S-CSCF that was assigned for the user and, therefore, cannot include any Route headers. Consequently, it will route the request on the basis of the host part of the request URI, which is "home1.fr" and can be resolved via DNS to one or more I-CSCF addresses of Tobias's home network. The I-CSCF then queries the HSS for the address of the S-CSCF that is assigned for the URI sip:tobias@home1.fr and sends the request to the S-CSCF.

Note that with this SUBSCRIBE request a new dialog is created, this time between the P-CSCF and the S-CSCF. This dialog has no relation to the UE's subscription to the very same registration-state information; therefore the S-CSCF will generate separate NOTIFY requests, including the registration-state information of Tobias, for the UE's and for the P-CSCF's subscription.

5.12.7 Elements of registration-state information

The S-CSCF generates a NOTIFY with Tobias's registration-state information immediately after a new subscription has been received and whenever the registration-state information changes (e.g., when a new public user identity becomes registered).

In this section we only look at the NOTIFY request and registration-state information that is received by Tobias's terminal immediately after the subscription. This information is identical to the information received by the P-CSCF at more or less the same time.

The NOTIFY request as received by Tobias's UE includes—among others—the following headers:

```
NOTIFY sip:[5555::1:2:3:4]:1357;comp=sigcomp SIP/2.0
Via: SIP/2.0/UDP scscf1.home1.fr;branch=nosctb
Via: SIP/2.0/UDP pcscf1.visited1.fi:7531;branch=nopctb
From: "Tobias" <sip:tobias@home1.fr>;tag=peruna
To: "Tobias" <sip:tobias@home1.fr>;tag=sipuli
Subscription-State: active;expires=599999
Event: reg
Content-Type: application/reginfo+xml
Contact: <sip:scscf1.home1.fr>
Content-Length: (...)
```

The things to note about this NOTIFY request are that:

- The To and From headers changed as this request was sent from the Notifier (S-CSCF) to the Subscriber (Tobias's UE). Although both headers have nearly identical content, their tags are different. The S-CSCF also has added a "To" tag ("peruna"), which now appears in the From header.

- A Subscription-State header has been added, which indicates that the subscription is active and will expire after 599,999 seconds.

5.12.8 Registration-state information in the body of the NOTIFY request

The registration-state information for the URIs associated with Tobias is included in the body of the NOTIFY request and shown in detail in Section 5.12.9. Registration-state information is a hierarchical list that consists of:

- The root element "reginfo", which includes registration-state information that is associated with one user.

- One or more "registration" sub-elements to the "reginfo" root element. A "registration" sub-element includes information about exactly one URI (i.e., one public user identity).

- Zero or more "contact" sub-elements to every "registration" sub-element. A "contact" sub-element includes information about an address that has been registered (or de-registered) for the URI in the "registration" sub-element.

Each registration sub-element can include the following attributes:

- The AOR (address of record) attribute, which is followed by the URI for the public user identity.

- The ID attribute, which uniquely identifies the registration sub-element from among all the others.

- The state attribute of the registration sub-element, which indicates whether the indicated URI is either:

    o "Active" (i.e., registered).

    o "Terminated" (i.e., de-registered).

    o "Init" (i.e., in the process of being registered, such as when an initial REGISTER request has been received, but authentication procedures have not yet been finished).

Each contact sub-element includes the registered contact address and can include the following attributes:

- The ID attribute, which uniquely identifies the contact sub-element from among all the others.

- The state attribute of the contact sub-element, which indicates whether the indicated contact—in relation to the URI of the registration sub-element—is either:

    o "Active" (i.e., the URI is registered with this contact information).

    o "Terminated" (i.e., the binding between the URI and this contact information has just been removed).

- The event attribute of the contact sub-element, which indicates the event that caused the latest change in the contact state attribute. The events can be:

- ○ Registered—this event switches the contact address from the "init" state to the "active" state and indicates that the AOR has been explicitly registered (i.e., a valid REGISTER request has been received for this AOR and the related contact information is bound to it).

- ○ Created—this event has the same meaning as the registered event, but indicates that the AOR has been implicitly registered (i.e., the binding was created automatically, such as when there is a received REGISTER request for another AOR).

- ○ Refreshed—this event occurs when re-registration for an AOR takes place and may also occur implicitly (i.e., when re-registration for an associated AOR is performed).

- ○ Shortened—this event occurs when the network shortens the expiration time of an AOR (e.g., to bring about network-initiated re-authentication, see Section 5.13.2).

- ○ Deactivated—this event occurs when the binding is removed by the network (e.g., due to a network-initiated de-registration), allowing the user to perform a new initial registration attempt afterwards.

- ○ Probation—with this event the network can de-register the user and request her to send a new initial registration after a certain time (dependent on the retry-after value).

- ○ Unregistered—this event occurs when the user has explicitly unregistered the contact.

- ○ And rejected—this event occurs when the network does not allow the user to register the specific contact.

- ● Additional attributes, such as:

- ○ The expires attribute—which indicates the remaining expiration time of the registration for the specific contact address (it must be set for the shortened event, but is optionally set for other events).

- ○ The retry-after attribute—which is only set for the probation event and indicates how long the UE should wait before it can try again to register.

5.12.9 Example registration-state information

Tobias's registration-state information is included in the body of the NOTIFY requests that the S-CSCF sends out to the UE and the P-CSCF. It includes first of all an XML document heading:

```
<?xml version="1.0"?>
<reginfo xmlns="urn:ietf:params:xml:ns:reginfo" version="0" state="full">
```

The heading indicates the XML version in use (1.0). The registration information then starts with the root element, named "reginfo", which includes a number of attributes:

- The xmlns attribute points to the uniform resource name (URN) that defines the XML document and the XML namespace.

- The version attribute always starts with the value "0" and is incremented by one every time a new (updated) version of the registration-state information is sent to the same recipient.

- The state attribute finally indicates that the following registration state information is a full list of all the AORs that relate to Tobias. The first version ("0") of a reginfo document always needs to be sent as a complete list ("full")—subsequent information (starting from "1") can be sent as "partial" and will only include information that has changed since the last notification.

All the public user identities that relate to Tobias and their registration states are now listed in the document:

```
<registration aor="sip:tobias@home1.fr" id="a1" state="active">
  <contact id="15" state="active" event="registered">
    sip:[5555::1:2:3:4]
  </contact>
</registration>
```

The first AOR or URI is "sip:tobias@home1.fr", which we already know from the above example. It is currently registered (state="active"). The content of this registration sub-element is one contact sub-element, which shows the binding that was created by the S-CSCF between sip:tobias@home1.fr and the contact information sip:[5555:1:2:3:4]. The event attribute is set to "registered", which indicates that this AOR was explicitly registered with this contact: this can be verified, because the registration procedures described in this chapter showed the AOR in the To header and the IP address in the Contact header:

```
<registration aor="tel:+44-123-456-789" id="a2" state="active">
  <contact id="16" state="active" event="created">
    sip:[5555::1:2:3:4]
  </contact>
</registration>
```

The next AOR is a tel URL that was implicitly registered (event="created") with the same IP address as the first AOR. This implicit registration was made by the S-CSCF, based on the user profile of Tobias. In this case the telephone number is directly related to the SIP URI sip:tobias@home1.fr:

```
<registration aor="sip:tobi@home1.fr" id="b1" state="terminated">
</registration>
<registration aor="tel:+44-123-456-111" id="b2" state="terminated">
</registration>
```

These two AORs are currently not registered (state="terminated") and, therefore, the registration sub-elements do not include any information at all.

Finally, Tobias is also the game master of an online role-playing game. He takes his job in this game very seriously and is therefore always registered from a gaming console that has the address sip:[5555::101:102:103:104]. The contact of the gaming console was explicitly registered (event="registered").

Nevertheless, Tobias also wants to stay informed about the ongoing status of the game when he is online with his IMS UE; therefore, this AOR was also implicitly registered (event="created") by the S-CSCF when the REGISTER request for sip:tobias@home1.fr was received.

```
<registration aor="sip:gameMaster@home1.fr" id="c1" state="active">
    <contact id="45" state="active" event="registered">
     sip:[5555::101:102:103:104]
    </contact>

    <contact id="19" state="active" event="created">
     sip:[5555::1:2:3:4]
    </contact>
  </registration>
</reginfo>
```

The last line of the registration-state information shows the   tag, which ends the XML document.

5.12.10 Multiple terminals and registration-state information

One or more public user identities can be registered from different terminals (i.e., different UE). In our example it could be that Tobias also owns a simple paging device that uses his public user identity sip:tobias@home1.fr. This device would also need to perform registration procedures before being able to use IMS services. The registration of this device could take place over a different P-CSCF, but would end up in the same S-CSCF as the first registration.

After this paging device has registered, Tobias's UE and his P-CSCF would receive another NOTIFY message indicating that additional contact information for the public user identity is now available: that is, information that relates to the first AOR in the body of the NOTIFY would include the following information:

```
<?xml version="1.0"?>
<reginfo xmlns="urn:ietf:params:xml:ns:reginfo" version="1" state="partial">
  <registration aor="sip:tobias@home1.fr" id="a1" state="active">
    <contact id="15" state="active" event="registered">
      sip:[5555::1:2:3:4]
    </contact>
    <contact id="20" state="active" event="registered">
      sip:[5555::171:171:172:173]
    </contact>
  </registration>
</reginfo>
```

The second lot of contact information in the registration information relates to the paging device. Note that this is now the second lot of registration-state information that the UE receives. To make sure that no registration-state information was lost, the "version" parameter is set to "1" (the first lot of information had version="0"). As the first NOTIFY included complete registration-state information, the UE will receive only information about changed registration elements, in this case for the AOR sip:tobias@home1.fr. Consequently, the state parameter is set to "partial" (in the first lot of information it was set to "full").

5.12.11 Related standards

Specifications relevant to Section 5.12 are:

* 3GPP TS 23.003   Numbering, addressing and identification.

* RFC3265   Session Initiation Protocol (SIP)-specific Event Notification.

* RFC3325   Private Extensions to the Session Initiation Protocol (SIP) for Asserted Identity within Trusted Networks.

* RFC3455   Private Header (P-Header) Extensions to the

Session Initiation Protocol (SIP) for the 3rd-Generation Partnership Project (3GPP).

• Draft-ietf-sipping-reg-event   A Session Initiation Protocol (SIP) Event Package for Registrations.

5.13 Re-registration and re-authentication

5.13.1 User-initiated re-registration

Tobias's UE can at any time perform re-registration by sending a new REGISTER request (see Section 5.4) to the network (Figure 5.13). This happens when, say, the registration needs to be refreshed due to expiration of the registration time. As the re-registration is handled in the same way as an initial SIP registration procedure, this is not further described here.

5.13.2 Network-initiated re-authentication

The IMS UE registers its contact information for a time of 600,000 seconds, which means that the binding of the registered public user identities and the physical IP address is kept for around 7 days in the S-CSCF. As user authentication procedures are directly coupled to registration procedures this would mean that the S-CSCF has no means of re-authenticating the user within this time period. Certain conditions may nevertheless make it necessary for the S-CSCF to re-authenticate the UE.

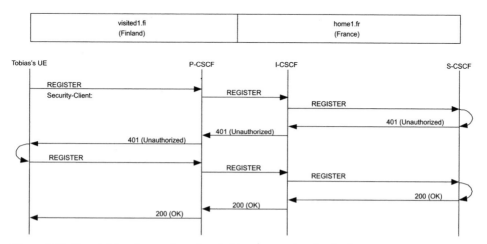

Figure 5.13 User-initiated re-registration (without re-authentication).

To achieve this, the S-CSCF can reduce the expiration time of the user's registration. Let us assume that Tobias has already been registered for 3 hours and his home operator wants to perform a random re-authentication. The S-CSCF assigned to Tobias will reduce the expiration time of Tobias's registration to 600 seconds (exactly 10 minutes). Up to that moment Tobias's UE is not aware of the reduced registration time and would therefore not perform a re-registration, which is needed for re-authentication. To inform the UE about this the S-CSCF makes use of the UE's subscription to the registration-state event package.

The S-CSCF generates a NOTIFY request for the registration-state event package, in which it indicates that it shortened the registration time and sends this NOTIFY request to Tobias's UE. On receiving this request the UE will immediately update the registration expiration time information.

Furthermore, all other subscribers to Tobias's registration-state information (e.g., the P-CSCF and the subscribed ASs) will receive a NOTIFY request from the S-CSCF with the updated state information.

After half the indicated time has elapsed (i.e., 300 seconds), the UE will send out another REGISTER request. From then on, the normal registration procedures as described in Section 5.4 will take place, during which the S-CSCF can authenticate the user again (see Section 5.6).

5.13.3 Network-initiated re-authentication notification

The NOTIFY message (Figure 5.14) that is sent from the S-CSCF to the UE will include the following information:

```
<?xml version="1.0"?>
<reginfo xmlns="urn:ietf:params:xml:ns:reginfo" version="2" state="partial">
  <registration aor="sip:tobias@home1.fr" id="a1" state="active">
    <contact id="15" state="active" event="shortened" expires="600">
      sip:[5555::1:2:3:4]
    </contact>
    <contact id="20" state="active" event="registered">
      sip:[5555::171:171:172:173]
    </contact>
  </registration>

  <registration aor="tel:+44-123-456-789" id="a2" state="active">
    <contact id="16" state="active" event="shortened" expires="600">
      sip:[5555::1:2:3:4]
    </contact>
  </registration>
```

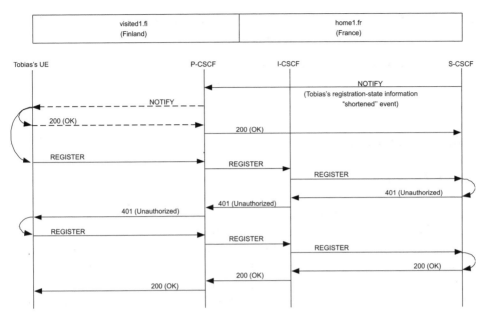

Figure 5.14 Network-initiated re-authentication.

```
<registration aor="sip:gameMaster@home1.fr" id="c1" state="active">
  <contact id="45" state="active" event="registered">
    sip:[5555::101:102:103:104]
  </contact>
  <contact id="19" state="active" event="shortened" expires="600">
    sip:[5555::1:2:3:4]
  </contact>
</registration>
</reginfo>
```

All registration and related contact states are still set to "active", but the latest event that occurred for contact is indicated as "shortened". The expires value shows that there are 10 minutes left for the UE to re-register.

In this document only partial registration-state information is delivered (state="partial" in the document header), as the rest of the registration-state information has not been changed.

5.13.4 Related standards

Specifications relevant to Section 5.13 are:

- RFC3265   Session Initiation Protocol (SIP)-specific Event Notification
- Draft-ietf-sipping-reg-event   A Session Initiation Protocol (SIP) Event Package for Registrations.

5.14 De-registration

5.14.1 Overview

All things come to an end at some point, and this is true of the registration of a user to the IMS. Tobias might want to be undisturbed after he called his sister and switches off his mobile phone. When doing so, his phone sends another REGISTER request to the S-CSCF, including all the information we have already seen, but indicating that this time it is for de-registration (Figure 5.15). The S-CSCF will then clear all the information it has stored for Tobias, update the data in the HSS and send a 200 (OK) response to Tobias's UE.

Sometimes the network sees the need to de-register the user (Figure 5.16): maybe the S-CSCF needs to be shut down or maybe Tobias is using a pre-paid card and has ran out of money. In these cases the S-CSCF would simply send another NOTIFY message with registration-state information to Tobias's UE, this time indicating that he has been de-registered.

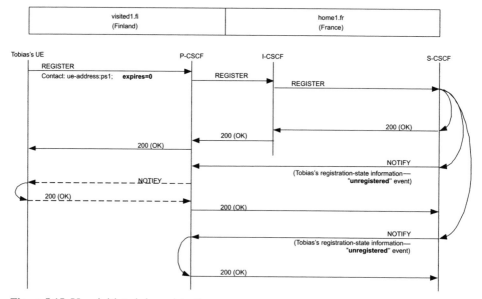

Figure 5.15 User-initiated de-registration.

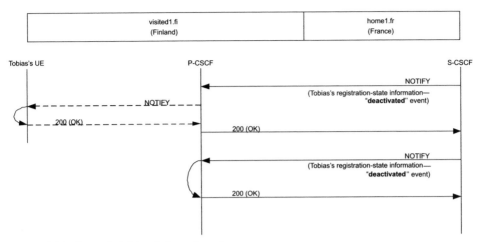

Figure 5.16 Network-initiated de-registration.

In both cases the S-CSCF will send NOTIFY requests to the P-CSCF and all other subscribers to Tobias's registration-state information, indicating that Tobias has been de-registered. By sending these NOTIFY requests the dialogs that were created during subscription to the registration-state event will also be terminated.

5.14.2 User-initiated de-registration

If Tobias decides to switch off his phone, the UE will send a REGISTER request to the network in order to de-register:

```
REGISTER sip:home1.fr SIP/2.0
Via: SIP/2.0/UDP [5555::1:2:3:4]:1357;comp=sigcomp;branch=99uetb
Route: sip:[5555::a:b:c:d]:7531;comp=sigcomp;lr
Max-Forwards: 70
From: <sip:tobias@home1.fr>;tag=ulkomaa
To: <sip:tobias@home1.fr>;tag=kotimaa
Authorization: Digest username="user1_private@home1.fr",
               realm="home1.fr",
               nonce=A34Cm+Fva37UYWpGNB34JP, algorithm=AKAv1-MD5,
               uri="sip:home1.fr",
               response="6629fae49393a05397450978507c4ef1",
               integrity-protected="yes"
               uri="sip:home1.fr",
Require: sec-agree
Proxy-Require: sec-agree
```

```
Security-Verify: tls ;q=0.2, IPsec-3gpp ;q=0.1
                 ;alg=hmac-sha-1-96
                 ;spi-c=98765434 ;spi-s=87654322
                 ;port-c=8644 ;port-s=7533
Security-Client: digest, IPsec-3gpp
                 ;alg=hmac-sha-1-96 ;spi-c=23456790 ;spi-s=12345679
                 ;port-c=2472 ;port-s=1357
Contact: <sip:[5555::1:2:3:4]:1357;comp=sigcomp>;expires=0
Call-ID: apb03a0s09dkjdfglkj49222
CSeq: 49 REGISTER
Content-Length: 0
```

This is principally the same information that we have already seen in the other REGISTER requests; the main difference is that the expires value is set to 0, which means that the user wants to de-register the binding between the public user identity (in the To header) and the IP address (in the Contact header).

This REGISTER request will be routed in exactly the same way as every other REGISTER request (i.e., it will not follow the stored Service-Route). Therefore, it will:

- Traverse the P-CSCF—which checks for integrity protection and adds the integrity-protected=yes flag to the Authorization header.

- Traverse the I-CSCF—which will ask the HSS for the S-CSCF address that was selected for the user.

- And finally be received at the S-CSCF—where de-registration will take place.

The S-CSCF will immediately send back a 200 (OK) response to the UE, which will also include the expires header set to the value 0.

Afterwards, the S-CSCF will generate NOTIFY requests to all subscribers to the registration-state information of Tobias, including Tobias's UE. Each of these NOTIFY requests will include the Subscription-State header set to the value "terminated", which indicates that the subscription to the registration-state information of that user has been terminated. For example:

```
NOTIFY sip:[5555::1:2:3:4]:1357;comp=sigcomp SIP/2.0
Subscription-State: terminated
```

The body of these NOTIFY requests will include Tobias's registration-state information:

```
<?xml version="1.0"?>
<reginfo xmlns="urn:ietf:params:xml:ns:reginfo" version="3" state="partial">
```

Once again, this XML document includes a "partial"-state notification, as it does not explicitly list those public user identities that have not been registered (see Section 5.12.7):

```
<registration aor="sip:tobias@home1.fr" id="a1" state="active">
  <contact id="15" state="terminated" event="unregistered">
    sip:[5555::1:2:3:4]
  </contact>

  <contact id="20" state="active" event="registered">
    sip:[5555::171:171:172:173]
  </contact>
</registration>
```

The public user identity sip:tobias@home1.fr is still active, as it was registered by Tobias's pager (see Section 5.12.10). Only the contact address of the mobile phone was set to terminated:

```
<registration aor="tel:+44-123-456-789" id="a2" state="terminated">
  <contact id="16" state="terminated" event="unregistered">
    sip:[5555::1:2:3:4]
  </contact
</registration>
```

Tobias's tel URL has been completely de-registered:

```
<registration aor="sip:gameMaster@home1.fr" id="c1" state="active">
  <contact id="45" state="active" event="refreshed">
    sip:[5555::101:102:103:104]
  </contact>

  <contact id="19" state="terminated" event="unregistered">
    sip:[5555::1:2:3:4]
  </contact>
</registration>
</reginfo>
```

Finally, the gaming URI sip:gameMaster@home1.fr also remains registered, as another UE is still actively using it. Only the contact that was explicitly de-registered ended up being removed.

5.14.3 Network-initiated de-registration

Whenever the network sees the need to de-register the user or some of the user's identities, the S-CSCF will generate NOTIFY requests in the same way as described in Section 5.14.2. Only the content of the XML document will look different:

```
<?xml version="1.0"?>
<reginfo xmlns="urn:ietf:params:xml:ns:reginfo" version="3" state="partial">
  <registration aor="sip:tobias@home1.fr" id="a1" state="terminated">
    <contact id="15" state="terminated" event="deactivated">
      sip:[5555::1:2:3:4]
    </contact>

    <contact id="20" state="terminated" event="deactivated">
      sip:[5555::171:171:172:173]
    </contact>
  </registration>

  <registration aor="tel:+44-123-456-789" id="a2" state="terminated">
    <contact id="16" state="terminated" event="deactivated">
      sip:[5555::1:2:3:4]
    </contact>
  </registration>

  <registration aor="sip:gameMaster@home1.fr" id="c1" state="terminated">
    <contact id="45" state="terminated" event="deactivated">
      sip:[5555::101:102:103:104]
    </contact>

    <contact id="19" state="terminated" event="deactivated">
      sip:[5555::1:2:3:4]
    </contact>
  </registration>
</reginfo>
```

All public user identities are now set to "terminated", as the network consequently de-registered every registration that was active for Tobias, even those from other terminals. The event has changed to "deactivated", which indicates that it was the network that de-registered, not the user.

5.14.4 Related standards

Specifications relevant to Section 5.14 are:

- RFC3265   Session Initiation Protocol (SIP)-specific Event
                                       Notification.

- Draft-ietf-sipping-reg-event   A Session Initiation Protocol (SIP) Event Package
                                       for Registrations.

6

An example IMS Session

6.1 Overview

This chapter shows an example session between Tobias and his sister Theresa, who are both registered in their home networks and are both currently roaming in different countries (see Section 4.1).

The Session Initiation Protocol (SIP) is facilitated by the IMS to ensure that Tobias and Theresa can talk to each other and even see each other on the screens of their user equipment (UE). In order to achieve this within the wireless environment certain steps have to be taken:

- Tobias's UE needs to construct an INVITE request that includes a registered public user identity of Theresa in order to reach her—Section 6.2.4.

- All SIP messages must traverse the Proxy Call Session Control Function (P-CSCFs) and the Serving-CSCF (S-CSCF) of both users—Section 6.3.

- All SIP messages are sent via the established IP Security (IPsec) security associations (ASs) between the UE and their P-CSCFs—Section 6.3.3.1.

- All SIP messages are sent compressed between the UE and their P-CSCFs— Section 6.4.

- The two items of UE agree on the media streams that they will exchange. In the example case they will exchange a bidirectional audio stream, so that brother and sister can talk to each other, and a bidirectional media stream, so that they can also see each other—Section 6.5.

- The two items of UE agree on a single codec for every media stream that they will exchange—Section 6.5.

- The networks will authorize the media for the session, so that the users can reserve the related resources—Section 6.7.2.

- Both items of UE perform resource reservation (i.e., they set up the necessary media PDP contexts over which the media streams to and from the network will be transported)—Section 6.6.

- Theresa's UE will not get any indication that her brother is calling her before the resources for media sessions (i.e., the media PDP contexts) have been reserved on both sides, in order to be sure that media sessions can really be established— Section 6.6.4.

- The network elements will exchange charging information, so that media sessions can be billed correctly—Section 6.8.

- The S-CSCFs may initiate advanced services for their served users—Section 6.3.8.

- Theresa's UE will finally start to ring and she will accept the session; this completes the session establishment phase.

After Tobias and Theresa have finished their call, they will hang up and one of their items of UE will send a BYE request to the other UE—Section 6.9.1.

The SIP message sequence for the example session will look like that shown in Figure 6.1.

6.2 Caller and callee identities

6.2.1 Overview

Section 5.12 described how an Internet Protocol Multimedia Subsystem (IMS) user becomes aware during registration of the public user identities he can use and which of them are currently registered. Subsequently, the users—in our example Theresa and Tobias—will use these identities for different purposes. For every kind of dialog within the IMS—in this example the INVITE dialog—two identities are essential:

- A registered and authenticated public user identity of the calling user (Tobias) needs to be indicated in the request, in order to guarantee the identification of the user within his home network and for execution rights for extended services as well. This is provided in the P-Asserted-Identity header within the INVITE request.

- A registered and authenticated public user identity of the called user (Theresa) needs to be indicated in the request, in order to be able to contact the user and to execute services for her. This is provided in the request URI (uniform resource identifier) of the INVITE request and in the P-Asserted-Identity header of the first response.

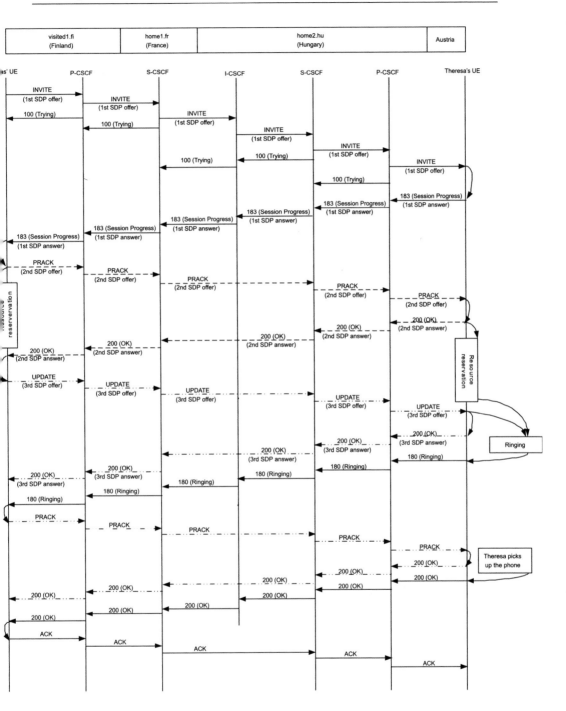

gure 6.1 IMS session establishment call flow.

6.2.2 From and To headers

The INVITE request that Tobias's UE sends toward Theresa includes the following headers that are related to either his or her identity:

```
INVITE sip:theresa@home2.hu SIP/2.0
From: "Your Brother" <sip:tobi@brother.com>;tag=veli
To: "My beloved Sister" <sip:Theresa@sister.com>
P-Preferred-Identity: <sip:tobias@home1.fr>
Privacy: None
```

Obviously, the From and To headers can be set to any value the sender likes. We chose the wording in this example to clearly indicate that the values of these two headers in any request (besides the REGISTER request) have no influence on any IMS routing or security procedures—they can be freely set. The only information that is needed by the protocol itself are the tag parameters in these two headers.

Tobias's home network operator may have certain restrictions to some values that the To header can be set at. Nevertheless, the home network can only reject the request if the setting of the From or To header does not fulfil the operator policy, because SIP does not allow any of these headers to be changed.

6.2.3 Identification of the calling user: P-Preferred-Identity and P-Asserted-Identity

6.2.3.1 Inclusion of the P-Preferred-Identity header by the originating UE

In the above example Tobias includes the P-Preferred-Identity header, which is optional. When present, it should include a registered public user identity of that user. In Section 5.12.5 we saw how Tobias became aware of all the public user identities that he can use. By means of the registration-state information to which his terminal subscribed, he also discovered which of these user identities he currently has registered.

If Tobias wanted to completely hide his identity from his sister, he would have needed to set the Privacy header to the "id" value. This value would force Theresa's P-CSCF to remove the P-Asserted-Identity header from the INVITE request, so that Theresa could only see the identity in the From header as the caller identification.

6.2.3.2 Originating P-CSCF includes the P-Asserted-Identity header

Tobias's UE will send out the INVITE request that is first received by the P-CSCF. The P-CSCF checks whether the request was received over a valid IPsec SA. If the request was received unprotected (i.e., not over an SA), the P-CSCF will reject the request.

Afterwards, the P-CSCF inserts a P-Asserted-Identity header in the INVITE request, which replaces the received P-Preferred-Identity header, if one was received. The P-Asserted-Identity header is the only identity within an IMS dialog that is guaranteed to include a registered and authenticated public user identity of the user.

If a P-Prefered-Identity header is present, the P-CSCF will check whether the URI in the header is a currently registered public user identity of the user who sent in the request. It will discover whether it is a registered public user identity from the registration-state information it is subscribed to (see Section 5.12.6). The P-CSCF can ensure that a certain request was sent in by a specific user based on the SA it was received over (see Section 5.7). If both checks are successful, the P-CSCF will replace the P-Preferred-Identity header with a P-Asserted-Identity header that includes the same content.

If the P-Preferred-Identity header did not include a currently registered public user identity, then the P-CSCF will remove the header. In this case or when no P-Preferred-Identity header was received at all, the P-CSCF will add a P-Asserted-Identity header that includes the default public user identity of the user. How the default public user identity of the user is determined is described in Section 5.12.4:

```
INVITE sip:theresa@home2.hu SIP/2.0
From: "Your Brother" <sip:tobi@brother.com>;tag=veli
To: "My beloved Sister" <sip:Theresa@sister.com>
P-Asserted-Identity: <sip:tobias@home1.fr>
Privacy: None
```

6.2.3.3 Originating S-CSCF and P-Asserted-Identity header

On receiving this INVITE request the S-CSCF of Tobias's home network operator will identify Tobias only by the information given in the P-Asserted-Identity header. This is why this header is so essential within the IMS. The S-CSCF will also check the authentication and registration state of the public user identity indicated in the header. Because of these checks, the header serves as the main identification of the user for the whole dialog. ASs (see Section 6.3.8) can base the identification and even the authentication of the user on this header as well.

Tobias's S-CSCF may add an additional URI to the P-Asserted-Identity header. In this example it adds the telephone universal resource locator (tel URL) of Tobias to the header:

```
INVITE sip:theresa@home2.hu SIP/2.0
From: "Your Brother" <sip:tobi@brother.com>;tag=veli
To: "My beloved Sister" <sip:Theresa@sister.com>
P-Asserted-Identity: <sip:tobias@home1.fr>, <tel:+44123456789>
Privacy: None
```

Before the S-CSCF of Tobias's home network routes the request toward Theresa's home network, it will check whether that network is within its trust domain (see Section 3.6.4.2). If the S-CSCF and the home network of Theresa do not share the same trust domain, the S-CSCF will remove the P-Asserted-Identity header from the request, as long as the Privacy header is set to the value "id". For this example we assume that the two networks have a trust relationship that allows the header to be sent on.

6.2.3.4 P-Asserted-Identity header at the terminating side

The P-CSCF of Theresa has to check the value of the Privacy header of the request. As it is not set to the value "id" it can send the P-Asserted-Identity header to Theresa's UE.

So, finally, the UE of Theresa receives the P-Asserted-Identity header. It can facilitate the information in the header by, say, displaying the "real name" of Theresa's caller.

6.2.4 Identification of the called user

6.2.4.1 The request URI

Let us look again at the INVITE message that Tobias sends. Its first line, the request URI, looks like:

```
INVITE sip:theresa@home2.hu SIP/2.0
```

The request URI is set to the final destination of the request (i.e., to Theresa's SIP URI). Section 6.3 explains how this URI is used for SIP and IMS routing procedures. But, this URI also identifies Theresa as the called user within her home network. This means that Theresa's S-CSCF will check whether this public user identity is currently registered and authenticated. If Theresa is currently not registered with this public user identity, the S-CSCF will return, say, a 404 (Not Found) response to the INVITE request and the call will fail or, based on the filter criteria for an unregistered user, will forward the INVITE to Theresa's voicemail box.

For our example we assume that Theresa has registered the public user identity that Tobias's UE put into the request URI.

6.2.4.2 The request URI and P-Called-Party-ID header

Another problem arises when this request is sent by Theresa's S-CSCF toward the terminating P-CSCF. The S-CSCF, which also acts as Theresa's SIP registrar, will rewrite the request URI with the registered contact address of Theresa, in order to route the request to the UE at which Theresa is currently registered. Therefore the public user identity in the request URI will be lost.

However, Theresa might have several public user identities and might want to know under which of them she receives a call: for example, she might have work-related user identities and others that relate to her private life. Maybe her UE even provides different ring tones for each of her user identities.

We already saw in Section 6.2.2 that Theresa cannot trust the To header in the request, as the originator can set it to any value—one that might be completely different from the public user identity in the request URI.

In order not to lose the information of the public user identity that is used by Tobias to call his sister, the S-CSCF, when rewriting the request URI with the registered contact address, will add a P-Called-Party-ID header to the INVITE request. This P-Called-Party-ID header includes the public user identity that was received in the request URI:

```
INVITE sip:[5555::5:6:7:8]:1006 SIP/2.0
P-Called-Party-ID: sip:theresa@home2.hu
```

6.2.4.3 P-Asserted-Identity header

After receiving the INVITE request, Theresa's UE will send back a P-Preferred-Identity header in the first response to the INVITE request—the 183 (Session in Progress) response—which will include one of Theresa's public user identities:

```
SIP/2.0 183 Session in Progress
From: "Your Brother" <sip:tobi@brother.com>;tag=veli
To: "My beloved Sister" <sip:Theresa@sister.com>;tag=schwester
P-Preferred-Identity: <sip:theresa@home2.hu>
Privacy: None
```

The P-CSCF of Theresa will perform the same checks as described before for Tobias's P-CSCF (see Section 6.2.3) and will replace it by a P-Asserted-Identity header:

```
SIP/2.0 183 Session in Progress
From: "Your Brother" <sip:tobi@brother.com>;tag=veli
To: "My beloved Sister" <sip:Theresa@sister.com>;tag=schwester
P-Asserted-Identity: <sip:theresa@home2.hu>
Privacy: None
```

6.2.5 Related standards

Specifications relevant to Section 6.2 are:

- RFC3323 A Privacy Mechanism for the Session Initiation Protocol (SIP).

- RFC3325 Private Extensions to the Session Initiation Protocol (SIP) for Asserted Identity within Trusted Networks.

- RFC3455 Private Header (P-Header) Extensions to the Session Initiation Protocol (SIP) for the 3rd-Generation Partnership Project (3GPP).

6.3 Routing

6.3.1 Overview

One of the most complex issues within the IMS is the routing of requests, especially the routing of initial requests. In our example, Tobias is sending the initial INVITE request to Theresa. Consequently, a SIP dialog is created within which several subsequent requests, such as ACK, PRACK, UPDATE and BYE, are sent.

Tobias's UE is not aware at the time of sending the INVITE request how Theresa's UE can be reached. All it can provide is:

- The final destination of the INVITE request—which is the SIP URI of Theresa (one of her public user identities) that Tobias had to provide (e.g., by selecting it from his phone book).

- The address of the P-CSCF—which is the outbound proxy of Tobias's UE and will be the first hop to route to. This address is obtained before SIP registration during the P-CSCF discovery procedures (see Section 5.3).

- The address of the S-CSCF—which was discovered during registration procedures by means of the Service-Route header (see Section 5.5.8).

Armed with this partial route information the INVITE request is sent on its way. It first traverses the P-CSCF and then the S-CSCF that have been selected for Tobias.

Tobias's S-CSCF now has no further routing information available for the request other than the final destination (i.e., the public user identity of Theresa, "sip:theresa@home2.hu"). As Tobias's S-CSCF does not act as a registrar for Theresa, it can only resolve the host part of the address: "home2.hu". This domain name is sent to the domain name system (DNS) server and the S-CSCF will receive back one or more Interrogating-CSCF (I-CSCF) addresses of Theresa's home network, will select one of them and will send the INVITE request to it.

The I-CSCF just acts as the entry point to Theresa's home network. It asks the local HSS for the address of the S-CSCF that was selected for Theresa and sends the INVITE further to the returned address.

Theresa's S-CSCF now acts as the registrar and replaces her SIP URI with the contact address that she has registered. It also sends the request indirectly to Theresa's UE, because it has not established an SA with it (see Section 5.7). The INVITE request therefore is first sent to Theresa's P-CSCF—the address the S-CSCF remembers from the Path header that it received during registration (see Section 5.5.9).

The P-CSCF finally forwards the INVITE request to Theresa's UE over the IPsec SA.

This shows that for the initial request the route from Tobias to Theresa is put together piece by piece, as the originating UE and the CSCFs have only information about the next one or two hops that have to be traversed. In order to make further routing within the dialog easier, SIP routing mechanisms (see Section 8.12) will be used:

- All CSCFs put their addresses on top of the Via header—this allows all responses to the INVITE request to be sent back over exactly the same route as the request.

- All CSCFs, other than Theresa's I-CSCF, put their addresses on top of the Record-Route header—this allows all subsequent requests in the dialog to be sent over the CSCFs that put themselves in the Record-Route header. The I-CSCF in Theresa's home network fulfilled its routing task when it discovered the addresses of Theresa's S-CSCF; so, it is no longer needed on the route.

When sending out subsequent requests the UEs will include a list of Route entries, which will force the request to follow the recorded route (Figure 6.2). Routing issues related to the provision of services are explained in Section 6.3.8.

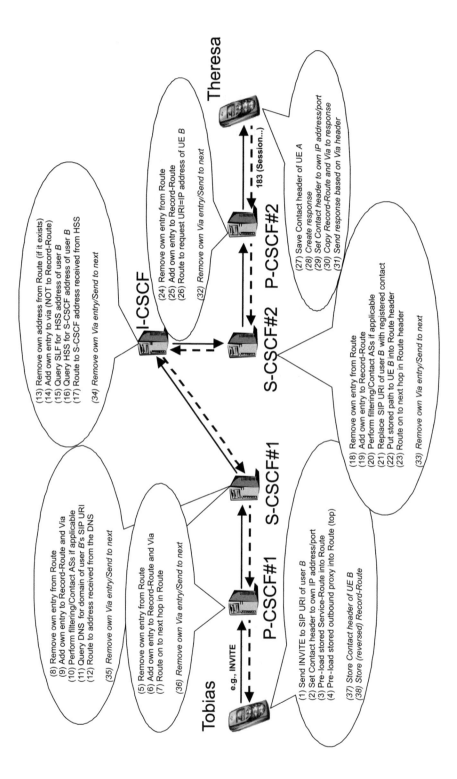

Figure 6.2 Routing an initial INVITE request and its responses.

Tobias

e.g., INVITE

(1) Send INVITE to SIP URI of user B
(2) Set Contact header to own IP address/port
(3) Pre-load stored Service-Route into Route
(4) Pre-load stored outbound proxy into Route (top)

(37) Store Contact header of UE B
(38) Store (reversed) Record-Route

P-CSCF#1

(5) Remove own entry from Route
(6) Add own entry to Record-Route and Via
(7) Route on to next hop in Route

(36) Remove own Via entry/Send to next

S-CSCF#1

(8) Remove own entry from Route
(9) Add own entry to Record-Route and Via
(10) Perform filtering/Contact ASs if applicable
(11) Query DNS for domain of user B's SIP URI
(12) Route to address received from the DNS

(35) Remove own Via entry/Send to next

I-CSCF

(13) Remove own address from Route (if it exists)
(14) Add own entry to via (NOT to Record-Route)
(15) Query SLF for HSS address of user B
(16) Query HSS for S-CSCF address of user B
(17) Route to S-CSCF address received from HSS

(34) Remove own Via entry/Send to next

S-CSCF#2

(18) Remove own entry from Route
(19) Add own entry to Record-Route
(20) Perform filtering/Contact ASs if applicable
(21) Replace SIP URI of user B with registered contact
(22) Put stored path to UE B into Route header
(23) Route on to next hop in Route header

(33) Remove own Via entry/Send to next

P-CSCF#2

(24) Remove own entry from Route
(25) Add own entry to Record-Route
(26) Route to request URI=IP address of UE B

(32) Remove own Via entry/Send to next

Theresa

183 (Session...)

(27) Save Contact header of UE A
(28) Create response
(29) Set Contact header to own IP address/port
(30) Copy Record-Route and Via to response
(31) Send response based on Via header

6.3.2 Session, dialog, transactions and branch

During session establishment and while the session is active as well, different types of signalling messages are exchanged and different kinds of relations between the two items of UE are established.

The term "session" describes the media connections between the two users. Tobias wants to exchange audio and video media streams with his sister. This exchange of media is done on the so-called "bearer level": this means that Real-time Transport Protocol (RTP) packets are sent from the two items of UE to their Gateway GPRS Support Nodes (GGSNs) and the GGSNs exchange these packets between each other directly over the backbone. This session is established on the basis of the SIP and Session Description Protocol (SDP) signalling that are exchanged via the "control plane".

A SIP dialog is the signalling relation between the two items of UE which is needed to establish, modify and release the multimedia session. The dialog will be first established (with the INVITE request) and will exist as long as the related session is active. Every SIP dialog is identified by the value of the Call-ID header and by the tags in the To and the From headers of the SIP requests, which in our example look like:

```
From: "Your Brother" <sip:tobi@brother.com>;tag=veli
To: "My beloved Sister" <sip:theresa@home2.hu>;tag=schwester
Call-ID: apb03a0s09dkjdfglkj49555
```

The SIP dialog for the multimedia session between Tobias and Theresa starts with the INVITE request and ends with the 200 (OK) response for the BYE request.

A SIP transaction comprises a single SIP request and all the responses related to it. In order to establish the session, Tobias's UE sends an INVITE request to Theresa's UE. At the very first it receives a 100 (Trying) response from the P-CSCF in response to the request. Afterwards, Theresa's UE responds with a 183 (Session in Progress), a 180 (Ringing) and finally with a 200 (OK) response. All these five messages belong to the same dialog and have the same CSeq number:

```
From: "Your Brother" <sip:tobi@brother.com>;tag=veli
To: "My beloved Sister"
<sip:theresa@home2.hu>;tag=schwester
Call-ID: apb03a0s09dkjdfglkj49555
CSeq: 1112 INVITE
```

Every subsequent request sent from the same side (in this case from Tobias's UE) will have a higher CSeq number than the preceding request: this means that, for example, the first PRACK request includes CSeq 1113, the following UPDATE request CSeq 1114 and so forth.

Every entity—either UE or CSCF—will correlate the responses that are received for a sent request on the basis of the branch parameter that it added as a parameter to its Via header entry: for example, the P-CSCF of Tobias adds the following Via header to the INVITE request:

```
INVITE sip:theresa@home2.hu SIP/2.0
Via: SIP/2.0/UDP pcscf1.visited1.fi;branch=9pctb
```

The branch parameter identifies the INVITE transaction (i.e., the INVITE request and the responses to it) at the P-CSCF. It is constructed from the tags in the To and From headers, the Call-ID, the CSeq number and the information in the topmost Via header of the request.

6.3.3 Routing of the INVITE request

6.3.3.1 From Tobias's UE to the P-CSCF

Tobias's UE will include the following routing-related headers into the initial INVITE request:

```
INVITE sip:theresa@home2.hu SIP/2.0
Via: SIP/2.0/UDP [5555::1:2:3:4]:1357;branch=8uetb
Route: <sip:[5555::a:b:c:d]:7531;lr>
Route: <sip:orig@scscf1.home1.fr;lr>
Contact: <sip:[5555::1:2:3:4]:1357>
```

The destination of the request is Theresa's SIP URI; hence its inclusion in the request URI.

During registration the route between Tobias's UE and its S-CSCF in the home network was discovered by the Service-Route header (see Section 5.5.8). The UE pre-loads this first part of the route into the Route header and puts the P-CSCF on top of it, because it always needs to contact its outbound proxy first.

Tobias's UE also puts its IP address into the Contact header of the request, so that the remote UE *B* can directly reach it. It also adds its IP address to the Via header in order to receive the responses to that request.

As the request is sent over established IPsec SAs (see Section 5.7), Tobias's UE puts:

• The protected server port of the UE (1357) as the port value in the Contact header, because it wants to receive all subsequent requests within this dialog via the established IPsec SA.

- The protected server port of the UE (1357) as the port value in the Via header, because it wants to receive all responses to the INVITE request via the established IPsec SA.

- The protected server port of the P-CSCF (7531) as the port value of the address of the P-CSCF in the Route header, because the P-CSCF must receive all requests from the UE via an established IPsec SA. The UE became aware of the P-CSCF's protected server port during SIP Security Mechanism Agreement procedures (see Sections 5.7.5 and 5.8).

The To and From headers are never used for routing purposes (see Section 6.2.2).

The INVITE request is now sent to the topmost entry in the Route header, which in this case is the P-CSCF that serves Tobias.

6.3.3.2 From Tobias's P-CSCF to the S-CSCF

When receiving this request the P-CSCF:

- Removes its own entry from the topmost Route header.

- Checks that the request includes further routing information in accordance with the routing information it saved during registration (i.e., that the UE does not try to deviate from the Service-Route).

- Puts its address at the top of the Via header, as it needs to receive all responses to the requests.

- Adds the first Record-Route header and puts its own address there. This guarantees that all subsequent requests within this dialog will traverse the P-CSCF.

- Does not include the protected server port number in both the Via and the Record-Route entry. The protected server port number identifies only the port over which the P-CSCF wants to receive SIP messages that are sent from the UE over the set of established IPsec SAs.

Having done that, the P-CSCF again routes toward the topmost entry of the Route header, which in this case is the S-CSCF that serves Tobias:

```
INVITE sip:theresa@home2.hu SIP/2.0
Via: SIP/2.0/UDP pcscf1.visited1.fi;branch=9pctb
Via: SIP/2.0/UDP [5555::1:2:3:4]:1357;branch=8uetb
```

```
Record-Route: <sip:pcscf1.visited1.fi;lr>
Route: <sip:orig@scscf1.home1.fr;lr>
Contact: <sip:[5555::1:2:3:4]:1357>
```

6.3.3.3 From Tobias's S-CSCF to Theresa's home network (I-CSCF)

Tobias's S-CSCF removes its entry from the topmost Route header, which afterwards is empty and can be removed. It then adds its address on top of the Record-Route and Via headers.

Afterwards, the S-CSCF will perform the procedures for service provisioning that are described in Section 6.3.8.

Having done that, it needs to route the request further. But, now there is a problem: there is no Route header left to point to the next hop. All the S-CSCF can do now is take the host part of the address of Theresa's public user identity that is indicated in the request URI (i.e., "home2.hu") and resolve a SIP server in that domain from DNS (see Chapter 12). In return, it receives one or more addresses of I-CSCFs that are located in the home network of Theresa. It takes one of them and sends the request there.

Note that the S-CSCF can only put the address of the I-CSCF into a Route header when it is aware that this I-CSCF is able to act as a loose router. As in the example case the S-CSCF and the I-CSCF are in different networks and it is not assumed that the S-CSCF knows about the routing capabilities of the I-CSCF. Therefore, it sends the UDP packet that transports the initial INVITE to the I-CSCF address.

```
INVITE sip:theresa@home2.hu SIP/2.0
Via: SIP/2.0/UDP scscf1.home1.fr;branch=asctb
Via: SIP/2.0/UDP pcscf1.visited1.fi;branch=9pctb
Via: SIP/2.0/UDP [5555::1:2:3:4]:1357;branch=8uetb
Record-Route: <sip:scscf1.home1.fr;lr>
Record-Route: <sip:pcscf1.visited1.fi;lr>
Contact: <sip:[5555::1:2:3:4]:1357>
```

6.3.3.4 From the I-CSCF to Theresa's S-CSCF

The I-CSCF in Theresa's home network now needs to discover the address of the S-CSCF that is allocated for Theresa. Even if Theresa is not currently registered, the I-CSCF may well be able to discover the address of a default S-CSCF as long as she is subscribed to some services as an unregistered user.

Information about the S-CSCF currently allocated for a user is stored in the Home Subscriber Server (HSS); as there are several HSSs within the network, the

I-CSCF first has to query the Subscription Locator Function (SLF) to discover which HSS holds the data for Theresa. After the SLF returns the address of the HSS, the I-CSCF queries that HSS, which finally returns the address of the S-CSCF that serves Theresa. The I-CSCF now adds a Route entry on top of the Route list and puts the received address of the S-CSCF into it. Furthermore, the I-CSCF:

- Removes its entry from the topmost Route header, if one is present (in this example this is not the case).

- Puts its address on top of the Via list, in order to receive all responses for the INVITE request.

- Does not put its address into the Record-Route, because it does not need to receive any subsequent requests in this dialog. The task of the I-CSCF is to find the S-CSCF of the called user, but as this was already done during initial request processing, there is no need for it to stay in the Route header.

The request once again goes toward the topmost entry in the Route header, which this time is set to Theresa's S-CSCF:

```
INVITE sip:theresa@home2.hu SIP/2.0
Via: SIP/2.0/UDP icscf1.home2.hu;branch=bicth
Via: SIP/2.0/UDP scscf1.home1.fr;branch=asctb
Via: SIP/2.0/UDP pcscf1.visited1.fi;branch=9pctb
Via: SIP/2.0/UDP [5555::1:2:3:4]:1357;branch=8uetb
Route: <sip:scscf2.home2.hu;lr>
Record-Route: <sip:scscf1.home1.fr;lr>
Record-Route: <sip:pcscf1.visited1.fi;lr>
Contact: <sip:[5555::1:2:3:4]:1357>
```

6.3.3.5 From Theresa's S-CSCF to the P-CSCF

Now Theresa's S-CSCF—her registrar—receives the INVITE request. Once again, it removes its entry from the Route header and puts itself into the Via and the Record-Route lists. Afterwards, it provides the services for Theresa as described in Section 6.3.8.

Having done that, the S-CSCF performs the actions of a registrar (i.e., it replaces the request URI, which is still set to Theresa's SIP URI, by her registered contact address). The registered contact address also includes the protected server port (1006) that is used to send requests from the P-CSCF to Theresa's UE via the established IPsec SA.

During Theresa's registration the S-CSCF received the Path header from the P-CSCF. It must now put the entries of the Path header into the Route header of the

INVITE request. Were this not done, the request would immediately be sent to Theresa's UE, which could not accept the request as it has not established an IPsec SA with the S-CSCF.

As there is no longer a Route header, the S-CSCF adds a new one, puts the P-CSCF address into it and, as this is now the topmost entry, sends the request to this address immediately:

```
INVITE sip:[5555::5:6:7:8]:1006 SIP/2.0
Via: SIP/2.0/UDP scscf2.home2.hu;branch=cscth
Via: SIP/2.0/UDP icscf1.home2.hu;branch=bicth
Via: SIP/2.0/UDP scscf1.home1.fr;branch=asctb
Via: SIP/2.0/UDP pcscf1.visited1.fi;branch=9pctb
Via: SIP/2.0/UDP [5555::1:2:3:4]:1357;branch=8uetb
Route: <sip:pcscf2.home2.hu;lr>
Record-Route: <sip:scscf2.home2.hu;lr>
Record-Route: <sip:scscf1.home1.fr;lr>
Record-Route: <sip:pcscf1.visited1.fi;lr>
Contact: <sip:[5555::1:2:3:4]:1357>
```

6.3.3.6 From the P-CSCF to Theresa's UE

The P-CSCF receives the request and does the usual: it removes the whole Route header, adds itself to the Record-Route and Via headers and then sends the request to the final destination indicated in the request URI—Theresa's UE:

```
INVITE sip:[5555::5:6:7:8]:1006 SIP/2.0
Via: SIP/2.0/UDP pcscf2.home2.hu:1511;branch=dpcth
Via: SIP/2.0/UDP scscf2.home2.hu;branch=cscth
Via: SIP/2.0/UDP icscf1.home2.hu;branch=bicth
Via: SIP/2.0/UDP scscf1.home1.fr;branch=asctb
Via: SIP/2.0/UDP pcscf1.visited1.fi;branch=9pctb
Via: SIP/2.0/UDP [5555::1:2:3:4]:1357;branch=8uetb
Route: <sip:pcscf2.home2.hu:1511;lr>
Record-Route: <sip:scscf2.home2.hu;lr>
Record-Route: <sip:scscf1.home1.fr;lr>
Record-Route: <sip:pcscf1.visited1.fi;lr>
Contact: <sip:[5555::1:2:3:4]:1357>
```

The entry of the P-CSCF in the Via header also includes the port number of the protected server port (1511), which was negotiated with Theresa's UE during the registration procedure in the same way as described for Tobias's registration in

Section 5.7.5. This entry forces Theresa's UE to send all responses to this request over the established IPsec SA.

The selfsame protected server port value (1511) is put into the Record-Route header entry of the P-CSCF, where it expects to receive all subsequent requests from Theresa's UE that are sent in this dialog.

After Theresa's UE has received the INVITE request, it stores the received Contact value and the Record-Route header list, as it will route subsequent requests in the dialog based on them.

6.3.4 Routing of the first response

6.3.4.1 From Theresa's UE to the P-CSCF

Theresa's UE now creates a response to the received INVITE request, which is due to the usage of preconditions (see Section 6.6) in a 183 (Session in Progress) response.

The UE puts its own IP address into the Contact header to indicate the address it wants to use to receive subsequent requests in this dialog. The contact address also includes the protected server port of Theresa's UE (1006), which guarantees that all subsequent requests will be received via the established IPsec SA as well.

The Record-Route and Via headers of the INVITE request also go into the response. After doing so, Theresa's UE sends the response to the address and port number of the topmost entry in the Via header, which is the protected server port of the P-CSCF:

```
SIP/2.0 183 Session in Progress
Via: SIP/2.0/UDP pcscf2.home2.hu:1511;branch=dpcth
Via: SIP/2.0/UDP scscf2.home2.hu;branch=cscth
Via: SIP/2.0/UDP icscf1.home2.hu;branch=bicth
Via: SIP/2.0/UDP scscf1.home1.fr;branch=asctb
Via: SIP/2.0/UDP pcscf1.visited1.fi;branch=9pctb
Via: SIP/2.0/UDP [5555::1:2:3:4]:1357;branch=8uetb
Record-Route: <sip:pcscf2.home2.hu:1511;lr>
Record-Route: <sip:scscf2.home2.hu;lr>
Record-Route: <sip:scscf1.home1.fr;lr>
Record-Route: <sip:pcscf1.visited1.fi;lr>
Contact: <sip:[5555::5:6:7:8]:1006>
```

All other responses that are sent from Theresa's UE to this INVITE request will include the same Via header entries as the 183 (Session in Progress) response.

6.3.4.2 From Theresa's P-CSCF onward to Tobias's P-CSCF

The P-CSCF identifies the INVITE transaction the request belongs to by the branch parameter that it set in its own entry in the Via header. It then manipulates the routing information in the 183 (Session in Progress) response in the following way:

- It removes its own address from the the Via header.

- It rewrites its own Record-Route entry.

- It sends the request to the topmost entry in the Via header, which is the S-CSCF in Theresa's home network.

Why does the P-CSCF re-write its own Record-Route entry? Well, it does this to ensure that no other entity than Theresa's UE sends messages to the P-CSCF's protected server port that is used for the IPsec SA with the UE. If Theresa's S-CSCF were to send the next request (the PRACK) to the P-CSCF's protected server port (1511), the request would be dropped by the IPsec layer in the P-CSCF's protocol stack, as it had not been sent integrity-protected via the IPsec SA:

```
SIP/2.0 183 Session in Progress
Via: SIP/2.0/UDP scscf2.home2.hu;branch=cscth
Via: SIP/2.0/UDP icscf1.home2.hu;branch=bicth
Via: SIP/2.0/UDP scscf1.home1.fr;branch=asctb
Via: SIP/2.0/UDP pcscf1.visited1.fi;branch=9pctb
Via: SIP/2.0/UDP [5555::1:2:3:4]:1357;branch=8uetb
Record-Route: <sip:pcscf2.home2.hu;lr>
Record-Route: <sip:scscf2.home2.hu;lr>
Record-Route: <sip:scscf1.home1.fr;lr>
Record-Route: <sip:pcscf1.visited1.fi;lr>
Contact: <sip:[5555::5:6:7:8]:1006>
```

From then on, nothing of consequence happens to the response until it reaches Tobias's P-CSCF—every hop simply removes its own Via entry and sends the message toward the next entry in the Via. The Record-Route stays untouched. Note that other servers on the way back are permitted to re-write their Record-Route entries in order to distinguish requests received from different directions; however, this is not shown in this example, as it is an implementation option for a CSCF to carry out.

6.3.4.3 From Tobias's P-CSCF to his UE

When receiving the 183 (Session in Progress) response, Tobias's P-CSCF performs similar actions to Theresa's P-CSCF. It also re-writes its entry in the Record-Route header; but, instead of removing the protected server port value in its entry (as Theresa's P-CSCF did during the handling of the same response), it adds this port (7531). Consequently, it forces Tobias's UE to send all subsequent requests via the established IPsec SA.

As the P-CSCF routes the response on the basis of the Via header, it will send it to the protected server port of Tobias's UE (1357) (i.e., via the IPsec SA):

```
SIP/2.0 183 Session in Progress
Via: SIP/2.0/UDP [5555::1:2:3:4]:1357;branch=8uetb
Record-Route: <sip:pcscf2.home2.hu;lr>
Record-Route: <sip:scscf2.home2.hu;lr>
Record-Route: <sip:scscf1.home1.fr;lr>
Record-Route: <sip:pcscf1.visited1.fi:7531;lr>
Contact: <sip:[5555::5:6:7:8]:1006>
```

After receiving the response, Tobias's UE:

- stores the IP address of Theresa's UE, as received in the Contact header; and

- stores the Record-Route list after reversing the order of all entries in it.

6.3.5 Re-transmission of the INVITE request and the 100 (Trying) response

After having sent out the INVITE request, Tobias's UE waits for responses from Theresa's UE. It will wait until its timer T1—in the IMS this is set to the value of 2 seconds—expires. Afterwards, it will re-transmit the INVITE request repeatedly until either a response to the request is received or until 128 (= 64*T1) seconds have elapsed; it will then indicate to Tobias that establishment of the session has failed.

As the INVITE request has to pass through several CSCFs all over Europe, it might take longer than 2 seconds for it to reach Theresa's UE, which has to construct the 183 (Session in Progress) response before once again travelling back to Finland.

To avoid frequent re-transmissions of the INVITE request from Tobias's UE, the P-CSCF sends back a 100 (Trying) response after it has received the INVITE request. This indicates that from now on the P-CSCF will take care of such re-transmissions.

The same is done by all other call-statefull SIP proxies on the route. The 100 (Trying) is always stopped at the SIP proxy that was the latest to take over responsibility for re-transmission.

For example, the S-CSCF of Theresa's home network sends back the 100 (Trying) response, which first reaches the I-CSCF. As the I-CSCF is not a call-statefull SIP proxy it just sends it on (based on the Via header). Next it reaches the S-CSCF of Tobias's home network. Tobias's S-CSCF has sent the 100 (Trying) response to the P-CSCF; consequently, it took over responsibility for the re-transmission of the INVITE request. Now the receipt of the 100 (Trying) response indicates that it no longer needs to re-transmit the INVITE request, as this responsibility is taken over by Theresa's S-CSCF.

6.3.6 Routing of subsequent requests in a dialog

When one of the two items of UE needs to send a subsequent request within a dialog, it copies the stored Record-Route entries into the Route header of the new requests and the remote UE's IP address into the request URI.

The request is routed toward the remote UE by strictly following the entries in the Route header (Figure 6.3). Every CSCF that is traversed puts itself into the Via header, in order to get all the responses to this request.

As the I-CSCF did not record any route in the beginning, it does not receive any subsequent request. For example, Tobias's UE has to send back a PRACK request to acknowledge the received 183 (Session in Progress) response (see Section 6.5.2). This PRACK request would include the following routing-related information:

```
PRACK sip:[5555::5:6:7:8]:1006 SIP/2.0
Via: SIP/2.0/UDP [5555::1:2:3:4]:1357;branch=82uetb
Route: <sip:pcscf1.visited1.fi:7531;lr>
Route: <sip:scscf1.home1.fr;lr>
Route: <sip:scscf2.home2.hu;lr>
Route: <sip:pcscf2.home2.hu;lr>
```

The PRACK request, therefore, will be routed:

- on the basis of the Route headers by means of Tobias's P-CSCF and his S-CSCF as well as Theresa's S-CSCF and her P-CSCF; and

- from Theresa's P-CSCF based on the address in the request URI—which Tobias's UE took from the received Contact header that was received in the 183 (Session in Progress) response—to Theresa's UE over the IPsec SA.

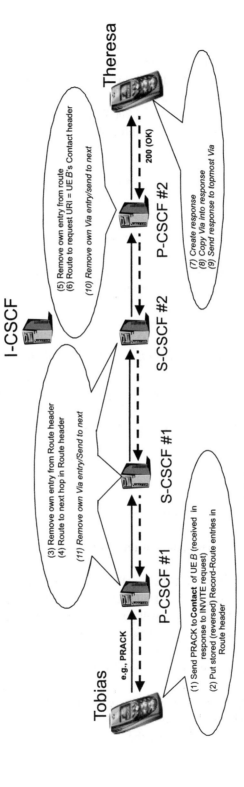

Figure 6.3 Routing of subsequent requests and their responses.

A subsequent request within a dialog does not include a Contact header, as the addresses of the two items of UE were already exchanged during the sending and receiving of the initial request and its first response. Furthermore, the CSCFs will not put any Record-Route headers in the request, because the route was already recorded during the initial request.

Theresa's UE will send back a 200 (OK) response to this PRACK request and will include the following routing information:

```
SIP/2.0 200 OK
Via: SIP/2.0/UDP scscf2.home2.hu;branch=c2scth
Via: SIP/2.0/UDP scscf1.home1.fr;branch=a2sctb
Via: SIP/2.0/UDP pcscf1.visited1.fi;branch=92pctb
Via: SIP/2.0/UDP [5555::1:2:3:4]:1357;branch=82uetb
```

This response will be routed back on the basis of the Via header entries. Record-Route headers are no longer returned.

6.3.7 Stand-alone transactions from one UE to another

For stand-alone transactions, such as MESSAGE or OPTIONS, the same routing procedures as those used for an initial request are performed, although record routing does not need to be done because a stand-alone transaction does not create a dialog.

6.3.8 Routing to and from ASs

6.3.8.1 Filter criteria evaluation in the S-CSCF

Service provisioning in the IMS is achieved by application servers (ASs), which are contacted on the basis of initial filter criteria. When Tobias's or Theresa's S-CSCF receives an initial request, they will go through these filter criteria one by one and, if one or more of them matches, they will send the request toward the indicated AS. Filter criteria are downloaded by the S-CSCF from the HSS during registration and are part of Tobias's and Theresa's service profile; this is further described in Section 3.12.

In this example we assume that there are three ASs that have set filter criteria for requests that originate from Tobias (see Table 6.1). Tobias's S-CSCF will check these filter criteria one by one against the information received in the INVITE request:

Table 6.1 Filter criteria in Tobias's S-CSCF.

Element of filter criteria	Filter criterion #1	Filter criterion #2	Filter criterion #3
SPT: session case	Originating	Originating	Terminating
SPT: public user identity	tel:+44-123-456-789	sip:tobias@hom1.fr tel:+44-123-456-789	sip:tobias@home1.fr
SPT: SIP method	*	INVITE	SUBSCRIBE
Further SPT	—	—	SIP header: event: pres
Application server	sip:as1.home1.fr;lr	sip:as2.home1.fr;lr	sip:as3.home1.fr;lr

The asterisk signifies the selector used in command line entries.

```
INVITE sip:theresa@home2.hu SIP/2.0
Via: SIP/2.0/UDP scscf1.home1.fr;branch=asctb
Via: SIP/2.0/UDP pcscf1.visited1.fi;branch=9pctb
Via: SIP/2.0/UDP [5555::1:2:3:4]:1357;branch=8uetb
Route: <sip:orig@scscf1.home1.fr;lr>
From: "Your Brother" <sip:tobi@brother.com>;tag=veli
To: "My beloved Sister" <sip:Theresa@sister.com>
P-Asserted-Identity: <sip:tobias@home1.fr>
Privacy: None
```

Filter criterion #1 does not match, because the P-Asserted-Identity header, which is checked against the Service Point Trigger (SPT) for the public user identity, does not include Tobias's tel URL.

Filter criterion #2 does match, because:

- The INVITE request is received from the originating user. The S-CSCF knows this from the user part it set in its Service-Route header entry (see Section 5.5.8) and which is now returned in the Route header;

- The P-Asserted-Identity is set to one of the public user identities that are filtered (sip:tobias@home1.fr).

- The SIP method is INVITE.

6.3.8.2 From the S-CSCF to the AS

Consequently, the S-CSCF now has to send the INVITE request to the AS (Figure 6.4) that is indicated in filter criterion #2. It also needs to take care that it receives the request again after the AS has fulfilled its actions, because the S-CSCF needs to evaluate filter criterion #3 and to send the request toward the home network

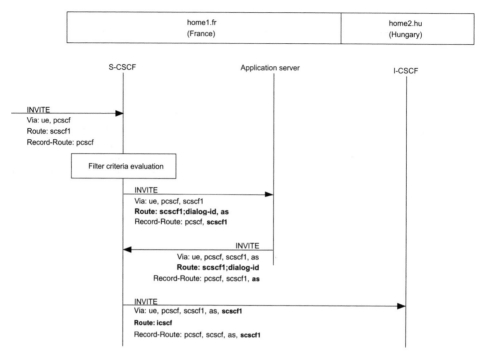

Figure 6.4 Routing to an AS.

of Theresa. To achieve this, the S-CSCF adds a set of routing-related headers by putting:

- Its own address on top of the Route headers, in order to receive the INVITE request back from the AS.

- The address of the AS on top of the Route headers, in order to route the INVITE request to the AS as the next hop.

- Its own address on top of the Record-Route headers, so that it stays on the route for subsequent requests as well.

- Its own address on top of the Via headers, so that it receives all responses to the request.

In addition to this, the S-CSCF will add an implementation-specific dialog identifier to its own Route header entry, which it has just added. It sets this dialog identifier to a value that allows it to identify the dialog that is created with this INVITE. But, what is the purpose of this?

The AS (as described in Section 3.12) could decide to act as a back-to-back user agent (B2BUA) and terminate the INVITE request locally. It would then send a new

INVITE request with a new Call-ID toward the S-CSCF. As this AS would use the URI that is included in the Route header for routing to the next hop, the S-CSCF would also get back the dialog identifier. Consequently, it recognizes that the new Call-ID is in fact related to the previously received INVITE request. The S-CSCF would then return to the point where it stopped after sending out the INVITE request to the AS.

We will not further consider the scenario of an AS acting as a B2BUA in this example:

```
INVITE sip:theresa@home2.hu SIP/2.0
Via: SIP/2.0/UDP sip:scscf1.home1.fr;branch=9sc2as2tb
Via: SIP/2.0/UDP pcscf1.visited1.fi;branch=9pctb
Via: SIP/2.0/UDP [5555::1:2:3:4]:1357;branch=8uetb
Route: <sip:as2.home1.fr;lr>
Route: <sip:scscf1.home1.fr;lr>;dia-id=6574839201
Record-Route: <sip:scscf1.home1.fr;lr>
Record-Route: <sip:pcscf1.visited1.fi;lr>
```

6.3.8.3 From the AS back to the S-CSCF

When receiving the INVITE request, the AS:

- Will store the topmost entry in the Route header that is pointing to the AS;

- Provide the service based on the information in the request.

- May modify the request in compliance with [RFC3261] (e.g., add another header).

- Put its own address at the top of the Via list.

- Decide whether it wants to receive subsequent requests within this dialog. If it wants to then it puts its own address at the top of the Record-Route list. In this example we assume that the AS wants to stay in the Route header.

- Route the INVITE request based on the topmost Route header back to the S-CSCF.

Our INVITE request now looks like:

```
INVITE sip:theresa@home2.hu SIP/2.0
Via: SIP/2.0/UDP sip:as2.home1.fr;branch=vas2tb
Via: SIP/2.0/UDP sip:scscf1.home1.fr;branch=9sc2as2tb
```

```
Via: SIP/2.0/UDP pcscf1.visited1.fi;branch=9pctb
Via: SIP/2.0/UDP [5555::1:2:3:4]:1357;branch=8uetb
Route: <sip:scscf1.home1.fr;lr>;dia-id=6574839201
```
Record-Route: <sip:as2.home1.fr;lr>
```
Record-Route: <sip:scscf1.home1.fr;lr>
Record-Route: <sip:pcscf1.visited1.fi;lr>
```

6.3.8.4 Evaluation of further filter criteria at the S-CSCF

When it receives the INVITE request again, the S-CSCF will then evaluate filter criterion #3; this does not match, because the SIP method is not SUBSCRIBE (as indicated in the SPT). Consequently, the S-CSCF will continue with its normal routing procedures, as described in Section 6.3.3.3 (i.e., it will send the INVITE request to the I-CSCF of Theresa's home network).

Because service provisioning further complicates the routing, no further attention is paid to it throughout this example; the Via, Route and Record-Route headers added here will likewise not be shown in the rest of this example.

6.3.9 Related standards

The IMS Service Provisioning Architecture is further described in 3GPP TS 23.218: IP Multimedia (IM) session handling; IM call model; Stage 2.

6.4 Compression negotiation

6.4.1 Overview

The basic compression capabilities of the UE and the P-CSCF have already been negotiated during the registration procedures (see Section 5.9). Consequently, all requests and responses that are sent between the two sets of UE and their P-CSCFs will be compressed.

In this example we only show how compression parameters are basically set during session establishment and concentrate only on the compression between Theresa's UE and her P-CSCF. The procedures for Tobias's side are identical.

6.4.2 Compression of the initial request

We assume that Theresa has registered a contact address that included the comp= SigComp parameter at her S-CSCF. Therefore, Theresa's S-CSCF will include this

parameter when it acts as a SIP registrar and re-writes the request URI of the
INVITE request (see Section 6.3.3.5).

INVITE sip:[5555::5:6:7:8]:1006;comp=SigComp SIP/2.0

When the P-CSCF receives this request, it will route it toward Theresa's UE based
on the request URI and, as the comp = SigComp parameter is included, it will send it
compressed. Furthermore, the P-CSCF will:

- Add the comp = SigComp parameter to its entry in the Via header, so that
  Theresa will send all responses to the INVITE request compressed.

- Add the comp = SigComp parameter to its entry in the Record-Route header, so
  that Theresa will send all subsequent requests in this dialog compressed.

Our INVITE request now looks like:

```
INVITE sip:[5555::5:6:7:8]:1006;comp=SigComp SIP/2.0
Via: SIP/2.0/UDP
pcscf2.home2.hu:1511;comp=SigComp;lr;branch=dpcth
Via: SIP/2.0/UDP scscf2.home2.hu;branch=cscth
Via: SIP/2.0/UDP icscf1.home2.hu;branch=bicth
Via: SIP/2.0/UDP scscf1.home1.fr;branch=asctb
Via: SIP/2.0/UDP pcscf1.visited1.fi;branch=9pctb
Via: SIP/2.0/UDP [5555::1:2:3:4]:1357;branch=8uetb
Record-Route: <sip:pcscf2.home2.hu:1511;comp=SigComp;lr>
Record-Route: <sip:scscf2.home2.hu;lr>
Record-Route: <sip:scscf1.home1.fr;lr>
Record-Route: <sip:pcscf1.visited1.fi;lr>
Contact: <sip:[5555::1:2:3:4]:1357;comp=SigComp>
```

6.4.3 Compression of responses

When Theresa's UE constructs the 183 (Session in Progress) response to the INVITE
request, it will add its IP address in the Contact header and will also include the
comp = SigComp parameter there. Based on this entry all subsequent requests will be
routed from Theresa's P-CSCF to her UE and, because Theresa wants them to be
sent compressed, it needs to add the parameter there.

The Record-Route headers are stored by Theresa's UE: whenever the UE sends
a subsequent request (e.g., the PRACK or BYE request) it will send it compressed
due to the compression parameter being set in the topmost entry.

Theresa's UE will send the 183 (Session in Progress) response to the P-CSCF and, as that shows the comp = SigComp parameter in the Via header, it will also send this response compressed:

```
SIP/2.0 183 Session in Progress
Via: SIP/2.0/UDP pcscf2.home2.hu:1511;comp=SigComp;lr
Via: SIP/2.0/UDP scscf2.home2.hu
Via: SIP/2.0/UDP icscf1.home2.hu
Via: SIP/2.0/UDP scscf1.home1.fr,
Via: SIP/2.0/UDP pcscf1.visited1.fi
Via: SIP/2.0/UDP [5555::1:2:3:4]:1357
Record-Route: <sip:pcscf2.home2.hu:1511;comp=SigComp;lr>
Record-Route: <sip:scscf2.home2.hu;lr>
Record-Route: <sip:scscf1.home1.fr;lr>
Record-Route: <sip:pcscf1.visited1.fi;lr>
Contact: <sip:[5555::5:6:7:8]:1006;comp=SigComp>
```

We saw in Section 6.3.4.2 that the P-CSCF re-writes its entry in the Record-Route header to remove its protected server port number from it. When doing so, it also removes the compression parameter from it, because it wants to receive compressed requests from the UE, and not from the S-CSCF:

```
SIP/2.0 183 Session in Progress
Via: SIP/2.0/UDP scscf2.home2.hu
Via: SIP/2.0/UDP icscf1.home2.hu
Via: SIP/2.0/UDP scscf1.home1.fr,
Via: SIP/2.0/UDP pcscf1.visited1.fi
Via: SIP/2.0/UDP [5555::1:2:3:4]:1357
Record-Route: <sip:pcscf2.home2.hu;lr>
Record-Route: <sip:scscf2.home2.hu;lr>
Record-Route: <sip:scscf1.home1.fr;lr>
Record-Route: <sip:pcscf1.visited1.fi;lr>
Contact: <sip:[5555::5:6:7:8]:1006;comp=SigComp>
```

6.4.4 Compression of subsequent requests

After the 183 (Session in Progress) request has reached Tobias's UE, it will send a PRACK request in the same dialog. The request URI of this PRACK request will be set to the address received in the Contact header of the 183 (Session Progress) response (see Section 6.3.5), which includes the compression parameter:

```
PRACK sip:[5555::5:6:7:8]:1006;comp=SigComp SIP/2.0
```

When this PRACK request is received at Theresa's P-CSCF, it will again be routed to Theresa's UE based on the request URI and can be sent compressed as it includes the compression parameter. Once again, the P-CSCF will add the comp = SigComp parameter to the Via header of the PRACK, so that Theresa can send the 200 (OK) response to it compressed.

Following these procedures, all requests and responses within the dialog will be sent compressed between the UE and their P-CSCFs.

6.4.5 Related standards

The comp parameter is defined in [RFC3486]: Compressing the Session Initiation Protocol (SIP).

6.5 Media negotiation

6.5.1 Overview

Media negotiation and the handling of preconditions, which will be described in Section 6.6.4, are closely related concepts in the IMS. Both are more concerned with the description of the session parameters in SDP. Nevertheless, they have a major influence on SIP signalling.

During media negotiation the two items of UE agree on the set of media they want to use for the session and which codecs will be used for the different media types. Therefore, the SDP offer/answer mechanism is used, which—in the IMS— basically works in the following way (Figure 6.5):

1. The calling UE sends a first SDP offer in the INVITE request to the called UE. This SDP lists all media types (e.g., audio, video or certain applications like whiteboard or chat) the caller wants to use for this session and lists the different codecs that the caller supports for encoding these different media types.

2. The called UE responds with a first SDP answer, in which it may reject some of the proposed media types. It also reduces the list of codecs by ignoring those that it does not support, such that only the codecs that are supported on both sides remain.

3. After receiving the first answer the caller has to make the final decision on the used codecs. It sends a second offer to the called user, which indicates a single codec for every media type that will be used during the session.

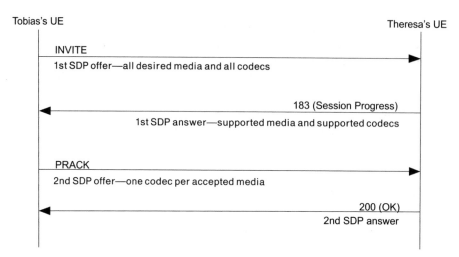

Figure 6.5 SDP offer/answer in IMS.

4. The called UE accepts the second offer and sends an answer back as confirmation.

SIP allows media connections to be set up after just one offer/answer exchange. In the IMS the selection of a single codec per media stream enforces a second exchange, if the first SDP answer includes more than one codec for any media type. This is done because both lots of UE must be prepared to receive any of the selected codecs and, therefore, would need to reserve resources on the air interface for the codec with the higher bandwidth, despite maybe using the codec with the lower bandwidth throughout the session.

Due to resource reservation, which is explained in Section 6.6, the two offer/answer exchanges must take place before the 200 (OK) for the INVITE is received. Consequently, the called UE needs to put the first answer into a 100-class response. We will also see in Section 6.6 that the first response is a 183 (Session in Progress) response. If this happens, two problems arise:

- The 183 (Session in Progress) response is—like all 100-class responses—a provisional response and, therefore, is not sent reliably, which means Theresa's UE cannot be sure that it will ever be received by the calling user.

- The calling side is no longer able to send a second offer back, as during a normal INVITE transaction there is no possibility for the calling UE to send any further SIP requests to the called UE besides the initial INVITE and the ACK at the very end of session establishment.

[RFC3262] solves both these problems by making the provisional 100-class responses reliable: this means that, when sending a provisional response back to Tobias's UE, Theresa's UE can indicate that it wants to send this response in a reliable way. Tobias's UE must then send back an acknowledgment (ACK) for the received provisional (PR) response: the PRACK request. As every request in SIP (as well as the ACK) must be answered by a final response, Theresa's UE will send a 200 (OK) response back, after receiving the PRACK request.

With this addition to SIP, the first SDP answer in the 183 (Session in Progress) response can be sent reliably and the second SDP offer/answer exchange can be done in the PRACK request and in its 200 (OK) response.

6.5.2 Reliability of provisional responses

The 100-class responses in SIP are provisional: that is, the terminal that sends them out does not get any indication back whether these responses were ever received by the other side. As shown above, there are some cases that require provisional responses to be sent reliably: that is, that the UE that receives the response can explicitly acknowledge it. One of these cases in the IMS is that the provisional response carries an SDP answer, which is obliged to be reliably delivered to the remote side.

The mechanism for sending provisional responses reliably is called, in short, "100rel" and its support is mandated for every UE that connects to the IMS.

In order to indicate that it supports the 100rel mechanism, Tobias's UE includes a Supported header in the INVITE request, indicating the "100rel" option tag:

```
INVITE sip:theresa@home2.hu SIP/2.0
From: "Your Brother" <sip:tobi@brother.com>;tag=veli
To: "My beloved Sister" <sip:theresa@home2.hu>
Supported: 100rel
CSeq: 1112 INVITE
Call-ID: apb03a0s09dkjdfglkj49555
```

After receiving this, Theresa's UE can start sending provisional responses reliably, as it knows that Tobias's terminal is going to acknowledge them. So, when Theresa's UE sends the 183 (Session in Progress) response, it inserts two additional headers:

```
SIP/2.0 183 Session in Progress
From: "Your Brother" <sip:tobi@brother.com>;tag=veli
To: "My beloved Sister"
<sip:theresa@home2.hu>;tag=schwester
Require: 100rel
```

```
RSeq: 1971
CSeq: 1112 INVITE
Call-ID: apb03a0s09dkjdfglkj49555
```

The Require header indicates that the terminal that receives the provisional response must send a PRACK request for it. In order to distinguish between multiple provisional responses, the RSeq header is included.

Tobias's UE is now requested to send a PRACK request back, in order to acknowledge the provisional 183 (Session in Progress) response:

```
PRACK <sip:[5555::5:6:7:8]:1006 SIP/2.0
From: "Your Brother" <sip:tobi@brother.com>;tag=veli
To: "My beloved Sister"
<sip:theresa@home2.hu>;tag=schwester
RAck: 1971 1112 INVITE
CSeq: 1113 PRACK
Call-ID: apb03a0s09dkjdfglkj49555
```

This request is now sent to the IP address of Theresa's terminal, which was returned in the Contact header of the 183 (Session in Progress) response. In addition, the CSeq number is incremented by one, as the PRACK is a subsequent request in the dialog that was created by the INVITE/183 (Session in Progress) exchange.

The provisional response that is explicitly acknowledged is identified in the RAck header, which includes the values of the RSeq and CSeq headers that were included in the received response. Both the RSeq and CSeq values are included in the RAck header in order to uniquely identify the 183 (Session in Progress) response.

As the PRACK is a normal SIP request, Theresa's terminal needs to return a final response to it:

```
SIP/2.0 200 OK
From: "Your Brother" <sip:tobi@brother.com>;tag=veli
To: "My beloved Sister"
<sip:theresa@home2.hu>;tag=schwester
CSeq: 1113 PRACK
Call-ID: apb03a0s09dkjdfglkj49555
```

The CSeq header in this 200 (OK) response indicates that this is a final response to the PRACK request (CSeq value "1113 PRACK") and not for the INVITE request (CSeq value "1112 INVITE"). Tobias's terminal is still waiting for the final response to the INVITE request and, as we are looking at an IMS case here, this will still take some time to be sent.

6.5.3 SDP offer/answer in IMS

The SDP offer/answer mechanism allows two users to agree on the media types and the codecs that they want to use for a specific session.

We assume that Tobias wants to use for his multimedia session with his sister the following media types:

- An audio stream—so that they can talk to each other.

- A first video stream—which is filmed by the camera built in to Tobias's UE, so that Theresa can see him and, if Theresa has a similar camera, he can see her.

- A second video stream—which is recorded by an external camera that is connected to the UE; this currently pictures a wooden house in Oulu in Finland.

Therefore the INVITE will include an SDP body that will look like this (note that not all SDP parameters are shown here):

```
v=0
o=- 2987933615 2987933615 IN IP6 IN IP6 5555::1:2:3:4
s=-
c=IN IP6 5555::1:2:3:4
t=907165275 0
m=audio 3458 RTP/AVP 0 96 97 98
a=rtpmap:0 PCMU
a=rtpmap:96 G726-32/8000
a=rtpmap:97 AMR-WB
a=rtpmap:98 telephone-event
m=video 3400 RTP/AVP 98 99
a=rtpmap:98 MPV a=rtpmap:99 H.261
m=video 3456 RTP/AVP 98 99
a=sendonly
a=rtpmap:98 H.261
a=rtpmap:99 MPV
```

6.5.3.1 General SDP parameters

Let us have a closer look at this SDP information: first, there is the heading with five lines:

```
v=0
o=- 2987933615 2987933615 IN IP6 5555::1:2:3:4
```

```
s=-
c=IN IP6 5555::1:2:3:4
t=907165275 0
```

The v-line indicates the protocol version and is always set to 0.

The o-line holds parameters related to the owner of the session, who is in this case Tobias:

- The first parameter should include the name that Tobias wants indicated to the receiver of SDP; however, as he is already identified by the SIP From header in the INVITE request, this can be left out.

- The second parameter is Tobias's session identifier, which is a number that allows his UE to make a link between the session description and the media he wants to set up.

- The third parameter is the version of the session information that Tobias sends; in this case it is initialized with the same value as the session identifier, but it could have any other value.

- The subsequent parameters tell us that Tobias's UE uses the Internet (IN) and IPv6 addressing and that he is located at a terminal that has a certain address.

The s-line may include a subject for the session; but, again, this is already handled by SIP.

The c-line contains information about the connection that has to be established for the multimedia session: that is, this indicates the addresses used for the real media streams and not those for signalling. In this case Tobias's UE just indicates that he wants to receive all media for this session on the IP address that he is also using for SIP signalling: that is, the address that was assigned to it during activation of the signalling Packet Data Protocol (PDP) context (see Section 5.2). Tobias's UE will also establish all secondary PDP contexts for media with the same IP address toward this access point.

Finally, there is the t-line, which indicates when the session was created and how long it is intended to last. There need be no time limitation to the session set in SDP, as SIP users already end a session by manually sending a BYE request. So the second parameter of the t-line can safely be given as zero.

The number given as the time at which the session starts is based on definitions from the Network Time Protocol (NTP): this is a 64-bit, unsigned, fixed-point number representation of the seconds that have elapsed since 00:00 o'clock on 1 January 1900.

6.5.3.2 Media lines

Individual media lines, or m-lines, represent the three different media streams that Tobias wants to send:

```
m=audio 3458 RTP/AVP 0 96 97 98
m=video 3400 RTP/AVP 98 99
m=video 3456 RTP/AVP 98 99
```

The first line indicates that Tobias intends to use audio for this session and that his UE will send this on the local port 3458. Real-time Transport Protocol/Audio Video Profile (RTP/AVP) will be used as the transport protocol for audio-related media and the terminal seems to be capable of coding the audio in four different ways, because the m-line gives four different formats (0, 96, 97 and 98). Later on, we will see that only three formats are supported, the last one of which (98) points to the DTMF (dual-tone multifrequency).

The last two m-lines represent two different video streams, which are intended to be sent in parallel: one shows Tobias and the other the wooden house. One is sent over the local port 3400 of Tobias's terminal and the other over port 3456. Both will be transported by RTP/AVP. The terminal holds two formats for each of them.

6.5.3.3 Audio and video formats

The m-lines in SDP include one or more formats that indicate the codecs in which the media streams are encoded. [RFC3551] includes up to 35 formats or codecs that have statically assigned RTP/AVP payload-type numbers (0 to 35). Since that assignment, many more codecs have been defined, which can also be transported via RTP/AVP. These newer codecs can be dynamically assigned to a payload-type number between 96 and 127.

In our example the first media line includes four RTP/AVP payload types, which are further explained in the subsequent lines of the SDP information. The lines between two SDP m-lines form a block that describes in more detail the media of the m-line under which they are placed:

```
m=audio 3458 RTP/AVP 0 96 97 98
a=rtpmap:0 PCMU
a=rtpmap:96 G726-32/8000
a=rtpmap:97 AMR-WB
a=rtpmap:98 telephone-event
```

The four format numbers that are indicated in the m-line for audio are now mapped to individual payload types or codecs.

Payload type 0 is statically assigned in [RFC3551] to the PCMU (pulse code modulation μ-law; ITU-T G.711) codec. The first a-line (attribute line) shows this relation again.

The following two attribute lines map (rtpmap) the next two dynamic RTP/AVP format numbers that are indicated in the media line (96 and 97) to specific codecs, in this case:

- Payload type 96 gets mapped to G.726.

- And payload type 97 gets mapped to adaptive multi-rate wideband (AMR-WB).

The last payload type gets mapped to the telephone event representation of DTMF tones, as described in [RFC2833]. This means that the terminal is able to generate standardized tones whenever the user presses a specific key. These tones are well known from normal (circuit-switched, or CS) phones when, say, an automatic voice response unit asks questions that the user has to answer with certain keys (e.g., "if you want information in English, press 1, if you want information in German, press 2"). The telephone event defines a text-based representation of these DTMF tones and other telephone-related tones that can be transported over RTP.

The first three payload types (PCMU, G.726 and AMR-WB) represent alternatives—each of which can be used to encode the media indicated in the m-line above them. The telephone event payload type cannot be seen as an alternative: it will always be possible for the UE to send this information whenever a tone needs to be generated.

Consequently, it should now be easy to interpret the last two media blocks in the SDP description:

```
m=video 3400 RTP/AVP 98 99
a=rtpmap:98 MPV
a=rtpmap:99 H.261
m=video 3456 RTP/AVP 98 99
a=sendonly
a=rtpmap:98 H.261
a=rtpmap:99 MPV
```

The terminal is able to encode the two video streams using either MPV (dynamically assigned RTP/AVP payload type 98) or H.261 (dynamically assigned payload type 99). The first video stream is sent from the camera that pictures Tobias; as nothing further is indicated this is deemed a send-and-receive stream, which means that Theresa can send video over the same stream toward Tobias. Video of the wooden house can be watched via the second video stream and, as it is assumed that Theresa will not be sending any video back, it is set to "sendonly".

6.5.3.4 The first SDP offer and answer exchange

The SDP information as shown above is the initial offer, and Tobias's UE sends it within the body of the INVITE request to Theresa. The SDP offer arrives at Theresa's UE due to SIP routing of the INVITE request. Theresa's UE afterwards generates an SDP answer for the received offer, which looks like this:

```
v=0
o=- 1357924 1357924 IN IP6 5555::5:6:7:8
s=-
c=IN IP6 5555::5:6:7:8
t=907165275 0
m=audio 4011 RTP/AVP 96 97 98
a=rtpmap:96 G726-32/8000
a=rtpmap:97 AMR-WB
a=rtpmap:98 telephone-event
m=video 4012 RTP/AVP 99
a=rtpmap:99 H.261
m=video 0 RTP/AVP 98
```

The SDP header now includes information about Theresa's UE: that is, its address is in the o-line along with UE session identification and version for this SDP information as well as in the c-line with the UE's IP address.

From the m-lines we see that the terminal is capable of handling the audio stream (on port 4011) and the first video stream (on port 4012), but cannot deliver the second video stream to the user; therefore, the port number in the third m-line is set to zero and all related a-lines are simply dropped.

The SDP answer must repeat all the m-lines that are included in the SDP offer; therefore, the terminal cannot drop the last m-line from here: the only way to indicate that this video cannot be handled is to set the port number to zero.

Furthermore, the UE does not support the MPV codec for video: this RTP/AVP payload-type value was dropped from the first video m-line, and the related a-lines are omitted in the answer.

For audio the UE does not support the PCMU payload type, but both the G.726 and AMR-WB would work with it. The UE does not make a decision here: it just sends both supported codecs back to the caller.

Additionally, DTMF signals can be represented as telephone events.

6.5.3.5 The second SDP offer and answer exchange

After receiving the first SDP answer, Tobias's UE has to make the final decision about the media streams and the codecs used to encode them. As already mentioned,

SIP and SDP do not insist on a user or a terminal choosing a single codec per media stream. As air interface resources are usually scarce, it is strongly recommended to make this decision when attached to the IMS via a wireless link.

The second SDP offer is therefore sent back in the PRACK request toward Theresa's terminal:

```
v=0
o=- 1357924 1357925 IN IP6 5555::1:2:3:4
s=-
c=IN IP6 5555::1:2:3:4
t=907165275 0
m=audio 3458 RTP/AVP 97 98
a=rtpmap:97 AMR-WB
a=rtpmap:98 telephone-event
m=video 3400 RTP/AVP 99
a=rtpmap:99 H.261
m=video 0 RTP/AVP 98
```

For the audio media the terminal chose AMR-WB as the desired codec. In parallel, DTMF signals can still be sent as telephone events. The first video stream, which was accepted by Theresa's terminal, will be sent in H.261 format, and, as the second video stream was not accepted by the remote side, Tobias's terminal also sets the port to the zero value.

As the second offer now includes all media-related information, the answer does not include any new information; that is, it is sent back including the port numbers used by Theresa's terminal and its address information:

```
v=0
o=- 1357924 1357924 IN IP6 5555::1:2:3:4
s=-
c=IN IP6 5555::1:2:3:4
t=907165275 0
m=audio 4011 RTP/AVP 97 98
a=rtpmap:97 AMR-WB
a=rtpmap:98 telephone-event
m=video 4012 RTP/AVP 99
a=rtpmap:99 H.261
m=video 0 RTP/AVP 98
```

6.5.4 Related standards

Specifications relevant to Section 6.5 are:

- RFC1305   Network Time Protocol (Version 3)
                                                  Specification, Implementation and
                                                  Analysis.

- RFC3550   RTP: A Transport Protocol for
                                                  Real-time Applications.

- RFC3551   RTP Profile for Audio and Video
                                                  Conferences with Minimal Control.

- RFC2327/Draft-ietf-mmusic-sdp-new   SDP: Session Description Protocol.

- RFC2833   RTP Payload for DTMF Digits,
                                                  Telephony Tones and Telephony
                                                  Signals.

- RFC3262   Reliability of Provisional Responses in
                                                  the Session Initiation Protocol (SIP).

- RFC3264   An Offer/Answer Model with SDP.

6.6 Resource reservation

6.6.1 Overview

The media sessions between Tobias and Theresa are negotiated via SIP and SDP signalling. As in our example both UEs make use of a dedicated signalling PDP context for the transport of SIP signalling, they will have to establish one or more media PDP contexts for the transport of the media streams.

The procedure for establishing media PDP contexts is called resource reservation. Both lots of UE perform these procedures completely independently of each other.

The establishment of media PDP contexts may consume some time and may even fail: when, say, insufficient resources are available over the wireless link. This means that, until the resources have been reserved, it cannot be guaranteed that the agreed media sessions can be established at all.

Therefore, Theresa's UE should not inform Theresa about the incoming session request (INVITE): that is, it should not start to ring until it has confirmation that resource reservation has succeeded locally as well as at the calling user's end.

In order to achieve this, both lots of UE exchange preconditions during the SDP offer/answer negotiation, which basically intstruct:

- Tobias's UE to send a SIP UPDATE request to the called UE, when resource reservation has succeeded at Tobias's UE; and

- Theresa's UE not to ring until it receives the SIP UPDATE request from the remote side and has also successfully reserved its resources.

Furthermore, the preconditions indicate what should happen with a session if a specific media stream could not be successfully reserved.

Figure 6.6 gives an overview of the relations between SIP, SDP offer/answer, resource reservation, preconditions, the SIP PRACK and the SIP UPDATE method during the establishment of a session.

6.6.2 The 183 (Session in Progress) response

As shown at the beginning of this chapter, the IMS session establishment attempt should only be indicated to Theresa (the called user) if resource reservation on the wireless link for both users has succeeded. Consequently, the SIP 180 (Ringing) response for the INVITE message will be sent by Theresa's terminal.

On the other hand, Tobias's UE expects an early response from the called terminal, especially as in the IMS the SDP offer/answer negotiation needs to be done before resource reservation can start. Therefore, after receipt of the INVITE message, the called UE will send back the 183 (Session in Progress) response, indicating that session establishment procedures have been started, but the called user has not been informed yet.

6.6.3 Are preconditions mandatorily supported?

Before sending out the first INVITE request, the calling terminal must also include a Require header with the value "precondition":

```
Require: sec-agree, precondition
```

This guarantees that the called terminal will reject this INVITE if it does not support the preconditions mechanism at all. But, this would mean that the session between Tobias and Theresa could not be established at all if Theresa's terminal does not support the precondition extension to SDP. Is that really what the IMS intended?

Let us assume that Theresa's terminal does not support preconditions but

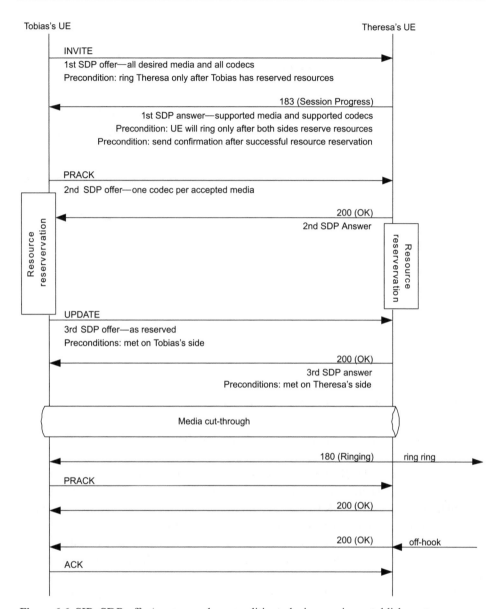

Figure 6.6 SIP, SDP offer/answer and preconditions during session establishment.

nevertheless goes ahead with session establishment (Figure 6.7). Assume further that resource reservation on Theresa's side is successful. Consequently, her terminal could immediately afterwards indicate the incoming call to her and start sending early media to Tobias.

At this point in time, Tobias would not even have started to reserve his resources over the wireless link, as he is still waiting for the SDP answer from Theresa's side.

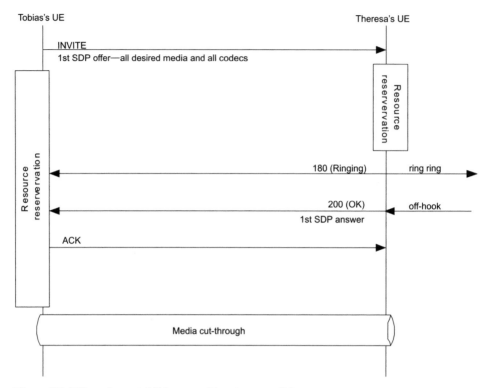

Figure 6.7 SIP session establishment without preconditions.

Even had he started reserving resources immediately after sending out the initial INVITE request, it cannot be guaranteed that this process would have finished when Tobias receives the 180 (Ringing) response from Theresa's UE.

Even worse, Theresa could pick up the phone—which would result in her terminal sending a 200 (OK) response—and start talking, while Tobias is still in the process of resource reservation. If resource reservation on Tobias's side fails in this scenario, Theresa will be left with a call that was indicated as successful on the signalling level, but never had a chance to connect through the session on the media plane.

Therefore, if Theresa's UE does not support preconditions, it will, due to the received Require header in the INVITE request, send back a 420 (Bad Extension) response that includes an Unsupported header:

```
SIP/2.0 420 Bad Extension
Unsupported: preconditions
```

After Tobias's UE terminal has received this response, it can decide whether to send the INVITE request again, this time without preconditions, or not to bother to call the other side at all.

In IMS Release 5 it is assumed that both terminals support the preconditions mechanism; this practically made calls between an IMS terminal and a terminal outside the IMS, which does not support preconditions, impossible. For Release 6, interworking procedures have been specified that allow such scenarios. Nevertheless, the support of preconditions in the IMS is mandated and necessary, due to resource reservation procedures over the wireless link.

6.6.4 Preconditions

[RFC3312] introduces the SDP preconditions mechanism that allows UE to delay completion of a SIP session establishment until one or both sides have successfully completed their resource reservation. This extension to SIP and SDP is mandatorily supported by every item of UE that connects to the IMS.

Up to now, the Internet Engineering Task Force (IETF) has only defined the "qos" precondition type, but there might be more to come. The "qos" tag means that the precondition is set due to certain quality requirements of the related media stream, which is indicated in the m-line. The quality of service (QoS) depends mostly on the bandwidth that is reserved over the wireless link and the priority by which the routers in the network handle the packets that transport the "chunks" of speech and video data.

6.6.4.1 Preconditions in the first SDP offer

As explained in Section 6.5, for the SDP offer/answer mechanism Tobias's UE proposes three different media types, which are offered in the first INVITE to Theresa's UE. In order to illustrate the preconditions mechanism, we take just one of these three media types as an example and extend it with precondition-specific indications (written in bold):

```
m=audio 3458 RTP/AVP 0 96 97 98
a=rtpmap:0 PCMU
a=rtpmap:96 G726-32/8000
a=rtpmap:97 AMR-WB
a=rtpmap:98 telephone-event
a=curr:qos local none
a=curr:qos remote none
a=des:qos mandatory local sendrecv
a=des:qos none remote sendrecv
```

Every m-line block in SDP needs to indicate a separate set of preconditions. Remember that the m = audio line is only taken as an example, the two remaining m-lines in the original SDP would also include precondition setting. First, we look at the penultimate line:

```
a=des:qos mandatory local sendrecv
```

This indicates the desired (des) quality of service (qos) precondition at the calling user (local) end. The resources for the calling user need to be reserved in both the sending and receiving directions (sendrecv), as the audio stream is bidirectional (i.e., both users can talk to each other and hear what the other says). It also states that the session will not take place if the indicated resources cannot be reserved by the calling user (mandatory).

The last line:

```
a=des:qos none remote sendrecv
```

indicates the desired (des) quality of service (qos)-related preconditions at the called user (remote) end. As the calling and the called terminal are not directly connected to each other, the calling terminal is not aware how the remote terminal is attached to the network. It might be that Theresa's terminal does not need to perform any resource reservation, as it is connected via a CS telephony network. Therefore the calling side can only indicate that, if the remote side needs to reserve resources, then they should be reserved in both the sending and receiving directions (sendrecv); however, the calling side is not currently aware that this is really required in order to get a media session established (none).

Up to now, the calling terminal has only expressed the desired (des) preconditions for each side, but it also needs to talk about the current status of resource reservation. This is the subject of the first two new lines:

```
a=curr:qos local none
a=curr:qos remote none
```

These two lines indicate that currently (curr) no (none) quality of service (qos)-related preconditions have been fulfilled by either the calling side (local) or the called side (remote).

The a = des lines make it possible to set preconditions for the local and the remote user, while the a = curr lines indicate the extent to which the set preconditions are already being fulfilled.

6.6.4.2 Preconditions in the first SDP answer

As we know, Theresa's terminal is also attached over a wireless link to the IMS and supports the preconditions mechanism. Therefore, it will respond to the received

SDP offer with a well-formed answer and will include its own preconditions. Once again, we only show here those preconditions for the first media type:

```
m=audio 4011 RTP/AVP 96 97 98
a=rtpmap:96 G726-32/8000
a=rtpmap:97 AMR-WB
a=rtpmap:98 telephone-event
a=curr:qos local none
a=curr:qos remote none
a=des:qos mandatory local sendrecv
a=des:qos mandatory remote sendrecv
a=conf:qos remote sendrecv
```

The important thing to note here is that the remote and local sides have changed, because from Theresa's point of view her terminal is local and Tobias's is remote. Her terminal now indicates its own preconditions line by line. Line 5:

```
a=curr:qos local none
```

it has currently not reserved any local qos-related resources. Line 6:

```
a=curr:qos remote none
```

it received information from the remote side (in the first offer) that no qos-related resources have been reserved at the moment. Line 7:

```
a=des:qos mandatory local sendrecv
```

it mandatorily requires that its own resources get reserved in both the sending and receiving directions, before the audio session can start. Note that the initial value "none", as set from the calling side, has changed to "mandatory" because Theresa's terminal is attached via the air interface and, therefore, is also mandated to reserve the resources locally before it can start sending media. Line 8:

```
a=des:qos mandatory remote sendrecv
```

it received information from the remote side (in the first offer) that resources are mandatorily reserved in both the sending and receiving directions. Line 9:

```
a=conf:qos remote sendrecv
```

the calling terminal (remote) should send a confirmation (conf) at the moment the resources (qos) have been reserved in the sending and receiving directions (sendrecv).

This is a new line that the called side adds to SDP. It is a necessary addition because the called terminal is not intended to ring the called user or to start sending media until *both* sides have reserved the resources.

This SDP answer is now sent in the 183 (Session in Progress) response to the calling terminal.

6.6.4.3 Start of resource reservation

Subsequently, the second SDP offer/answer exchange takes place: that is, Tobias's terminal sends a PRACK request with the final selection of media types and codecs, which for the audio stream will look like:

```
m=audio 3458 RTP/AVP 97 98
a=rtpmap:97 AMR-WB
a=rtpmap:98 telephone-event
a=curr:qos local none
a=curr:qos remote none
a=des:qos mandatory local sendrecv
a=des:qos mandatory remote sendrecv
```

The received a = conf line is not repeated because the terminal knows that it needs to send a confirmation after it has reserved the resources.

At the moment Tobias's UE sends the PRACK request it can start resource reservation, because the QoS requirements for the different media streams and the specific codecs, which will be used to encode these media streams, are known.

It is also possible for Tobias's terminal to have already started resource reservation after the initial INVITE request was sent. All it would need do is to modify the reserved resources (i.e., the characteristics of the established media PDP context in the case of the GPRS) when it sends the PRACK request. This modification would be necessary because the initial reservation would have been made for media streams and codecs with the highest QoS requirements, while the agreed media and codecs could well have different QoS requirements.

Theresa's UE can start resource reservation after sending the 183 (Session in Progress) response, but it is assumed here that it starts after sending the 200 (OK) response for the PRACK request. This response will include the second SDP answer, which for the audio stream will look like:

```
m=audio 4011 RTP/AVP 97 98
a=rtpmap:97 AMR-WB
a=rtpmap:98 telephone-event
a=curr:qos local none
```

```
a=curr:qos remote none
a=des:qos mandatory local sendrecv
a=des:qos mandatory remote sendrecv
```

Once again, there is no need for the a = conf line to be repeated. As the status of the resource reservation has not changed, the precondition-related information in the second answer does not include anything new.

6.6.4.4 Successful resource reservation

Both lots of UE now try to reserve the requested resources. In the General Packet Radio Service (GPRS) case this would mean that they establish one or more media PDP contexts as secondary PDP contexts (see Chapter 13). It is possible for either Tobias's or Theresa's UE to finish the reservation first. In either case Theresa's terminal must not start to indicate the incoming call to Theresa (i.e., it must not ring) before knowing that both sides have successfully reserved resources.

Let us assume that the called side finishes the reservation first: in this case it will simply wait until it gets the confirmation that it requested from the calling side in the 183 (Session in Progress) response answer by setting the a = conf line in the first SDP answer.

As soon as the calling side has reserved the required resources, it will confirm this to the called terminal by sending a SIP UPDATE request as a subsequent request in the dialog that was established with the INVITE request. This UPDATE request will include a third SDP offer in its body, which shows that the resources at the calling side have been reserved:

```
m=audio 3458 RTP/AVP 97 98
a=rtpmap:97 AMR-WB
a=rtpmap:98 telephone-event
a=curr:qos local sendrecv
a=curr:qos remote none
a=des:qos mandatory local sendrecv
a=des:qos mandatory remote sendrecv
```

The only difference here is that the first a = curr line has changed from "none" to "sendrecv". Consequently, Tobias's terminal indicates that the status of the local QoS-related resources for the audio stream has changed. They have now been successfully reserved in both the sending and receiving directions.

As Theresa's terminal was the first to secure resource reservation, it can immediately start to ring after sending the 200 (OK) response for the UPDATE request, because it is now sure that both sides have sufficient resources reserved to send and receive the audio stream. When it starts to ring, it sends a 180 (Ringing) response for the INVITE request in parallel.

The 200 (OK) for the UPDATE will include a third SDP answer with the following SDP information about the audio stream:

```
m=audio 4011 RTP/AVP 97 98
a=rtpmap:97 AMR-WB
a=rtpmap:98 telephone-event
a=curr:qos local sendrecv
a=curr:qos remote sendrecv
a=des:qos mandatory local sendrecv
a=des:qos mandatory remote sendrecv
```

We can see from this that all the current states for resource reservation match the desired states; so, the preconditions negotiation has been successful and has finished.

Let us assume that Theresa's resource reservation takes longer than Tobias's: in this case Theresa's UE will receive the UPDATE request while it is still reserving resources. Although it will not start to ring, it will send back a 200 (OK) response with the following SDP information about the audio stream:

```
m=audio 4011 RTP/AVP 97 98
a=rtpmap:97 AMR-WB
a=rtpmap:98 telephone-event
a=curr:qos local none
a=curr:qos remote sendrecv
a=des:qos mandatory local sendrecv
a=des:qos mandatory remote sendrecv
```

The only difference here is that the status of local resource reservation is still set to "none". After local resources have been reserved, Theresa's UE will start to ring and send a 180 (Ringing) response that will *not* carry another SDP body. Nevertheless, Tobias's UE can interpret the 180 (Ringing) response as an indication that the resources on the remote side (i.e., Theresa's) have been reserved successfully: this is the reason why the 180 (Ringing) response also has to be sent reliably, obliging Tobias's UE to answer it with a PRACK request (see Section 6.5.2).

Once both sides have reserved their resources, media can be exchanged between the two lots of UE.

6.6.5 Related standards

Specifications relevant to Section 6.6 are:

- RFC3311   The Session Initiation Protocol (SIP) UPDATE Method.

- RFC3312   Integration of Resource Management and Session Initiation Protocol.

6.7 Controlling the media

6.7.1 Overview

As each UE is usually attached to the IMS over a wireless link, bandwidth resources are scarce and need to be handled with care. In Section 6.5 we saw that both sides exchange their preferences on the media and codecs used during a session until they agree on a specific set of media streams, each media stream using a single codec.

As media streams will not be routed through any CSCF, the IMS network needs to authorize reservation of the agreed resources. Consequently, the P-CSCF requests the policy decision function (PDF) to generate a media authorization token (Figure 6.8) whenever the first SIP message for the dialog is sent toward the UE (see Section 3.9.1). This token is added to the INVITE request and sent to Theresa's UE. Tobias' UE will receive its token in the 183 (Session in Progress) response.

During the SDP offer/answer process, P-CSCFs can tell the UE to group specific media streams together within one media PDP context or to keep certain media streams in separate media PDP contexts: this is achieved by the SDP media line grouping mechanism.

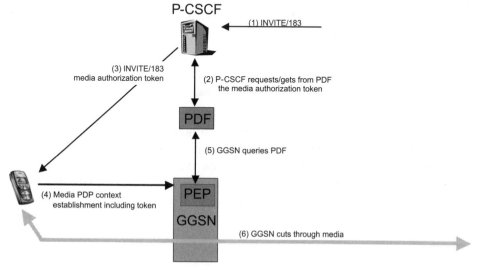

Figure 6.8 Transport of media authorization information.

In addition to all this, network operators might want to restrict the media types and codecs that each UE can use. To do so, the P-CSCFs and S-CSCFs check the SDP that is included in the INVITE request and send a rejection to Tobias's UE, if a certain media type or codec should not be used. In our example we do not assume that such a situation occurs. Nevertheless, the general procedure for media policing is described in Section 6.7.6.

After completion of the SDP offer/answer process, each UE can begin resource reservation for the agreed media. In the case of the GPRS and when using a dedicated signalling PDP context, this means that each UE will now try to establish one or more secondary PDP contexts for the media.

Within every request for media PDP context establishment, the UE will include the media authorization token that it received from the P-CSCF.

The GGSN receives the request for a new PDP context and first sends the token back to the PDF that created it. It also includes the requested session parameters in the message to the PDF. The PDF and the policy enforcement point (PEP) perform checks on the resource reservation request. Subsequently, the GGSN establishes the requested media PDP context with the UE.

Similar procedures will exist for other types of access networks in the future: the GPRS is used here as an example.

6.7.2 Media authorization

When Theresa's P-CSCF receives the INVITE request it will request the PDF to generate a media authorization token. The P-CSCF will add a P-Media-Authorization header to the INVITE request, include the received token and send the request to Theresa's UE:

```
INVITE sip:[5555::5:6:7:8]:1006 SIP/2.0
P-Media-Authorization: example-auth-token2
```

Theresa's UE will use this token when establishing the media PDP context or contexts for the media streams that are negotiated between the two lots of UE.

Shortly after, Tobias's P-CSCF will receive the 183 (Session in Progress) response and will also request a media authorization token from its PDF. It includes the token in a P-Media-Authorization header and sends it with the response to Tobias's UE:

```
SIP/2.0 183 Session in Progress
P-Media-Authorization: example-auth-token1
```

As the reservation and authorization of resources is a local matter for each UE, the two tokens will likely be different.

Each UE will receive only one token for every session that was created by an INVITE request. Regardless of how many PDP contexts the UE establishes, it can always blindly use the same token when requesting the resource. Consequently, the UE does not need to be at all aware of the content or the coding of the token, it just receives it in the SIP message and puts it into the ACTIVATE PDP CONTEXT REQUEST.

6.7.3 Grouping of media lines

The section on authorizing QoS resources (p. 78) already includes an example of the grouping of media lines. For the example in this chapter we will assume the grouping is different on both sides.

6.7.4 A single reservation flow

Theresa's P-CSCF will send to her UE the following information related to media line grouping in the second SDP offer in the PRACK request:

```
v=0
o=- 1357924 1357924 IN IP6 5555::1:2:3:4
s=-
c=IN IP6 5555::1:2:3:4
t=907165275 0
a=group:SRF 1
m=audio 3458 RTP/AVP 97 98
a=mid: 1
a=rtpmap:97 AMR-WB
a=rtpmap:98 telephone-event
m=video 3400 RTP/AVP 99
a=mid: 1
a=rtpmap:99 H.261
m=video 0 RTP/AVP 98
```

The a-line in the SDP heading defines a single reservation flow (SRF) group with the number "1": that is, all media streams in this group should go into a single PDP context.

Both media lines—one for audio and one for video—are followed by an a-line that indicates a media stream identification (MID) that is set in both cases to the

above-defined group number "1". This means that the P-CSCF instructs Theresa's
UE to group the two media streams together into one PDP context.

6.7.5 Separated flows

The P-CSCF in Finland sees the need for separated media streams. Therefore, it adds
the following media line grouping-related information to the SDP answer that is sent
to Tobias's UE in the 200 (OK) response to the PRACK request:

```
v=0
o=- 1357924 1357925 IN IP6 5555::5:6:7:8
s=-
c=IN IP6 5555::5:6:7:8
t=907165275 0
a=group:SRF 1
a=group:SRF 2
m=audio 4011 RTP/AVP 97 98
a=mid: 1
a=rtpmap:97 AMR-WB
a=rtpmap:98 telephone-event
m=video 4012 RTP/AVP 99
a=mid: 2
a=rtpmap:99 H.261
m=video 0 RTP/AVP 98
```

The a = group lines create two individual SRF groups with the numbers "1" and "2".
As the audio stream indicates SRF group "1" and the video stream SRF group "2",
Tobias's UE will have to reserve separate media PDP contexts for each of the two
media streams (Figure 6.9).

6.7.6 Media policing

The P-CSCF and S-CSCF can reject certain media types or codecs that are offered in
SDP. This might be due to operator policy. One reason for this could be that an
operator does not allow the use of any unknown media types or unknown codecs, as
the network would not be able to charge for such media.

If a CSCF detects that an unsupported media type or codec is included in the
SDP offer, it will reject the request with 488 (Not Acceptable Here) and will indicate
the supported media types in the body of the response.

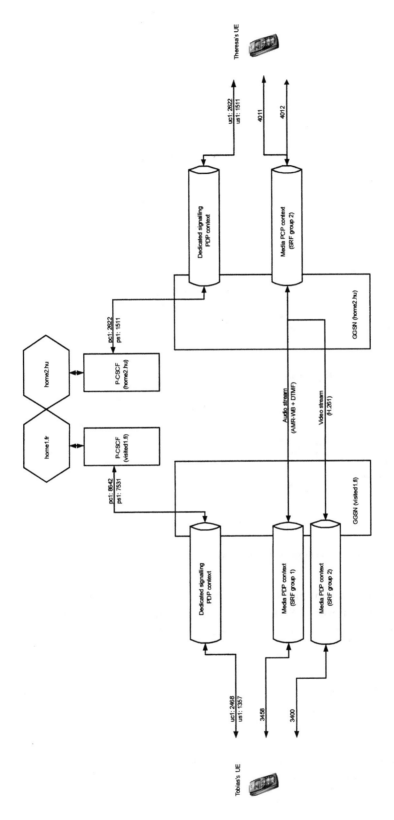

Figure 6.9 Media streams and transport in the example scenario.

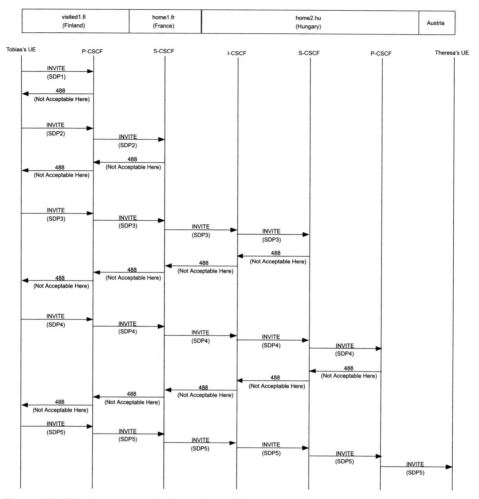

Figure 6.10 Worst case scenario for media policing.

In the example scenario this is not assumed to happen. Figure 6.10 shows a worst case scenario when every CSCF on the route does not like certain media elements. Such extreme scenarios are unlikely to occur in reality.

6.7.7 Related standards

Specifications relevant to Section 6.7 are:

- RFC3313   Private Session Initiation Protocol (SIP) Extensions for Media Authorization.

- RFC3388   Grouping of Media Lines in the Session Description Protocol (SDP).

- RFC3524   Mapping of Media Streams to Resource Reservation Flows.

6.8 Charging-related information for sessions

6.8.1 Overview

In the case of a GPRS access network (chosen for this example) charging is usually done on the basis of the amount of data that is sent via a PDP context. The network operator can decide to charge specific PDP contexts differently: for example, an MMS could be charged differently from "normal" Internet traffic.

For the media session Tobias's UE needs to establish several media PDP contexts with the GGSN. The GGSN will start generating charging records whenever data (i.e., media) are sent over that PDP context. These charging records are based on the GPRS charging ID (GCID) that is generated by the access network.

As described in Section 5.11, the IMS charging ID (ICID) is created during initial registration by the P-CSCF. During a session the P-CSCF creates an additional ICID for the charging of media streams that are transported over established media PDP contexts.

The GGSN sends GCIDs of these media PDP contexts via the Go interface to the P-CSCF, which then creates an ICID and sends it toward the home network of the served user.

These procedures will take place on both sides (i.e., for Tobias and for Theresa). Which of them will get charged what will be decided by the charging entities in their home networks.

The S-CSCFs of the users will furthermore distribute within their home networks the addresses of the Charging Collection Function (CCF) and the Event Charging Function (ECF) that will collect the charging records for the two users. This information is needed by all entities that create charging records for the session.

6.8.2 Exchange of ICID for a media session

When Tobias's P-CSCF receives the INVITE request for the session it will create a new ICID and include it in a P-Charging-Vector header that it adds to the INVITE request:

```
INVITE sip:theresa@home2.hu SIP/2.0
P-Charging-Vector: icid-value= "AyretyU0dm+6O2IrT5tAFrbHLso=023551025"
```

Tobias's S-CSCF will add the originating interoperator identifier (IOI) to the P-Charging-Vector header. The IOI parameter identifies Tobias's home network to Theresa's home network for charging purposes and is usually set to the home domain of the user:

```
INVITE sip:theresa@home2.hu SIP/2.0
P-Charging-Vector: icid-value= "AyretyU0dm+6O2IrT5tAFrbHLso=023551025"
 ;orig-ioi=home1.fr
```

The S-CSCF of Theresa's home network will store and remove the received orig-ioi parameter for the P-Charging-Vector header.

This P-Charging-Vector header will be sent further with the INVITE request until it reaches the P-CSCF of Theresa, where it will be removed. Charging-related headers are never sent toward a UE. All entities on the way that are involved in charging for this session will store the ICID.

When receiving the first response for the INVITE request—i.e., the 183 (Session in Progress) response—Theresa's P-CSCF will once again add the P-Charging-Vector header, including the same ICID value as received in the INVITE request:

```
SIP/2.0 183 Session in Progress
P-Charging-Vector: icid-value= "AyretyU0dm+6O2IrT5tAFrbHLso=023551025"
```

Theresa's S-CSCF will add the terminating IOI information to the response, before sending it further:

```
SIP/2.0 183 Session in Progress
P-Charging-Vector: icid-value= "AyretyU0dm+6O2IrT5tAFrbHLso=023551025"
 ;term-ioi=home2.hu
```

The terminating IOI information will again be stored and removed by Tobias's S-CSCF, and his P-CSCF will remove the P-Charging-Vector from the response.

6.8.3 Correlation of GCID and ICID

The media PDP context on Tobias's side is established after successful resource reservation, and, as we saw in Section 6.6, Tobias's UE will then immediately send an UPDATE request toward Theresa. As this is the first request that is received by Tobias's P-CSCF after the media PDP context has been established, it will include in the UPDATE request the P-Charging-Vector header with the following information:

- The ICID that it created for this media session.

- The charging parameters for every media PDP context that were established for this session:

  o   The GCID that was received from the GGSN.

  o   The address of the related GGSN.

  o   An indication ("pdp-sig") of whether the related PDP context is a signalling PDP context or not (in this case it is not a signalling PDP context).

  o   The flow identifier of the media stream.

  o   And the token (auth-token) that was used for media authorization (see Section 6.7.2).

The UPDATE request now looks like:

```
UPDATE sip:[5555::5:6:7:8]:1006 SIP/2.0
P-Charging-Vector: icid-value= "AyretyU0dm+6O2IrT5tAFrbHLso=023551024"
  ;ggsn=[5555::4b4:3c3:2d2:1e1];
   ;pdp-sig=no; gcid=723084371
   ;auth-token=example-auth-token1
   ;flow-id=1
  ;ggsn=[5555::4b4:3c3:2d2:1e1]
   ;pdp-sig=no; gcid=723084372
   ;auth-token=example-auth-token1
   ;flow-id=2
```

As we saw in Section 6.7.3, Tobias has established two media PDP contexts; hence, two charging parameter lists are also included.

Tobias's S-CSCF will store and remove all data from the P-Charging-Vector header and add the originating and terminating IOI parameters, before sending the UPDATE request to Theresa's S-CSCF:

```
UPDATE sip:[5555::5:6:7:8]:1006 SIP/2.0
P-Charging-Vector: icid-value= "AyretyU0dm+6O2IrT5tAFrbHLso=023551024"
  ;orig-ioi=home1.fr
  ;term-ioi=home2.hu
```

Once again, these two parameters (orig-ioi and term-ioi) will be removed by Theresa's S-CSCF before sending it to the P-CSCF, which removes the header before sending the UPDATE request toward Theresa's UE.

In our example Theresa's UE has already finished resource reservation at the moment the UPDATE request is received from Tobias's side (see Section 6.6.4.4). Therefore, Theresa's P-CSCF will include the P-Charging-Vector header again in the 200 (OK) response to the UPDATE request, this time with all the information related to the reserved media PDP context:

```
SIP/2.0 200 OK
P-Charging-Vector: icid-value= "AyretyU0dm+6O2IrT5tAFrbHLso=023551024"
  ;ggsn=[5555::802:53:58:6]
    ;pdp-sig=no
    ;gcid=306908949
    ;auth-token=example-auth-token2
    ;flow-id=2
```

As Theresa established only one media PDP context (see Section 6.7.3), her P-CSCF includes exactly one charging parameter list.

The S-CSCFs now behave in the same way as before:

- Theresa's S-CSCF will remove PDP context-related information from the P-Charging-Vector header and will add the saved orig-ioi and term-ioi parameters.

- Tobias's S-CSCF will remove the orig-ioi and term-ioi parameters.

- Finally, the P-CSCF will again remove the P-Charging-Vector header from the 200 (OK) response before sending it to Tobias's UE.

6.8.4 Distribution of charging function addresses

The addresses of the CCF and ECF are distributed within the home network of the S-CSCFs.

Every S-CSCF adds a P-Charging-Function-Address header to the INVITE request, which is then sent along the route until it reaches the border of the home network. The CSCF at the border of the home network will remove this header.

In the case of Tobias's home network this means that his S-CSCF:

- Adds the P-Charging-Function-Address header when it first receives the INVITE request and then sends it to all ASs that belong to the home network and are contacted due to the filter criteria in Tobias's user profile (see Section 6.3.8). As the S-CSCF is the last entity in Tobias's home network, it will remove the P-Charging-Function-Address header before sending it toward Theresa's home network.

- Adds the P-Charging-Function-Address header again when it receives the 183 (Session in Progress) response, sends it again to all ASs that belong to the home network and are contacted due to the filter criteria and removes the header before sending the response toward the P-CSCF that is located in the visited network.

The P-CSCF in Tobias's visited network will need to discover its local charging function addresses by other means.

On the other hand Theresa's S-CSCF:

- Adds the P-Charging-Function-Address when it receives the INVITE request and then sends it toward the ASs, as stated above. However, in this case the S-CSCF does not remove the header from the request before sending it to the P-CSCF, as Theresa's P-CSCF is located in her home network. She uses GPRS roaming, not IMS roaming, to access her home network's P-CSCF (see Section 4.1). The P-Charging-Function-Address header will therefore be removed from the INVITE request by the P-CSCF.

- Adds the P-Charging-Function-Address header again when it receives the 183 (Session in Progress) response and sends it again with the response. However, this time the header will be removed by the I-CSCF of Theresa's home network, which will also receive the response.

The P-Charging-Function-Address header that is added by Tobias's S-CSCF in the initial INVITE request would look something like this:

```
INVITE sip:theresa@home2.hu SIP/2.0
P-Charging-Function-Addresses: ccf=[5555::b99:c88:d77:e66]
  ;ccf=[5555::a55:b44:c33:d22]
  ;ecf=[5555::1:2ee:3dd:4cc]
  ;ecf=[5555::6aa:7bb:8cc:9dd]
```

6.8.5 Related standards

IMS-specific SIP headers for these charging procedures are specified in [RFC3455]: Private Header (P-Header) Extensions to the Session Initiation Protocol (SIP) for the 3rd-Generation Partnership Project (3GPP).

6.9 Release of a session

6.9.1 User-initiated session release

Of course, Tobias and Theresa will at one point stop their conversation. Let us say that Theresa meets one of her friends at Stephansdom in Vienna and has to say goodbye to her brother (Figure 6.11). She will be the one who drops the call by pressing the red button on her mobile phone.

Consequently, her UE will generate a BYE request, which is sent to Tobias's UE in the same way as any other subsequent request. In parallel to this, her UE will also drop the media PDP context that was established for the session:

```
BYE sip:[5555:1:2:3:4]:1357 SIP/2.0
Route: <sip:pcscf2.home2.hu;lr>
Route: <sip:scscf2.home2.hu;lr>
Route: <sip:scscf1.home1.fr;lr>
Route: <sip:pcscf1.visited1.fi;lr>
To: "Your Brother" <sip:tobi@brother.com>;tag=veli
From: "My beloved Sister" <sip:Theresa@sister.com>;tag=schwester
```

We can see from this BYE request that the information in the To and From headers has been swapped, as this request is now sent from Theresa's side.

Tobias's UE will also drop its PDP context immediately after it receives the BYE request. It will also respond to the request with a 200 (OK) response, which will be sent back toward Theresa. The four CSCFs and all ASs on the route will clear all dialog states and information related to the session.

6.9.2 P-CSCF performing network-initiated session release

There might be situations in which it is necessary for one of the CSCFs to release the session, rather than the user.

For example, Theresa's P-CSCF would need to release an ongoing session when it realizes that Theresa's UE has lost radio coverage and is no longer connected to the access network (Figure 6.12). In that case the P-CSCF would need to send a BYE request on behalf of Theresa:

```
BYE sip:[5555:1:2:3:4]:1357 SIP/2.0
Route: <sip:scscf2.home2.hu;lr>
Route: <sip:scscf1.home1.fr;lr>
Route: <sip:pcscf1.visited1.fi;lr>
To: "Your Brother" <sip:tobi@brother.com>;tag=veli
From: "My beloved Sister" <sip:Theresa@sister.com>;tag=schwester
```

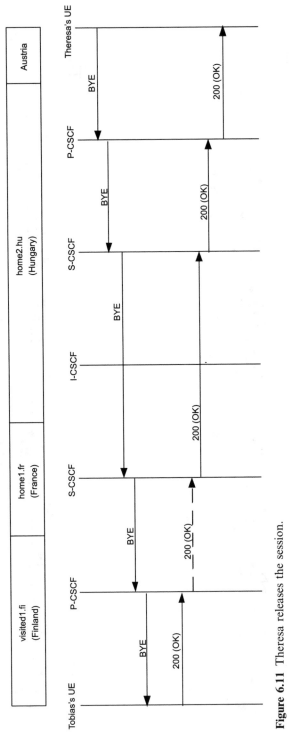

Figure 6.11 Theresa releases the session.

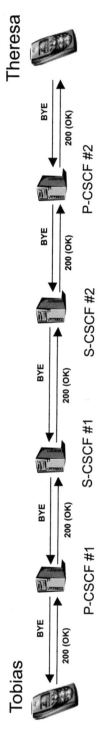

Figure 6.12 P-CSCF terminates a session.

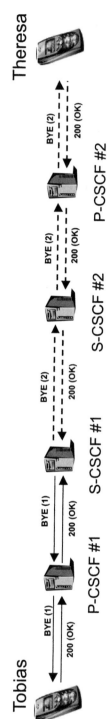

Figure 6.13 S-CSCF terminates a session.

6.9.3 S-CSCF performing network-initiated session release

There may be occasions when Tobias's S-CSCF needs to be shut down: Tobias may be using a prepaid card and runs out of money. In this case Tobias's S-CSCF would release the session (Figure 6.13) by issuing one BYE request toward Tobias's UE:

```
BYE sip:[5555:1:2:3:4]:1357 SIP/2.0
Route: <sip:pcscf1.visited1.fi;lr>
To: "Your Brother" <sip:tobi@brother.com>;tag=veli
From: "My beloved Sister" <sip:Theresa@sister.com>;tag=schwester
```

and another BYE request toward Theresa's UE:

```
BYE sip:[5555:1:2:3:4]:1006 SIP/2.0
Route: <sip:scscf2.home2.hu;lr>
Route: <sip:pcscf2.home2.hu;lr>
From: "Your Brother" <sip:tobi@brother.com>;tag=veli
To: "My beloved Sister" <sip:Theresa@sister.com>;tag=schwester
```

To generate the BYE request with the correct set of Route headers, the P-CSCFs and S-CSCFs need to keep track of all routing information that is collected during the establishment of any dialog.

7

Routing of PSIs

The concept of a public service identity (PSI: i.e., a URI that is not related to a user but to a service) is explained in Section 3.4.2.

This chapter is a basic introduction to the routing principles of PSIs, as they are quite different from those that are applied between two Internet Protocol Multimedia Subsystem (IMS) users.

As PSIs are not registered, requests to and from them do not need to traverse any Serving-Call Session Control Function (S-CSCF).

There are three scenarios for PSI routing.

7.1 Scenario 1: routing from a user to a PSI

This occurs, for example, when a user calls into a conference (Figure 7.1). In this case the request needs first to traverse the user's S-CSCF, which then can:

1. Either, resolve the PSI immediately and route it directly to the application server (AS).

2. Or, is able to resolve the address of an Interrogating-CSCF (I-CSCF) in the home network of the PSI. The I-CSCF will then query the Home Subscriber Server (HSS)—where routing information about the PSIs will be stored—and route the request directly to the hosting AS. Note that an S-CSCF can also be assigned for the PSI, in which case this would be contacted first. This scenario is not shown here.

7.2 Scenario 2: routing from a PSI to a user

This occurs, for example, when a conference server (the focus) invites a user to a conference (Figure 7.2). In this case the AS sends the request:

The IMS. Miikka Poikselkä, Georg Mayer, Hisham Khartabil and Aki Niemi
Copyright 2004 by John Wiley & Sons, Ltd. ISBN 0-470-87113-X

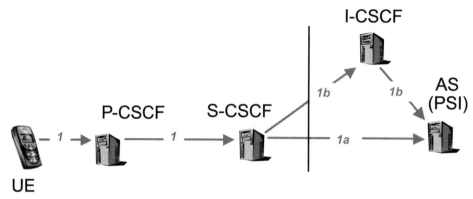

Figure 7.1 Routing from a user to a PSI.

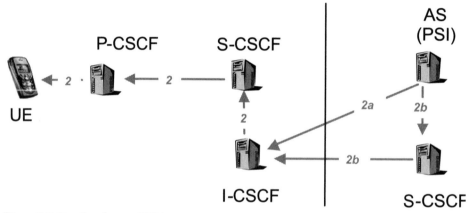

Figure 7.2 Routing from a PSI to a user.

1. Either, directly to the I-CSCF of the terminating user's home network, if the AS can resolve this address on its own.

2. Or, to an S-CSCF in the ASs home network, which then resolves the address of the I-CSCF of the terminating user's home network.

7.3 Scenario 3: routing from a PSI to another PSI

This occurs, for example, when two conferences are interconnected such that one conference server (focus) sends an INVITE request to another focus (Figure 7.3).

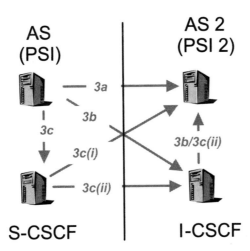

Figure 7.3 Routing from an AS to a PSI.

For this case several routing possibilities exist. The originating AS can route the request:

1. Either, directly to the terminating AS that is hosting the second PSI, if the originating AS is able to resolve the PSI directly.

2. Or, if it cannot resolve the address of the second PSI directly, to an I-CSCF of the terminating AS, which then queries the HSS with the second PSI and sends the request directly to the terminating AS. Note that an S-CSCF can also be assigned for the PSI, in which case this would be contacted first. This scenario is not shown here.

3. Or, if it cannot resolve any part of the terminating address, to an S-CSCF in its own home network, which then will:

   o   Either, resolve the address of the second PSI and send the request to the terminating AS directly.

   o   Or, cannot resolve the address of the second PSI directly and therefore sends the request to an I-CSCF in the home network of the second PSI. This I-CSCF will then act in the same way as item (2) in Scenario 3.

Part III

Protocols

8

SIP

This chapter does not provide a full Session Initiation Protocol (SIP) specification. Instead, it tries to point out the important aspects of SIP as they apply to the Internet Protocol Multimedia Subsystem (IMS). In particular, this chapter does not discuss how a SIP entity should behave using the maddr parameter in URIs (uniform resource identifiers) nor does it explain how the SIP entity should behave in certain error conditions. For a full SIP specification, please refer to [RFC3261].

8.1 Background

SIP is an application layer protocol that is used for establishing, modifying and terminating multimedia sessions in an IP network. It is part of the multimedia architecture whose protocols are continuously being standardized by the Internet Engineering Task Force (IETF). Its applications include, but are not limited to, voice, video, gaming, messaging, call control and presence.

The idea of a session signalling protocol over IP dates back to 1992 where multicast conferencing was in mind. SIP itself originated in late 1996 as a component of the IETF Mbone (multicast backbone), an experimental multicast network on top of the public Internet. It was used by the IETF for the distribution of multimedia content, including IETF meetings, seminars and conferences. Due to its simplicity and extensibility, SIP was later adopted as a Voice Over Internet Protocol (VoIP) signalling protocol, finally becoming an IETF-proposed standard in 1999 as [RFC2543]. SIP was further enhanced to take into account interoperability issues, better design and new features. The actual document was re-written entirely for clarity. The protocol remains mostly backward compatible with [RFC2543]. This newly created document became the proposed standard as [RFC3261] in 2002, making [RFC2543] obsolete.

The IMS. Miikka Poikselkä, Georg Mayer, Hisham Khartabil and Aki Niemi
Copyright 2004 by John Wiley & Sons, Ltd. ISBN 0-470-87113-X

8.2 Design principles

SIP, as part of the IETF process, is based on the Hyper Text Transfer Protocol (HTTP) and the Simple Network Management Protocol (SNMP). Figure 8.1 shows where SIP fits into a protocol stack.

SIP was created with the following design goals in mind:

- Transport protocol neutrality—able to run over reliable (TCP, SCTP) and unreliable (UDP) protocols.

- Request routing—direct (performance) or proxy-routed (control).

- Separation of signalling and media description—can add new applications or media.

- Extensibility.

- Personal mobility.

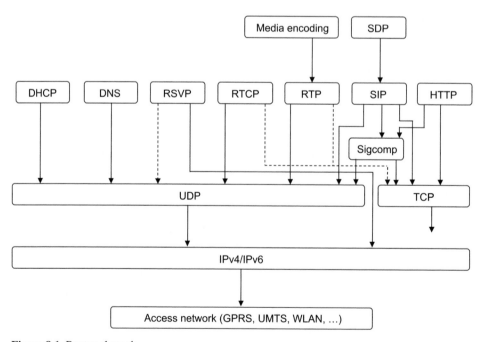

Figure 8.1 Protocol stack.

8.3 SIP architecture

Elements in SIP can be classified into user agents (UAs) and intermediaries (servers). In an ideal world, communications between two end points (or UAs) happen without the need for intermediaries. But, this is not always the case as network administrators and service providers would like to keep track of traffic in their network.

Figure 8.2 depicts a typical network set-up, which is referred to as the "SIP trapezoid".

A SIP UA or terminal is the end point of dialogs: it sends and receives SIP requests and responses, it is the end point of multimedia streams, and it is usually the user equipment (UE), which is an application in a terminal or a dedicated hardware appliance. The UA consists of two parts:

- User Agent Client (UAC)—the caller application that initiates requests.

- User Agent Server (UAS)—accepts, redirects, rejects requests. Sends responses for incoming requests on behalf of the user.

Gateways are special cases of USs.

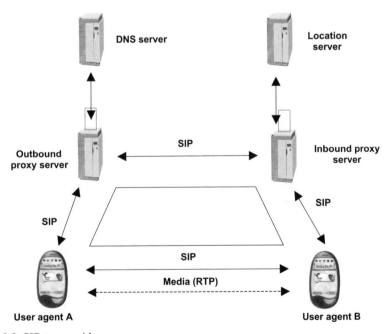

Figure 8.2 SIP trapezoids.

SIP intermediaries are logical entities where SIP messages pass through on their way to their final destination. These intermediaries are used to route and redirect requests. These servers include:

- *Proxy server*—receives and forwards SIP requests. It can interpret or rewrite certain parts of SIP messages that do not disturb the state of a request or dialog at the end points, including the body. A proxy server can also send a request to a number of locations at the same time. This entity is labelled as a forking proxy. Forking can be parallel or sequential. There are three proxy server types:

  o Dialog-statefull proxy—a proxy is dialog-statefull if it retains the state for a dialog from the initiating request (INVITE request) to the terminating request (BYE request).

  o Transaction-statefull proxy—maintains client and server transaction-state machines during the processing of a request.

  o Stateless proxy—forwards every request it receives downstream and every response it receives upstream.

- *Redirect servers*—maps the address of requests into new addresses. It redirects requests but does not participate in the transaction.

- *Location server*—keeps track of the location of users.

- *Registrars*—a server that accepts REGISTER requests. These servers are used to store explicit binding between a user's address of record (SIP address) and the address of the host where the user is currently residing or wishing to receive requests on.

Two more elements that are used to provide services for SIP users:

- *Application server*—an AS is an entity in the network that provides end users with a service. Typical examples of such servers are presence and conferencing servers.

- *Back-to-back-user-agent*—as the name depicts, a B2BUA is where a UAS and a UAC are glued together. The UAS terminates the request as a normal UAS. The UAC initiates a new request that is somehow related to requests received at the UAS side, but not in any protocol-specific link. This entity is almost like a proxy, but it breaks all the rules that govern the way a proxy can modify a request.

8.4 Message format

As shown in Figure 8.3, the SIP message is made up of 3 parts: the start line, message headers and body.

The start line contents vary depending on whether the SIP message is a request or a response. For requests it is referred to as a request line and for responses it is referred to as a status line.

An example SIP request looks like:

```
INVITE sip:bob.smith@nokia.com SIP/2.0
Via: SIP/2.0/UDP cscf1.example.com:5060;branch=z9hG4bK8542.1
Via: SIP/2.0/UDP [5555::1:2:3:4]:5060;branch=z9hG4bK45a35h76
Max-Forwards: 69
From: Alice <sip:alice@nokia.com>;tag=312345
To: Bob Smith <sip:bob.smith@nokia.com>
Call-ID: 105637921
CSeq: 1 INVITE Contact: sip:alice@[5555::1:2:3:4]
Content-Type: application/sdp
Content-Length: 159

[body]
```

8.4.1 Requests

SIP requests are distinguished from responses using the start line. As indicated earlier, the start line in the request is often referred to as the request line. It has three components: a method name, a request-URI and the protocol version. They

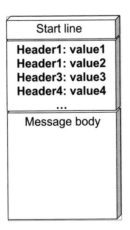

Figure 8.3 SIP message format.

appear in that order and are separated by a single space character. The request line itself terminates with a carriage return–line feed pair (CRLF):

- *Method*—the method indicates the type of request. Six are defined in the base SIP specification [RFC3261]: the INVITE request, CANCEL request, ACK request and BYE request are used for session creation, modification and termination; the REGISTER request is used to register a certain user's contact information; and the OPTIONS request is used as a poll for querying servers and their capabilities. Other methods have been created as an extension to [RFC3261].

- *Request-URI*—the request-URI is a SIP or a SIPS URI that identifies a resource that the request is addressed to.

- *Protocol version*—the current SIP version is 2.0. All requests complaint with [RFC3261] must include this version in the request, in the form "SIP/2.0".

8.4.2 Response

SIP responses are distinguished from requests using the start line. As indicated earlier, the start line in the response is often referred to as the status line. It has three components: the protocol version, status code and reason phrase. They appear in that order and are separated by a single space character. The status line itself terminates with a CRLF pair:

- *Protocol version*—this is identical to the protocol version in the request line.

- *Status code*—the status code is a three-digit code that identifies the nature of the response. It indicates the outcome of the request.

- *Reason phrase*—this is a free text field providing a short description of the status code. It is mainly aimed at human users.

Status codes are classified into six classes (classes 2xx to 6xx are final responses):

- *1xx*—provisional/informational responses. They indicate that the request was received and the recipient is continuing to process the request.

- *2xx*—success responses. The request was successfully received, understood and accepted.

- *3xx*—redirection responses. Further action needs to be taken by the requester in order to complete the request.

- *4xx*—client error responses. The request contains a syntax error. It can also indicate that the server cannot fulfil the request.

- *5xx*—server error responses. The server failed to fulfil a valid request. It is the fault of the server.

- *6xx*—global failure responses. The request cannot be fulfilled at any server. The server responding with this response class needs to have definitive information about the user.

The "xx" are two digits that indicate the exact nature of the response: for example, a "180" provisional response indicates ringing by the remote end, while a "181" provisional response indicates that a call is being forwarded.

8.4.3 Header fields

Header fields contain information related to the request: for example, the initiator of the request, the recipient of the request and call identifier. Header fields also indicate message body characteristics.

Header fields end with a CRLF pair. The headers section of a SIP message terminates with a CRLF.

The format of the header fields is as follows:

```
Header-name: header-value
```

Some headers are mandatory in every SIP request and response. Those headers and their formats are listed below:

- To header   To: SIP-URI(;parameters)

- From header   From: SIP-URI(;parameters)

- Call-ID header   Call-ID: unique-id

- CSeq header   CSeq: digit method

- Via header   Via: SIP/2.0/[transport-protocol] sent-by(;parameters)

- Max-Forwards header   Max-Forwards: digit

- Contact header   Contact: SIP-URI(;parameters)

The Contact header is mandatory for requests that create dialogs, the Max-Forwards header is typically set to 70. Note that the brackets around parameters indicate that they are optional. The brackets are not part of the header syntax.

Whenever (;parameters) appears it indicates that multiple parameters can appear in a header and that semicolons separate the parameters. The transport protocol for the Via header is the User Datagram Protocol (UDP), Transmission Control Protocol (TCP) or Transport Layer Security (TLS).

8.4.4 Body

The message body (payload) can carry any text-based information, while the request method and the response status code determine how the body should be interpreted.

When describing a session the SIP message body is typically a Session Description Protocol (SDP) message.

8.5 The SIP URI

The SIP URI follows the same form as email addresses: `user@domain`. There are two URI schemes:

- sip:bob.smith@nokia.com is a SIP URI. This is the most common form and was introduced in [RFC2543].

- sips:bob.smith@nokia.com is a Secure SIP (SIPS) URI. This new scheme was introduced in [RFC3261] and requires TLS over TCP as transport for security.

There are two types of SIP and SIPS URIs:

- Address of record (AOR)—this is a SIP address that identifies a user. This address can be handed out to people much like a phone number is: for example, sip:bob.smith@nokia.com (needs DNS SRV records to locate SIP Servers for the nokia.com domain).

- Fully qualified domain name (FQDN) or IP address (identifies a device) of the host—for example, sip:bob.smith@127.233.4.12 or sip:bob.smith@pc2.nmp. nokia.com (which needs no resolution for routing).

The SIP URI has the form: sip:userinfo@hostport[parameters][headers]. The SIPS URI follows exactly the same syntax as the SIP URI:

- Userinfo—a user name or a telephone number.

- Hostport—the domain name or numeric network address and port.

- Parameters—defines specific URI parameters, such as transport, time to live, etc.

- Headers—another rarely used form that passes on extra information.

Below are some examples of SIP URIs:

- sip:bob.smith@nokia.com

- sip:bob@nokia.com; transport = tcp

- sip: + 1-212-555-1234@gw.com;user = phone

- sip:root@136.16.20.100:8001

- sip:bob.smith@registrar.com;method = REGISTER

8.6 The tel URI

The telephone URI (tel URI) is used to identify resources using a telephone number. SIP allows requests to be sent to a tel URI. This means that the request-URI of a SIP request can contain a tel URI.

   The tel URI can contain a global number or a local number. A global number follows the rules of E.164 numbers and starts with a "+", while a local number follows the rules of local private numbering plans. Local numbers need to have the phone-context parameter, which identifies the context (owner) of the local number and, therefore, the scope of the number. This makes the number globally unique. The context can be represented by a global number or a domain name: the former must contain a valid global number that is owned by the local number distributor and the latter must contain a valid domain name that is under the authority of the owner distributing the local numbers. Here are some tel URI examples:

- a global number—tel: + 358-9-123-45678

- a local number with a domain context—tel:45678;phone-context = example.com

- a local number with a global number context: tel:45678;phone-context = + 358-9-123

Notice that the tel URI allows visual separators like hyphens "-" in the number to improve readability and the tel URI parameters are separated by semicolons ";". The full tel URI syntax can be found in [Draft-ietf-iptel-rfc2806bis].

8.7 SIP structure

SIP is a layered protocol that allows different modules within it to function independently with just a loose coupling between each layer. Figure 8.4 visualizes the layered approach taken.

8.7.1 Syntax and encoding layer

The first layer in the protocol is the syntax and encoding layer. Encoding makes use of the augmented Backus-Naur Form grammar (BNF), the complete description of which can be found in [RFC3261].

8.7.2 Transport layer

The second layer is the transport layer. As the name indicates, this is the layer that dictates how clients send requests and receive responses and how servers receive requests and send responses. The transport layer is closely related to the sockets layer of a SIP entity.

8.7.3 Transaction layer

The third layer is the transaction layer. A transaction, in SIP terms, is a request that is sent by a client to a server, along with all responses to that request sent from the server back to the client. The transaction layer handles the matching of responses to requests. Application-layer retransmissions and application-layer transaction time-outs are also handled in this layer and are dependent on the transport protocol used. A client transaction sends requests and receives responses, while a server transaction receives requests and sends responses. The transaction layer uses the transport layer for sending and receiving requests and responses.

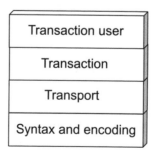

Figure 8.4 SIP protocol layers.

The transaction layer has four transaction-state machines. Each transaction-state machine has its own timers, retransmission rules and termination rules:

- INVITE client transaction.

- Non-INVITE client transaction.

- INVITE server transaction.

- Non-INVITE server transaction.

8.7.4 TU layer

The fourth layer is the transaction user (TU) layer. This is the layer that creates client and server transactions. When a TU wishes to send a SIP request it creates a client transaction instance and sends the request along with the destination IP address, port, and transport protocol to use. TUs are defined to be UAC core and UAS core, or simply UAC and UAS. UACs create and send requests and receive responses using the transaction layer, while UASs receive requests and create and send responses using the transaction layer.

There are two factors that can affect TU behaviour: one is the method name in the SIP message and the other is the state of the request with regard to dialogs (dialogs are discussed in Section 8.9).

Other than these two factors, the TU behaves in a standard way. This is described in the following sections.

8.7.4.1 UAC behaviour

For requests that arrive outside a dialog, the steps that a UAC needs to take include populating the request-URI, the To header, the From header, the Call-ID header, the CSeq header and the Via header. Other headers like the Require header and Supported header that indicate any extension that the UAC requires or supports may also be added. A Contact header must be added if the request creates a dialog or if a registration binding is required. Any additional components can also be populated at this stage; this includes the message body. In the presence of a message body in the SIP request, the Content-type header and Content-length header must also be populated:

- The To header is populated with the target's AOR (an AOR is similar to an address from a business card).

- The From header is populated with the sender's AOR. It is also populated with a tag parameter. The tag is one of the components used to identify a dialog.

- The Call-ID header is populated with an identifier that is unique.

- The CSeq header is used to identify the order of transactions. The CSeq number is arbitrary for requests outside a dialog. It contains two parts, a CSeq and a method name separated by a space. The method part is populated with the same method as the one in the request line.

- The Max-Forwards header is used to limit the number of hops a request traverses and is used to avoid loops. It is typically set to 70 (indicating the number of hops). Each hop decrements the value by 1.

- The Via header contains two vital pieces of information: the transport protocol and the address where the response is to be sent. The protocol name and value are always set to the SIP and 2.0, respectively. The Via header contains a branch parameter that identifies transactions and is used to match requests to responses. It must be unique. The branch inserted by an element compliant with this specification always begins with the characters "z9hG4bK".

- The Contact header is populated with a URI that is typically the address of the host where the request originated.

- The request-URI is normally populated with the value in the To header. REGISTER requests are special cases in which the request-URI is populated with the registrar address.

A UAC may have a pre-existing route set, which is a set of intermediaries that the UAC wants requests to propagate to before reaching their final destination, including an outbound proxy. This route set is represented in the request as Route headers. The request-URI population may differ in this case depending on whether the URI in the topmost Route header contains a loose-route parameter. Section 8.12.2 explains the concept of loose routing and how a remote-target should be populated in case the next hop is a loose router. The procedures in Section 8.12.2 are followed for populating the request-URI as the remote target.

The UAC must then route the request according to the rules defined in Section 8.12.

The UAC also handles the responses for requests it sends. These responses can be timeout error responses or SIP success or failure responses, including redirection responses (3xx).

8.7.4.2 UAS behaviour

For requests that arrive outside a dialog, a UAS inspects the request method for recognition and inspects the request-URI and To header to determine whether this request is destined to it. If either of the two inspections fail, an error response is returned.

The UAS then decides whether any extensions are required and returns an error if it cannot satisfy them. If it can, the UAS continues processing the request by examining and processing the contents of the request (the message body).

If all the above is successful, the UAS can then apply any extensions that are supported by the UAC (as indicated in the Supported header). Processing of the request from this point on is method-specific: for REGISTER requests see Section 8.8 and for INVITE requests see Section 8.10.

Once the UAS has processed the request it generates a response, which can be provisional or final. Multiple provisional responses can be sent for one request, but only one final response must be sent. Typically, a provisional response is only sent in response to an INVITE request.

When generating a response, the From header, the Call-ID header, the CSeq-header, the Via header and the To header in the response are all copied from the request. The sequence of Via headers in the request must be maintained.

If a request contained a To header tag parameter in the request, then a new tag must not be created. However, if the To header in the request did not contain a tag, then the UAS must add a tag to the To header in the response. For "100" provisional responses a To header tag is not mandated. This tag is used as one of the components that identify a dialog. It is also used by a forking proxy to identify the UAS.

8.8 Registration

SIP supports the concept of user mobility and discovery. A user can make herself available for communication by explicitly binding her AOR with a certain host address. This allows user mobility since the user can register from any device that supports SIP, including personal computers, wireless devices and cellular phones.

The discovery of the intended recipient of a SIP request is typically the function of SIP intermediary servers: for example, the user creates a binding with the registrar, which acts as a front end to a location server where all the bindings are stored, and then a proxy server, receiving a request that is destined to a domain it is responsible for, contacts that location server to retrieve the exact location of that intended recipient.

A user creates a binding by placing her AOR in the To header and the host address in the Contact header.

A user can be registered from many devices simultaneously by sending a REGISTER request from each device. Similarly, a user can create multiple bindings from the same device; this can be achieved by sending one REGISTER request with multiple bindings to the AOR. To do this, a user adds multiple contact headers in the REGISTER request.

A user can discover all the current bindings to her AOR using a process called registration fetching. Registration fetching is accomplished by sending a REGISTER request without a contact header. The registrar returns all the current bindings in the register response. Each binding has its own contact header in the response.

SIP registrations, by nature, are soft-state: this means that registration bindings must be periodically refreshed (updated). The expiration time of a binding is indicated by the registering entity using the expires parameter in a Contact header. If this parameter is not present, the registrar assumes an expiration time of 1 hour. If the UA does not refresh or otherwise explicitly remove the binding, the registrar silently removes it when the expiration time lapses. A UA can explicitly remove a binding by sending a REGISTER request and adding a Contact header for the binding to be removed. This Contact header contains the expires parameter with a value of 0.

A registrar can be discussed using the procedures in Chapter 20.

8.9 Dialogs

A dialog is a SIP relationship between two collaborators. The dialog provides the necessary states required for the routing and sequencing of messages between those collaborators.

Dialogs are identified using Dialog-IDs and UAs use them to track messages sent within a dialog. A Dialog-ID consists of a Call-ID, a local tag and a remote tag. For a UAC the local tag is the tag that appears in the From header of the initial dialog-creating request and the remote tag is the tag that appears in the To header of the dialog-creating response. For a UAS the local tag is the tag that appears in the To header of the dialog-creating response and the remote tag is the tag that appears in the From header of the initial dialog-creating request. For subsequent requests using dialogs sent from either end, the local tag is placed in the From header and the remote tag is placed in the To header.

Note that a UAS needs to be prepared to receive a request without a tag in the From header, in which case the tag is considered to have a null value.

A dialog state is needed for creating, sending, receiving and processing of messages within a dialog. This state consists of the dialog-ID, a local sequence number, a remote sequence number, a local URI, a remote URI, a remote target, a Boolean flag called a "secure" flag and a route set.

When a dialog is in an "early" state it is referred to as an "early dialog". This occurs when a provisional response arrives at the UAC to an initial request, thus creating a dialog. A dialog moves to a "confirmed" state when a "2xx" success response arrives. If a final response other than a "2xx"-class response arrives or if no response arrives at all, the early dialog terminates.

A UAS responding to a request with a final response indicating success must copy all Record-Route headers that appear in the request into the response, maintaining the order they appear in. The UAS then stores those URIs in the Record-Route headers as the route set, maintaining the order. If no Record-Route headers are present, the route set is left empty. This route set, even if empty, is retained for the remainder of the dialog. This means that other Record-Route headers appearing in requests within dialogs do not override the already-existing route set.

Record-Route headers are added to a request by intermediaries that wish to remain on the signalling path of any subsequent requests sent from the UAC to the UAS, or vice versa, within a dialog.

A UAS must also add a Contact header in the response that indicates the address where subsequent requests within the dialog should be targeted.

The dialog state at the UAS is constructed as follows:

- If the request arrived over the TLS and the request-URI contained a SIPS URI, the "secure" flag is set to true; otherwise, it is set to false.

- The remote target is set to the URI from the Contact header of the request.

- The remote sequence number is set to the value of the sequence number in the CSeq header of the request.

- The local sequence number remains empty at this stage. It is populated when the remote end sends a request within a dialog.

- The remote URI is set to the URI in the From header.

- The local URI is set to the URI in the To header.

- The Dialog-ID is created as indicated above.

- The route set is set as indicated above.

The UAC must provide a Contact header in an initial request that creates a dialog. When the UAC receives a response that creates a dialog, it creates the dialog state at its end as follows:

- If the request was sent over TLS and the request-URI contained a SIPS URI, the "secure" flag is set to true; otherwise, it is set to false.

- The remote target is set to the URI from the Contact header of the response.

- The remote sequence number is set to empty at this stage. It is populated when the remote end sends a request within a dialog.

- The local sequence number is set to the value of the sequence number in the CSeq header of the request.

- The local URI is set to the URI in the From header.

- The remote URI is set to the URI in the To header.

- The Dialog-ID is created as indicated above.

- The route set is set using URIs in the Record-Route headers, but the order is reversed. If no Record-Route headers are present, the route set is left empty. This route set, even if empty, is retained for the remainder of the dialog. This means that other Record-Route headers appearing in requests within dialogs do not override the already-existing route set. If the route set was created using Record-Route headers in a provisional response, then the 2xx final response that confirms the dialog re-sets the route set using URIs in the Record-Route headers, but once again the order is reversed.

Requests within dialogs are populated using dialog states. The local CSeq header value is incremented by 1 for every new request created within a dialog. Requests within dialogs may update the remote target if they are target refresh requests: examples of target refresh requests are INVITE requests and UPDATE requests.

An early dialog is terminated when a non-2xx final response is returned to the request. Confirmed dialogs are terminated uniquely, depending on the method used.

8.10 Sessions

A multimedia session consists of a set of multimedia senders and receivers and the data streams that flow between them. Sessions use SIP dialogs and follow SIP rules for sending requests within dialogs.

The role SIP plays in establishing a multimedia session revolves around its ability to carry SDP media descriptions in its message bodies. SDP is used to describe the session and the offer/answer model is employed [RFC3264]. SDP and the offer/answer model are described in Chapter 10. Section 8.10.1 describes the restrictions SIP has on such a model.

The session is initiated using the INVITE method, the request line and headers of which are populated by the UAC (see Section 8.7.4.1). The body is populated with an SDP offer. The answer may arrive in a provisional response or in the 2xx response.

INVITE requests follow a three-way handshake model: this means that the UAC, after receiving a final response to an INVITE request, must send an ACK request. The ACK request does not require a response; in fact, a response must never be sent to an ACK request.

If the UAC wants to cancel an invitation to a session after it sent the INVITE request, it sends a CANCEL request. The CANCEL request is constructed in a similar way to how the request-URI, the To header, the From header, the Call-ID header and the numeric part of the CSeq header are copied from the INVITE request. The method part of the CSeq header holds the CANCEL method. A UAS receiving a CANCEL responds to it with a 200 response and then follows it up with a "487 Request Terminated" response to the INVITE request. It is important to remember that all transactions must complete independently of each other: the reason the UAS must respond to the INVITE request.

If the UAC is not satisfied with the SDP answer that arrives in the 2xx response, it sends an ACK request followed by a BYE request to terminate the session. If the UAS is not satisfied with the SDP offer, it rejects the request with a 488 response.

INVITE requests can also be sent within dialogs to renegotiate the session description.

A session is terminated with a BYE request. The BYE request is sent like any other request within a dialog.

8.10.1 The SDP offer/answer model with SIP

Using the basic SIP, offers and answers can only appear in INVITE requests, reliable responses to INVITE requests and ACK requests. However, Sections 8.13.4 and 8.13.5 describe further opportunities for offer/answer exchanges using SIP extensions.

If an INVITE request results in multiple dialogs, each dialog has its own offer/answer exchange.

The general rule for the offer/answer exchange is that an offer must not be sent unless a previously sent (received) offer, if any, has received (been sent) an answer. This rule restricts the basic SIP to the following scenarios when an offer/answer exchange is possible:

- If the offer is in the INVITE request, then the answer must appear in the 2xx response of the INVITE.

- If the INVITE request did not contain an offer, then the 2xx response contains the offer and the ACK contains the answer.

8.11 Security

8.11.1 Threat models

SIP is susceptible to the following threats and attacks:

- *Denial of service*—the consequence of a DOS attack is that the entity attacked becomes unavailable. This includes scenarios like targeting a certain UA or proxy and flooding them with requests. Multicast requests are further examples.

- *Eavesdropping*—if messages are sent in clear text, any malicious user can eavesdrop and get session information, making it easy for them to launch a variety of hijacking-style attacks.

- *Tearing down sessions*—an attacker can insert messages like a CANCEL request to stop a caller from communicating with someone else. It can also send a BYE request to terminate the session.

- *Registration hijacking*—an attacker can register on a user's behalf and direct all traffic destined to that user toward the attacker's machine.

- *Session hijacking*—an attacker can send an INVITE request within dialog requests to modify requests en route to change session descriptions and direct media elsewhere. A session hijacker can also reply to a caller with a 3xx-class response, thereby redirecting a session establishment request to the hijacker's machine.

- *Impersonating a server*—someone else pretends to be the server and forges a response. The original message could be misrouted.

- *Man in the middle*—this attack is where attackers tamper with a message on its way to a recipient.

8.11.2 Security framework

There are six aspects to the SIP security framework:

- *Authentication*—this is a means of identifying another entity or user and making sure that the user is really who she claims to be. Typical methods involve user IDs passwords or digitally signing a set of bytes using a keyed hash.

- *Authorization*—once the user is authenticated, she must be authorized. Authorization involves deciding whether the user with the authenticated identity should be granted access to the requested services. This is often achieved using access control lists (ACLs).

- *Confidentiality*—this is where messages must remain confidential and only the intended recipient is allowed to see the contents of a message. This is usually achieved by means of encryption.

- *Integrity*—a user needs to be assured that a message was not tampered with en route. A message integrity check is a means of ensuring that.

- *Privacy*—anonymity of users is the key. Users do not want others to know who they are, what they are communicating or whom they are communicating with.

- *Non-repudiation*—reverse protection.

8.11.3 Mechanisms and protocols

8.11.3.1 Hop-by-hop mechanism

Hop-by-hop authentication provides the user with total confidentiality. It involves a complex security infrastructure that requires each proxy to decrypt the message and, therefore, relies on trust relationships between hops. Two protocols are in use for SIP: IP Security (IPsec: Chapter 18) and the Transport Layer Security protocol, or TLS (Chapter 14).

SIP, TLS and SIPS URI

TLS provides authentication, integrity and confidentiality. As mentioned earlier, the use of TLS in SIP messages requires all SIP entities to use the SIPS URI. A UAC wishing to communicate securely places a SIPS URI in the To header. If the next hop URI or the request-URI of a SIP request contains a SIPS URI, the UAC must place a SIPS URI in the Contact header. If the request-URI contains a SIPS URI, any alternative destinations to the request must be contacted using TLS.

TLS-secured requests must be sent using a reliable transport protocol like TCP or the Stream Control Transmission Protocol (SCTP). The default port for sending TLS-secured requests and for sending TLS-secured responses is 5061.

When registering a binding, a UAC must create a SIPS URI in the Contact header unless it can guarantee that the host represented in the Contact header has other means of security.

A UAS responding with a request that creates a dialog with a dialog-creating response (non-failure response) places a SIPS URI in the Contact header of the response if the request-URI, top Record-Route header (if there is one) or the

Contact header (if there are no Record-Route headers) has a SIPS URI. UASs must send responses using TLS if the request arrived on TLS (the Via header shows TLS as the transport).

UACs and UASs examine the "secure" flag in the dialog state when sending requests within dialogs. A "secure" flag value of true requires those entities to place a SIPS URI in the Contact headers.

For proxies inserting a Record-Route header, they must place a SIPS URI in the header if the request-URI or the topmost Route header (after post-processing the request) has a SIPS URI.

All SIP entities must use TLS if the next hop URI is a SIPS URI. Entities sending new requests using Contact headers in 3xx responses should not send a new request to a non-SIPS URI if the request-URI in the initial request contained a SIPS URI. Independently of which URI is being used as input to the procedures of discovering the next hop (Section 8.12), if the request-URI specifies a SIPS resource, the SIP entity making the discovery must follow the same procedures just as if the input URI was a SIPS URI.

A proxy that is processing responses changes the URI it placed in a Record-Route header from a SIPS URI to a SIP URI if it received the request from a non-TLS connection and forwarded it to a TLS connection. Similarly, a proxy that is processing responses changes the URI it placed in a Record-Route header from a SIP URI to a SIPS URI if it received the request from a TLS connection and forwarded it to a non-TLS connection.

The format of the SIPS URI is identical to the SIP URI except for the scheme: for SIPS URIs it is "sips", while for SIP URIs it is "sip". A SIP URI and a SIPS URI are not equivalent.

IPsec

IPsec provides authentication, integrity and confidentiality by securing SIP messages at the IP layer. It supports both TCP and UDP (see Chapter 18 for more details).

8.11.3.2 User-to-user and proxy-to-user mechanisms

User-to-user (or end-to-end) and proxy-to-user security can be regarded as a more secure mechanism since only two entities are available for attack. Two protocols are used in SIP for this mechanism: SIP digest and Secure Multipurpose Internet Mail Extension, or S/MIME [RFC2633]. In addition, there is an extension to the digest framework (namely, digest AKA) which is used in the Third Generation Partnership Project (3GPP) IMS.

Digest authentication

SIP digest authentication mostly makes use of the HTTP digest [RFC2617] authentication mechanism with a few minor modifications. Although digest only provides limited integrity protection, it does provide client authentication and replay protection. It also provides a form of mutual authentication that enables clients to authenticate servers.

Digest authentication requires a shared secret: this means that there is a need for a pre-existing relationship between all users and between all users and proxies. This is very problematic for public services.

Digest AKA Authentication

As described in Section 3.6.2, IMS authentication utilizes the UMTS Authentication and Key Agreement (AKA) protocol. This protocol needs to be transported within SIP signalling, which is the idea behind digest AKA: it integrates the AKA protocol and the digest authentication framework.

In practice, this means that the AKA authentication request is encapsulated in the WWW-Authenticate header field or the Proxy-Authenticate header field carried in the 401 "Unauthorized" and 407 "Proxy Authentication Required" responses, respectively. Similarly, the client's authentication response is encapsulated in the Authorization header field or the Proxy-Authorization header field of the request.

The actual AKA parameters (namely, the random challenge, or RAND, and the network authentication token, or AUTN) are concatenated and appended to the server nonce parameter. The response (RES) is calculated in the request digest by simply treating the RES parameter as the digest password. The normal authentication flow using digest AKA is illustrated in Figure 8.5.

The only exception to fully following the digest framework occurs when there is an AKA synchronization failure. Then, the synchronization failure parameter AUTS is included in an extension digest parameter, encoded in Base64: the reason for this is simply that there exists no other proper protocol element to carry the parameter. (Figure 8.6).

S/MIME

S/MIME provides message integrity, confidentiality and authentication, which is achieved by protecting SIP headers in an encrypted and/or signed S/MIME SIP message body. It does not require a shared secret.

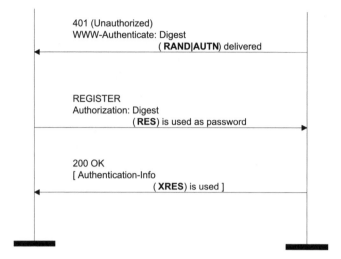

Figure 8.5 Normal digest AKA message flow. Note that [ ] indicates the element is optional and the message syntax is figurative.

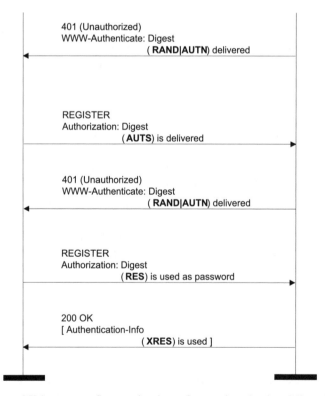

Figure 8.6 Digest AKA message flow at the time of a synchronization failure. Note that [ ] indicates the element is optional and the message syntax is figurative.

8.12 Routing requests and responses

8.12.1 Server discovery

The procedures indicated in this section are minimal and only show steps needed in the most basic scenarios (please refer to [RFC3263] for full procedures).

8.12.1.1 Sending requests

The TU layer performs the next hop discovery by using the URI it chose to be the next hop URI (see Section 8.12.5). The basic steps are as follows:

- If the URI contains a transport parameter, then that transport is used.

- If the URI contains a numeric IP address, but no transport parameter, then the response is sent using that IP address and the port indicated (or to the default port if no port is specified). The transport used is the UDP for SIP URIs and the TCP for SIPS URIs.

- If the sent-by field contains a domain name and a port, then an A (IPv4) or AAAA (IPv6) query is made. The resulting IP address and the available port are used to send the response. The transport used is UDP for SIP URIs and TCP for SIPS URIs.

- For URIs without numeric IP addresses or ports, a domain name system (DNS) server is queried for naming authority pointer (NAPTR) records for the domain in the URI, which are then examined.

- The TU chooses a transport protocol, if more than one record different transport protocol is available, depending on external factors (e.g., configuration).

- Using the chosen transport protocol and the NAPTR records, the TU then queries the DNS server for service (SRV) records. The server may return one or more SRV records.

- Those records are then tested in sequence, one by one, by performing an A or AAAA DNS query on each, so that an IP address and port can be found to send on the request. Note that an A or AAAA query on one SRV record might result in more than one IP address being returned. Each one of these IP addresses should be tried first and it is only after all have failed that the next A or AAAA query should be performed on the next SRV record.

8.12.1.2 Sending responses

This procedure is followed when a UAS fails to send a request using the source IP address and the port in the "sent-by" field. The procedure is as follows:

- If the sent-by field contains a numeric IP address, then the response is sent there using the port indicated (or to the default port if no port is specified).

- If the sent-by field contains a domain name and a port, then an A or AAAA query is made. The resulting IP address and the available port are used to send the response.

- If the sent-by field contains a domain name, but no port, then an SRV record query is performed on the domain name. These records are then tested in sequence, one by one, by performing an A or AAAA DNS query on each, so that an IP address and port can be found to send on the request. Note that an A or AAAA query on one SRV record might result in more than one IP address being returned. Each one of these IP addresses should be tried first and it is only after all have failed that the next A or AAAA query should be performed on the next SRV record.

8.12.2 The loose routing concept

Loose routing was introduced in [RFC3261]. It offers a more robust way of forwarding messages to hops and provides the means for the request-URI to remain unchanged throughout the request's journey to the proxy that is responsible for servicing it (i.e., the proxy responsible for the domain present in the request-URI).

A SIP URI carrying the loose router parameter indicates that the owner (typically an intermediary) of this SIP URI is [RFC3261]-compliant and supports loose routing. An example of such a URI is: sip:proxy.example.com;lr. The lr parameter identifies the entity as a loose router. The absence of such a parameter indicates that the next hop is a strict router.

8.12.3 Proxy behaviour

For all new requests, including those with unknown methods, a statefull proxy performs the following:

- validate the request;
- pre-process routing information;

- determine target(s) for the request;

- forward the request to each target;

- process all responses.

[RFC3261] describes each step in detail. For the purpose of completing the routing analysis, the steps taken by a proxy to route a request are listed below:

- If the request-URI contains a URI that the proxy has previously placed in the Record-Route header of that request, then the proxy replaces it with the URI in the last Route header. The proxy then removes that Route header.

- If the first Route header contains a URI representing this proxy, then the proxy removes that Route header.

- If the domain name of the request-URI indicates a domain that this proxy is not responsible for, then the proxy proceeds with the task of forwarding the request and only places the request-URI in the target set of addresses where the request will be proxied.

- If the domain name of the request-URI indicates a domain that this proxy is responsible for, then the proxy uses any mechanism it has been configured with in order to determine the target set. Taking the first target in the target set, the proxy places it in the request-URI. The proxy places a Record-Route header as the topmost Record-Route header, if it wishes to remain on the path of any subsequent requests within a dialog. The URI in the Record-Route header must contain an lr parameter and must be a SIP or a SIPS URI.

- If a proxy has a set of entities that it wants a request to pass before it arrives at its final destination, it needs to insert Route headers with the addresses of those entities on top of any other Route headers, if present. The proxy needs to ensure that those entities are loose routers.

- If the request contains Route headers, then the proxy examines the topmost Route header and if the topmost Route header does not contain the lr parameter, then the proxy places the request-URI in a Route header as the last Route header. It then places the URI in the topmost Route header into the request-URI and removes that Route header. The proxy then sends the request using the steps in Section 8.12.5 and using the calculated first target in the target set as the entry in the route set.

The proxy server may try each address in the target set serially or in parallel: a concept referred to as forking.

8.12.4 Populating the request-URI

The UAC uses the remote target and the route set to populate the request-URI as follows:

- If the route set is empty, then the remote target is placed in the request-URI.

- If the route set is not empty and the topmost URI is a loose router, then the remote target is placed in the request-URI. Route headers are then built using the route set.

- If the route set is not empty and the topmost URI is a strict router, then the topmost URI is placed in the request-URI. Route headers are then built using the route set, but excluding the topmost URI in that route set. The remote target is then placed as the last Route header.

8.12.5 Sending requests and receiving responses

The procedures outlined in Section 8.12.1.1 are used by the TU layer to send the request as follows:

- If the topmost URI in the route set indicated that the next hop is a strict router and resulted in forming the request as described in Section 8.12.4, then the procedures are applied to the request-URI.

- If the topmost URI in the route set indicates that the next hop is a loose router, then the procedures are applied to the topmost URI in the route set.

- If there is no route set, then the procedures are applied to the request-URI.

The TU then creates a transaction instance and passes the request, the IP address and the port to it and indicates the transport protocol to use. The transaction layer passes this information to the transport layer, which sends the request as follows:

- If the request is within 200 bytes of the maximum transfer unit (MTU) en route to the destination, then the request must be sent using a congestion-safe protocol (TCP or SCTP).

- If the MTU is unknown and the request is greater than 1,300 bytes, then the request must be sent using a congestion-safe protocol (TCP or SCTP).

- If the transport protocol indicated in the Via header needs changing after the above steps, then it is changed. The Via header's sent-by field is populated by an IP address and a port (or preferably an FQDN).

Received responses are matched to requests using the branch parameter in the Via header.

8.12.6 Receiving requests and sending responses

If the request was received from a different IP address than the one indicated in the sent-by field of the Via header in the request, then the transaction layer adds a "received" parameter in the Via header and populates it with the IP address from which it received the request. The request is then matched to a server transaction or (if no match is found) is passed to the TU, which may choose to create a new server transaction instance.

Once the TU has completed processing the request and has generated a response, it passes the response to the transaction instance from which it received the request. The transaction layer forwards the response to the transport layer, which performs the following:

- If the request was received on a connection-oriented protocol, then the response is sent on the same connection.

- If the connection is no longer open, then the received parameter and port in the sent-by field (or the default port, if no port is specified) are used to open a new connection and send the request.

- If no received parameter is present, then the procedures in Section 8.12.1.2 are followed.

8.13 SIP extensions

8.13.1 Event notification framework

SIP has been the extension used for the purpose of event notification. A user or resource subscribes to another resource that has an event of interest and receives notifications of the state and any changes in such an event.

The SIP SUBSCRIBE method is used for subscription while the NOTIFY method is used to deliver notifications of any changes to an event.

[RFC3265] is the IETF paper that documents this extension. It is a framework that describes subscriptions and notifications in a generic manner and provides rules for creating SUBSCRIBE requests and NOTIFY requests. It also describes the behaviour of subscribers when sending and receiving subscription requests, as well as the notifiers' behaviour when receiving subscription requests and sending notifications.

The event notification framework also introduces new SIP headers and response codes, along with the SUBSCRIBE and NOTIFY methods:

- *Event header*—this identifies the event to which a subscriber is subscribing for notifications.

- *Allow-Events header*—this indicates to the receiver that the sender of the header understands the event notification framework. The tokens present in the header indicate the event packages that it supports.

- *Subscription-State header*—this indicates the status of a subscription. "Active", "pending" and "terminated" are the three defined subscription states. This header also carries the reason for a subscription state: "deactivated", "probation", "rejected", "timeout", "giveup" and "noresource". Extensions are possible for subscription-state and reason values.

- *"202 Accepted" response*—this indicates that the subscription request has been preliminarily accepted, but is still pending a final decision, which will be indicated in the NOTIFY request.

- *"489 Bad Event" response*—this response is returned when the notifier does not understand an event as described in the Event header.

SUBSCRIBE requests are dialog-establishing requests. A dialog is created when a 2xx response or a NOTIFY request arrives for the SUBSCRIBE request. Subsequent SUBSCRIBE and NOTIFY requests are sent within the created dialog.

The request-URI in an initial SUBSCRIBE request addresses the resource that the subscriber wishes to receive state information about. The Event header identifies the event related to the subscription.

Much like registration, subscriptions are in soft state and need refreshing. The duration of a subscription is indicated in an Expires header. The default value is 1 hour if the header is not present in the SUBSCRIBE request. A subscription is terminated when not refreshed and can be explicitly terminated by sending a SUBSCRIBE request within the dialog and setting the Expires header value to 0.

The event notification framework also introduces the concept of an event package: an extension to the framework. Each event package created introduces a new use case for the event notification framework.

The NOTIFY request payload (body) is used to carry the state information. Each event package defines its own MIME type for carrying such information.

The event template package is a special event package and is associated with other event packages, including itself. Template packages define states that can be applied to other event packages. A subscription to a template package is indicated in the Event header by appending a period (full stop) to an event package, followed by the template package name: for example, Event: presence.winfo.

8.13.2 State publication (the PUBLISH method)

The event notification framework specifies how to subscribe to the state of an event and how to get notification of changes to the state of an event. It does not specify how the state can be published. However, the SIP extension for state publication specification [Draft-ietf-sip-publish] allows a client to publish its event state to the state agent, which acts as the compiler of such state and generating notifications. This is achieved using the PUBLISH method.

PUBLISH requests are in soft state and need to be refreshed. The Event header defined in [RFC3265] (and in Section 8.13.1) is used by the publisher to identify the event whose state it is publishing. The request-URI is used to identify the resource whose state is being published. An entity tag is used by the client and is supplied by the server to enable the client to update a state using the PUBLISH method. The state of a resource is carried in the body of the PUBLISH request.

8.13.3 SIP for instant messaging

For a more detailed description of instant messaging, please refer to Chapter 24.

SIP is extended for instant messaging by the introduction of the MESSAGE method in [RFC3428].

There are two modes of instant message exchange: page mode and session mode. The MESSAGE method is used in page mode. Page mode is a one-shot instant message where a subsequent instant message is not related, at the protocol level, to the preceding one. It is used when a conversation or interaction is not unexpected.

The request-URI in a MESSAGE request carries the resource where the request will be sent. The MESSAGE request body carries the actual contents of an instant message, which, again, is a MIME type. The most common MIME body uses the "text/plain" MIME type. For interoperability with non-SIP instant-messaging clients using an IETF standard, the MIME type "message/cpim" as defined in [Draft-ietf-impp-cpim-msgfmt] "Common Presence and Instant Messaging Message Format" is used.

Session-based instant messaging uses SIP for signalling and the Message Session Relay Protocol (MSRP) for carrying the data (instant messages) after the session has been established (see Section 24.4 for details about how an instant-messaging session can be established using SIP and SDP, as well as how the MSRP operates).

8.13.4 Reliability of provisional responses

In the basic SIP, provisional responses are transmitted unreliably, unlike the 2xx responses for INVITE requests. It was later discovered that reliability in the transmission of provisional responses in some cases was both important and useful: therefore, [RFC3262] was created. In the 3GPP, reliable provisional responses and their acknowledgments are used to exchange additional SDP offer/answer messages.

The reliability of provisional responses extension only applies to INVITE requests.

A UAC generating an INVITE request and wishing to indicate its support for the reliable provisional responses extension includes the Supported header in the INVITE request with the option tag "100rel". A UAC requiring the UAS to support the reliable provisional responses extension indicates this by placing the option tag "100rel" in a Require header of the INVITE request.

A UAS receiving an INVITE request with a Require header carrying the option tag "100rel" responds with an error response "420 Bad Extension" if it does not support such extension. If the option tag appears in a Supported header, the UAS may ignore it.

A UAS supporting this extension must transmit the provisional response reliably if the "100rel" option tag was present in a Require header and may transmit it reliably if the option tag was present in a Supported header.

A UAS sending a reliable provisional response indicates this by placing the extension header RSeq in the provisional response and by placing the option tag "100rel" in a Require header. The RSeq carries an integer value between 1 and $2^{31} - 1$.

Multiple reliable provisional responses can be sent for the same INVITE request. Any subsequent reliable provisional response must carry an RSeq value that is 1 higher than the previous value in the same transaction space (within the same INVITE transaction).

A UAC receiving a reliable provisional response responds to it with a provisional response acknowledgment (PRACK) request. The PRACK request carries a RACK header. The RACK header reflects the value in the RSeq header and that in the CSeq header, in that order.

Just like any other request, the PRACK must be responded to. The UAS holds off sending a 2xx to an INVITE request until it has received PRACK requests for all reliable provisional responses it has sent. Examples:

* Require: 100rel.

* RSeq: 12345.

* RACK: 12345 1 INVITE.

"100 Trying" responses are not sent reliably.

8.13.4.1 The SDP offer/answer model with PRACK request

Section 8.10.1 described the offer/answer restrictions that occur when using the basic SIP. This section offers additional opportunities for the offer/answer exchange:

- If the INVITE contained an offer, then an answer may be present in a reliable provisional response.

- If the INVITE did not carry an offer, then the offer must be present in the first reliable provisional response.

- If the reliable provisional response carries the first offer, then the answer must be in the PRACK.

- If a reliable provisional response carries an answer, then the PRACK may carry an additional offer.

- If the PRACK carried an offer, then the 2xx of the PRACK must carry an answer.

8.13.5 The UPDATE method

The UPDATE method is an extension to SIP that enables UAs to update a session description without having any impact on a dialog. The UPDATE method can be sent within early and confirmed dialogs, but must not be sent if a dialog is not created (i.e., before a dialog-creating provisional response is sent/received). It is constructed like any other request within a dialog.

An UPDATE request can be sent by the caller (the INVITE UAC) as well as the callee (the INVITE UAS) and is only used for sessions: this means UPDATE requests are only used for dialogs created using INVITE requests.

A UAC wishing to indicate its support for the UPDATE method extension does so by including an Allow header in the INVITE request, listing UPDATE as a method. A UAS wishing to indicate its support for the UPDATE method does so by including an Allow header, using the UPDATE method to deliver a provisional response.

The UPDATE request is a target refresh request: that is, it can update the remote target of a dialog.

8.13.5.1 The SDP offer/answer model using an UPDATE request

The UPDATE request always carries the offer, while the 2xx response carries the answer.

UPDATE requests sent within early or confirmed dialogs can only carry an offer if answers have been returned for all the offers communicated by either side. Of course, an UPDATE with an offer cannot be sent if no offer/answer exchanges have yet taken place.

8.13.6 Integration of resource management and SIP (preconditions)

The inability to reserve network resources for a session is a serious drawback to session establishment. To minimize session failure once a session has started, it is necessary to reserve resources before the callee is alerted (i.e., before the phone rings).

To reserve resources the network needs to know the callee's IP address, port and session parameters. This is not possible without offer/answer exchanges. The problem is that a session is established after the offer/answer exchange and, typically, a user is only alerted after session establishment has occured. To solve this problem the concept of preconditions was introduced. In this concept the caller indicates by means of an SDP offer a set of constraints about the session. The answerer responds to the offer with an SDP answer; however, it neither establishes a session nor does it alert the user. Session establishment occurs when the caller and callee learn that the preconditions have been met by local events and by the caller sending a new offer confirming that the preconditions have been met and the answerer sending an answer confirming the same. In the IMS this new offer/answer exchange is carried in an UPDATE request and response.

This SDP extension is defined in [RFC3312], where preconditions for quality of service (QoS) are also defined. Three new SDP attributes are introduced:

- *Current status*—carries the current resource reservation status in the network for a particular media stream.

- *Desired status*—carries the preconditions for resource reservation in the network in order to proceed with session establishment.

- *Confirmation status*—the confirmation status attribute carries threshold conditions for a media stream. It is used so that one end point can indicate to the other end point that it needs a confirmation that the preconditions at the remote end have been met.

Each of these three attributes have the following four fields, which appear in the same order (the current status attribute is an exception as it does not have a strength tag):

- *Precondition type*—at the moment only the QoS precondition type has been defined (extensions are possible).

- *Strength tag*—this indicates the strength of the preconditions. Defined values are "mandatory", "optional", "none", "failure" and "unknown".

- *Status type*—the end-to-end (or "e2e") status type indicates the status of end-to-end resource reservation. The "local" and "remove" status types indicate that a preconditions status is needed for local and remove networks, respectively. They are used when each side performs its own resource reservation locally.

- *Direction tag*—this indicates the current direction of resource reservation: desired or confirmed.

Here are some a-line attribute examples:

```
a=curr:qos e2e send
a=des:qos optional e2e send
a=conf:qos e2e recv
```

A new SIP extension option tag "precondition" was introduced to deal with preconditions. It appears in a Require header if one of the strength tags that appear in an offer is "mandatory". If all the strength tags are either "optional" or "none", then the option tag can appear in a Supported header. Here is an example of how the option tag is used:

```
Require: 100rel, precondition
```

8.13.7 The SIP REFER method

The REFER method is standardized in [RFC3515].

The sender of a REFER request refers the recipient to a resource that is identified in the REFER request itself. A REFER request also implicitly creates a subscription (implicit subscription means creating a subscription state as described in Section 8.13.1 without explicitly sending a SUBSCRIBE request) where senders of the REFER request can receive notifications about the outcome of such a referral (see Section 8.13.1 for details about event notification). A REFER request, therefore, is a dialog-creating request.

One application of the REFER method is call transfer. For instance, a secretary receives a call from an associate asking for the manager. The secretary then transfers the call to the manager by sending the associate a REFER request with the manager's contact information. The associate's UA then calls the manager by sending an INVITE.

The REFER standard also introduces a SIP extension header, the Refer-To header, which provides a URI. Its syntax follows that of the Contact header (with the exception of STAR "*", which cannot appear in a Refer-to header). The Refer-to header can only appear in a REFER request, but must not appear more than once.

The REFER request is constructed in the same way as any other request outside a dialog (defined in Section 8.7.4.1).

The receiver of a REFER request typically responds to it with a "202 Accepted" response. Its recipient also creates a subscription and sends notifications of the outcome. According to the event notification framework, a NOTIFY request is generated and sent immediately after accepting the REFER request. NOTIFY requests are constructed in the same way as any other request within a dialog. Every NOTIFY request contains the Event header with the tag "refer".

Unlike a SUBSCRIBE request, the REFER request does not carry an expiration time. The implicit subscription duration is determined by the referred party and is indicated to the referrer in the first NOTIFY request, using the Subscription-State header. A referrer wanting to extend the subscription duration can do so by sending a SUBSCRIBE request within a dialog for that event.

The NOTIFY request also contains a body of MIME type "message/sipfrag" (described in Section 8.13.8). The body, at minimum, contains a SIP response status line.

8.13.8 The "message/sipfrag" MIME type

The "message/sipfrag" MIME type allows a fragment of a SIP message to be represented. If a start line, header or body of a SIP message is represented, then it must be complete according to the BNF of such a representation. If a SIP message body is represented, then the fragment must also carry appropriate SIP message headers that describe the body.

8.13.9 SIP extension header for registering non-adjacent contacts (the Path header)

Intermediaries, such as SIP proxy servers, could exist in a network topology between a UA and a registrar. Because of this scenario a SIP extension introduced the Path header, which is used to record the path taken (proxies traversed) by a REGISTER request from the UA to the registrar. This Path header is similar to the Record-Route header used for dialog-creating requests, but is only used by the home network of a user to send requests destined to that user.

The Path header syntax is similar to that of the Record-Route header. A UA sending a REGISTER request includes a Supported header in the request with the "path" option tag. Path header values returned in the 2xx response of a REGISTER request can be ignored by the UA.

An intermediate proxy wanting to remain on the path of any requests sent to the UA inserts a Path header value with a URI pointing to itself as long as the REGISTER request contained a Supported header with an option tag "path". The Path header value or a new Path-header, must appear as the topmost header. If a proxy requires the registrar to support the path extension, then it inserts a Require header with an option tag "path".

Proxies may add a Path header value with a URI pointing to other proxies, if it is aware of the network topology.

If the UA does not indicate support for the path extension and a Path header appears in the REGISTER request, it is recommended the registrar reject the registration request with a "420 Bad extension" response.

If the registrar accepts the request, then it stores the Path header values preserving the order and associates them with the AOR and the Contact header. It then copies the Path header values into the 2xx response.

Typically, a home proxy is associated with the registrar of a user. If a home proxy receives a request that is destined to a user whose AOR is registered, then the home proxy determines whether there are Path headers stored with that registration. If there are, then the home proxy inserts those Path headers in the request as Route headers. The Path extension is defined in [RFC3327].

8.13.10 Private SIP extensions for asserted identity within trusted networks

[RFC3325] provides a private extension to SIP that enables a network of trusted SIP servers to assert the identity of end users or end systems. It also provides a mechanism whereby an end user can indicate requested privacy. Entities are responsible for withholding asserted identity information outside the trusted domain when privacy is requested.

If proxy server receiving a SIP request from an untrusted entity authenticates a user, it then inserts the P-Asserted-Identity extension header into a request before forwarding it to a trusted next hop. If any P-Asserted-Identity headers were already present in the request when it arrived from the untrusted entity, the proxy removes or replaces them. If a proxy receives a SIP request from a trusted entity, then it can safely rely on the identity information in already-existing P-Asserted-Identity headers.

In the case of multiple public user identities the user can give a proxy an idea of the preferred public Identity to be inserted into the request. This is done using the P-Preferred-Identity extension header. A proxy receiving the request from an untrusted entity uses that header as an idea of what to insert in the P-Asserted-Identity header after it has authenticated the user. If no such identity exists, then the proxy may either reject the request or insert a P-Asserted-Identity header and ignore the P-Preferred-Identity header.

A proxy forwarding a request to an untrusted next hop removes all P-Asserted-Identity headers if the end user has requested privacy. A user wishing to request privacy inserts the privacy token "id" into a Privacy header (defined in Section 3.6).

The syntax of the P-Asserted-Identity header and the P-Preferred-Identity header are similar to that of the From header.

8.13.11 Security mechanism agreement for SIP

8.13.11.1 Introduction

The SIP Security Agreement (SecAgree) is an extension to SIP that allows a UA and its first hop server to exchange and agree security mechanisms that they will use in future communications. The most important aspect of SecAgree is that this negotiation is secure itself. The protocol has mechanisms in place that make it impossible for an attacker to bid down on the exchanged security mechanism in an attempt to force an intentionally weak security mechanism on them. This makes systems future-proof: new stronger security mechanisms can be added and weaker ones deleted at a later date.

8.13.11.2 Operation

The SecAgree handshake contains the following stages (see Figure 8.7):

1. On receiving a request from the UA the first hop SIP server initiates the security agreement by challenging the UA. This challenge contains a list of security mechanisms supported by the server. Note that the client may pre-emptively include its supported security mechanisms in any request. However, the server always challenges with a full list of its supported security mechanisms, irrespective of the contents of the client list.[1]

2. The UA inspects the server's security mechanism list, which is an ordered list of mechanisms, ordered by preference, and chooses the security mechanism with the highest preference (i.e., most commonly supported). It then proceeds to switch on the chosen security mechanism (e.g., it establishes the manually

---

[1] The reason for this behaviour is simply that there is no security applied to the client list; it cannot be trusted. It is merely a tool to relay certain parameters or ideas to the server.

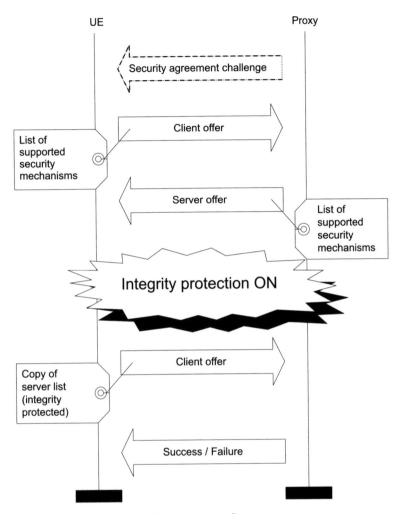

Figure 8.7 Security agreement handshake message flow.

keyed IPsec SAs or starts a TLS handshake with the server). The UA then resubmits its original request, which is now protected by the security mechanism used. It also returns the server's list of supported security mechanisms.

3. The server receives a security mechanism list that is supposed to be a mirror image of the list it originally sent with the SecAgree challenge. The server verifies that the mirrored list is in fact identical to the list it had previously sent out. If the lists were not identical, then it would mean that an attack was in progress,[2] in

---

[2] This aspect of SecAgree requires that any mechanism negotiated using SecAgree must at a minimum provide data integrity protection.

which case the server would immediately abort and reject the request. If everything checks out, then the request is forwarded to the next hop.

8.13.12 Private SIP extensions for media authorization

This SIP extension links the QoS applied to media in the bearer network with session signalling, which helps to guard against DOS attacks. The use of this extension is only feasible in administrative domains with trust relationships. The SIP intermediary (typically a proxy server) authorizes the QoS and the policy control function (or policy decision function, or PDF) provides the QoS. This mechanism prohibits any end-to-end encryption of message bodies that describe media sessions; it must only be used in specialized SIP networks like the 3GPP IMS.

This extension assumes that a UA wishing to obtain QoS for a session is connected to a QoS-enabled proxy and a policy enforcement point (PEP). A QoS-enabled proxy is a proxy that has a link to a PDF.

The solution utilizes something called a token (i.e., a string of hexadecimal digits). The UA acquires a media authorization token from the QoS-enabled proxy server, which itself acquires a media authorization token from the PDF and then passes it to the UA. The proxy server passes this token to the UA by means of a SIP extension header. The UA then presents this token to the PEP in order to get the bearer with the allocated QoS.

The extension header is named P-Media-Authorization; it can carry multiple media authorization tokens separated by commas. The P-Media-Authorization header itself can only be carried in requests and responses that carry an SDP offer or answer. As already mentioned, media authorization token is a string of hexadecimal digits.

When a SIP UA sends an INVITE request to a QoS-enabled proxy server (typically referred to as the originating proxy server), the proxy includes one or more media authorization tokens in all unreliable provisional responses (except 100), the first reliable 1xx or 2xx response and all re-transmissions of that reliable response. This is done for each dialog that is created as a result of the INVITE request. When the UA requests QoS from the bearer network it includes media authorization tokens with the resource reservation request.

When a SIP UA receives an INVITE request from the QoS-enabled proxy (typically referred to as the destination proxy), it examines the request to see what media authorization tokens the proxy may have included. When the UA requests QoS from the bearer network it includes media authorization tokens with the resource reservation request.

The media authorization extension is defined in [RFC3313].

8.13.13 SIP extension header for service route discovery during registration

The UA is free to include a set of Route headers—the preloaded route—in an initial request to force that request to visit and, potentially, be serviced by one or more proxies. This allows a UA to request services from a specific set of proxy servers, which are typically located in the user's home network.

[RFC3608] defines a header called "Service-Route", which is used by the UA to learn the service route (or preloaded route). A registrar wanting to inform a UA of a service route set uses this Service-Route header in a 2xx response to a REGISTER request. If used by the UA, the service route set will provide services from a set of one or more proxies associated with that registrar.

The UA choosing to use the service route set provided by the registrar places the contents that appear in the Service-Route headers into the Route headers, maintaining the order they appear in.

The mechanism that the registrar uses to construct the header value is not described in [RFC3608]. It is a registrar-local policy.

8.13.14 Private header extensions to SIP for 3GPP

[RFC3455] defines the following 3GPP IMS-specific headers for SIP:

- *P-Charging-Vector*—this transports the IMS charging ID (ICID) and correlated access network (e.g., GPRS)-related charging information between IMS network entities.

- *P-Charging-Function-Address*—this transports the addresses of the charging functions between the IMS network entities of the user's home network.

- *P-Visited-Network-ID*—this transports the identification string of the visited network to the home network of the user during registration, thus allowing the home network to discover details about the roaming agreements between the two networks.

- *P-Access-Network-Info*—this transports information about the access network technology and the user's location (GPRS Cell-ID) from the visited network to the home network.

- *P-Called-Party-ID*—this is included in an initial request by the registrar of the terminating user. The registrar rewrites the request URI with the registered contact address of the terminating user. Consequently, the URI originally indicated in the request URI would be lost; therefore, the URI originally indicated in the request URI is saved to the P-Called-Party-ID header and sent along with the request.

- *P-Associated-URI*—this is included in the 200 (OK) response of a REGISTER request. It includes additional URIs that are associated with the user. The registrar may implicitly register some of these additional URIs.

8.13.15 Compressing SIP

[RFC3486] provides a SIP mechanism to indicate whether signalling compression is supported. Two parameters are defined to enable a SIP entity to show that it supports signalling compression. The parameters, when present, also indicate the entity's willingness to receive compressed messages.

The first parameter is defined as a SIP URI parameter and is named "comp". Currently, only one value is defined for this parameter: 'SigComp". When sending requests the SIP entity (a UA or an intermediary) inspects the next hop URI, and if the URI carries the parameter "comp = SigComp", the UAC sends the request compressed according to the signalling compression specified in [RFC3320].

The second parameter is defined as a Via header parameter. It is identical to that of the URI, but is used by the SIP entity forwarding the response to learn the upstream entity's ability and willingness to receive compressed responses. Hence, if that parameter is present in the Via header of a request, then the response is sent compressed.

Typically, a UA sending a compressed request inserts the "comp = SigComp" parameter into the Via header. If a Contact header is present, the parameter is also added to the Contact header URI. Proxy servers forwarding the compressed request also insert the parameter into their Via header. If a proxy is record routing, then it adds the parameter into the URI of the inserted Record-Route header.

9

SDP

The Session Description Protocol (SDP) is an application-layer protocol intended to describe multimedia sessions. It is a text-based protocol. When describing a session the caller and callee indicate their respective "receive" capabilities, media formats and receive address/port.

Capability exchange can be performed during session set-up or during the session itself (while the session is in progress).

At the time of writing a new SDP specification is in the process of being finalized. The current version is [Draft-ietf-mmusic-sdp-new] and the earlier version was specified in [RFC2327].

9.1 SDP message contents

An SDP message contains three levels of information:

- Session-level description—this includes the session identifier and other session-level parameters, such as the IP address, subject, contact info about the session and/or creator.

- Timing description—start and stop times, repeat times, one or more media-level descriptions.

- Media type and format—transport protocol and port number, other media-level parameters. Note that the media address may be different from the signalling address.

The three levels of information must appear in the order described above. The SDP message is a collection of SDP lines.

The IMS. Miikka Poikselkä, Georg Mayer, Hisham Khartabil and Aki Niemi
Copyright 2004 by John Wiley & Sons, Ltd. ISBN 0-470-87113-X

Table 9.1 Session-level description SDP lines.

Field	Description	Mandate
v	Protocol version	m
o	Origin and session ID	m
s	Session name	m
i	Session information	o
u	URI for sesssion	o
e	Email address	o
p	Phone number	o
c	Connection information	m*
b	Bandwidth information	m
z	Time zone correction	o
k	Encryption key	o
a	Attribute lines	o

* Not required if present in every media line.

9.1.1 Session description

Table 9.1 lists all the session-level description lines and indicates their mandate and the letter used as the line name.

9.1.2 Time description

Table 9.2 lists all the time description lines and indicates their mandate and the letter used as the line name.

9.1.3 Media description

Table 9.3 lists all the media-level description lines and indicates their mandate and the letter used as the line name.

Table 9.2 Time-level description SDP lines.

Field	Description	Mandate
t	Time session is active	m
r	Repeat times	o

Table 9.3 Media-level description SDP lines.

Field	Description	Mandate
m	Media and transport	o
i	Media title	o
c	Connection information	o*
b	Bandwidth information	o
k	Encryption key	o
a	Media attributes	o

*Not required if present at the session level.

9.2 SDP message format

The SDP syntax is very strict and all lines follow the same format. Every SDP line has the format:

```
<character>=<value>
```

No spaces are allowed on either side of the "=" sign. The <value> part of the line contains one or more parameters and there is exactly one space between each parameter:

```
value=parameter1 parameter2 ... parameterN
```

Each SDP line ends with a carriage return line feed (CRLF) and each line has a defined number of parameters.

9.3 Selected SDP lines

9.3.1 Protocol version line

The SDP protocol version is 0 and, therefore, the v-line in an SDP message must always be set to 0:

```
v=0
```

9.3.2 Connection information line

The c-line must be either present at the session level or media level. It must be present at the media level if it is not present at the session level. If it is present at both levels, then media-level connection information overrides session-level information:

```
c=<network type> <address type> <network address>
```

The c-line has three parameters:

- Network type—the only currently defined network type is the Internet. The value appears as "IN".

- Address type—there are two address types, IP4 or IP6.

- Network address—this parameter identifies the Internet Protocol (IP) address or domain name where media are received.

9.3.3 Media line

The m-line carries information about the media, including transport information. The syntax is as follows:

```
m=<media> <port> <transport> <format-list>
```

The m-line has four parameters:

- *Media*—the type of media (e.g., audio, video, game).

- Port—contains the port number where these media can be received.

- Transport—the transport protocol to use, either the User Datagram Protocol (UDP) or Real-time Transport Protocol Audio and Video Profile (RTP/AVP) (RTP is explained in Chapter 11).

- Format-list—contains more information about the media, usually payload types defined in RTP/AVPs (see Section 11.3 for details).

If the transport is RTP/AVP, then the port number for the RTP Control Protocol (RTCP) = RTP port + 1. RTCP is assumed to be sent whenever RTP is carrying the media. The RTP port number must always be an even number and, therefore, the RTCP port is an odd number.

9.3.4 Attribute line

The a-line defines the attributes of the media. It is used to extend SDP: in fact it is the only way of doing so. Attributes can be session-level attributes, media-level attributes or both. Attribute interpretation depends on the media tool being invoked. Its syntax is as follows:

```
a=<attribute field> [":"<attribute value>]
```

The attribute field contains the name of the attribute. The attribute value is optional, but if present is separated from the attribute field by a colon.

Table 9.4 shows a list of the most commonly used attributes that are defined in [Draft-ietf-mmusic-sdp-new]. This list is not exhaustive since extensions have also been defined in separate Internet Engineering Task Force (IETF) documents (see Section 8.13.6 on preconditions as an example of how SDP can be extended).

9.3.5 The rtpmap attribute

For RTP-transported media, SDP can be used to bind a media-encoding codec to the media's RTP payload type. This is done using a payload-type number.

For static payload types (see Section 11.4.1 for the definitions of static and dynamic payload types) the payload-type number is sufficient for the binding, but for dynamic payload types the payload-type number is not sufficient and additional encoding information is needed. This is achieved using the rtpmap attribute. The payload-type number is carried in the format-list parameter of the media line. The syntax for the rtpmap attribute is:

```
a=rtpmap:<payload type> <encoding name>/<clock rate>[/<encoding parameters>]
```

The rtpmap attribute consists of four parameters:

- Payload type—carries the payload-type number as indicated in the m-line.

- Encoding name—the codec name.

- Clock rate—bits per second.

- Encoding parameters—media-specific parameters, including the number of channels, but not codec-specific parameters.

Table 9.4 Most common SDP attribute lines.

Attribute name	Syntax	Description	Session or media-level attribute
Packet time	`a=ptime:<packet time>`	It represents how long media are allowed to remain in one packet, in milliseconds. It is just a recommendation and is not necessary to be used when decoding RTP	Media
Maximum packet time	`A=maxptime:<max packet time>`	The maximum number of milliseconds media are allowed to be included in a packet	Media
RTP map	`a=rtpmap:<payload type> <encoding name>/<clock rate> [/<encoding parameters>]`	See Section 9.3.5 for details	Both
Receive only	`a=recvonly`	The sender of the media description only wants to receive the media stream (or streams, if in session level)	Both
Send only	`a=sendonly`	The sender of the media description only wants to send the media stream (or streams, if in session level)	Both
Send and receive	`a=sendrecv`	The sender of the media description wants to send and/or receive the media stream (or streams, if in session level). This is the default value if none of "sendonly", "recvonly", "sendrecv" or "inactive" are present	Both
Inactive	`a=inactive`	In this mode the media are neither sent nor received. It can be used to put media steams on hold. Note that in the case of RTP, RTCP continues to be sent even if the stream is set to inactive	Both
Frame rate	`a=framerate:<frame rate>`	It indicates the maximum video frame rate in frames per second	Media
Formats	`a=fmtp:<format-specific parameters>`	Allows parameters for specific formats to be conveyed. The formats are the encoding parameters in the rtpmap attribute	Media

10

The Offer/Answer Model with SDP

The Session Description Protocol's (SDP) Offer/Answer Model is used by two entities to reach agreement on session description, such as which media streams are in the session, the codecs, etc. The offerer indicates the desired session description in the offer. The answerer replies to the offer by indicating the desired session description from the answerer's viewpoint. The Offer/Answer Model can be used to create sessions or to modify an existing session. There are, however, restrictions on what can appear in an offer and answer. The following sections describe how offers and answers are created (the Offer/Answer Model is described in more detail in [RFC3264]).

10.1 The offer

An entity wanting to create a session or currently in a session can generate an SDP session description that constitutes the offer. This offer contains the set of media streams and the set of codecs the offerer wishes to use, as well as the Internet Protocol (IP) addresses and ports the offerer would like to use to receive the media.

The session line, or s-line, is always set to "-" or a single space in the offer/answer.

Since signalling protocols, such as SIP, are used to create unicast sessions, the start and stop times in an SDP offer/answer exchange is set to 0 and 0, respectively.

10.2 The answer

The answerer generates an answer, which is an SDP answer that responds to an offer provided to it. The answer indicates:

The IMS. Miikka Poikselkä, Georg Mayer, Hisham Khartabil and Aki Niemi
Copyright 2004 by John Wiley & Sons, Ltd. ISBN 0-470-87113-X

- Whether a media stream is accepted.

- The codecs that will be used.

- The IP addresses and ports that the answerer wants to use to receive media.

10.3 Offer/answer processing

There are certain restrictions to when an offer or answer can be generated:

- The client that received an offer cannot generate a new offer until it has responded with an answer to the offer or until it has rejected that offer.

- The client that generated an offer cannot generate a new offer until it has received an answer to the original offer or until that original offer is rejected.

There is also a constraint on the number of m-lines an answer can have. The number of m-lines in an answer must equal the number of m-lines in the offer.

If multiple media streams of different types are present, this means that the offerer wishes to use those streams at the same time. Rejecting media can be achieved by setting the port in a media line to 0, but not by removing the media line because, as stated earlier, the number of m-lines in an answer must equal that in the offer.

If the offerer wishes to send (and/or receive) multiple streams of the same type at the same time, then it places multiple media streams of that type in an offer. Each stream may use different encoding.

The offerer sends media after receiving an answer (it cannot do so before it receives an answer since it does not know the IP address and port to direct media to). It uses the media format listed in the answer and, typically, uses the first media format listed in the answer: it is recommended that formats appear in an SDP offer/answer in an order in which the most preferred appear at the top of the list and the least preferred appears last.

10.3.1 Modifying a session description

At any point during the session, either participant can issue a new offer to modify characteristics of the session. Each stream in the newly created offer must have a matching media stream in the previous offer/answer exchange, and the order of the media streams must not change. As a result of these two requirements, the number of m-lines in a stream never decreases: it either stays the same or increases. Below are

some examples to illustrate this (note that the syntax is not shown correctly): First
offer:

```
m=a
m=b
```

First answer:

```
m=a
m=b
```

Correct second offer:

```
m=a
m=b
m=c
```

Correct second answer:

```
m=a
m=b
m=c
```

Incorrect second offer (order changed):

```
m=a
m=c
m=b
```

Incorrect second offer (number of lines decreased):

```
m=a
```

As happens when creating an initial offer/answer, by setting the port value to 0 we
can disable a media stream. Also, new media streams that are added can reuse a
media line whose stream was disabled with a port 0 value by an earlier offer/answer
exchange.

10.3.2 Putting the media stream on hold

A client that wishes to instruct the remote end to stop sending media on that stream,
but does not wish to completely disable the stream, can do so in two different ways
which depend on the stream direction that was negotiated in an earlier offer/answer
exchange. A stream that was earlier set with a sendrecv attribute (media for that
stream is sent and received) is placed on hold by assigning it an attribute with
sendonly in the new offer/answer. A stream that was earlier set with a recvonly
attribute is placed on hold by assigning it an attribute with inactive in the new
offer/answer.

11

RTP

[RFC3550] defines the Real-time Transport Protocol (RTP) as a protocol for end-to-end delivery for real-time data. It also contains end-to-end delivery services for real-time data: payload-type (codec) identification, sequence numbering, time stamping and delivery monitoring. RTP does not provide quality of service (QoS); it does, however provide QoS monitoring using the RTP Control Protocol (RTCP). RTCP also conveys information about media session participants.

11.1 RTP for real-time data delivery

11.1.1 RTP fixed header fields

The RTP fields that are present in an RTP packet (Figure 11.1) represent the following:

- *Version (V)*—the version of RTP. It is always set to "2" if the implementation is [RFC3550]-compliant.

- *Padding (P)*—if set, then the RTP packet contains padding octets. Padding may be needed by some encryption algorithms with fixed block sizes or for carrying multiple RTP packets in a lower layer protocol packet.

- *Extension (X)*—if set, then there is an RTP header extension.

- *CSRC count (CC)*—the number of contributing source (CSRC) IDs that follow the fixed header.

- *Marker (M)*—its interpretation is defined by a profile.

- *Payload type (PT)*—Identifies the payload format (codec).

- *Sequence number*—this increments by 1 for each RTP data packet. This sequence

The IMS. Miikka Poikselkä, Georg Mayer, Hisham Khartabil and Aki Niemi
Copyright 2004 by John Wiley & Sons, Ltd. ISBN 0-470-87113-X

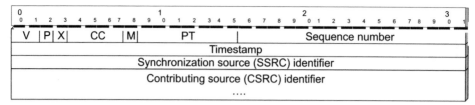

Figure 11.1 RTP packet format.

number is used by the receiver to re-order packets that arrive out of sequence. It also allows the receiver to determine whether packets have been lost.

- *Timestamp*—indicates the time when the first octet in the payload was sampled. This field is used along with the sequence number to remove jitter (see Section 11.1.2).

- *Synchronization source (SSRC)*—RTP is not dependent on the underlying Internet Protocol (IP) and, therefore, uses this field to identify the source of the RTP packets.

- *Contributing source (CSRC)*—this field carries a list of SSRCs indicating the sources that have contributed to a mixed media stream, if a mixer has been involved. It allows between 0 and 15 SSRCs. If there are more than 15 sources of media, then only the first 15 are identified. An audio conference mixer is a good example.

11.1.2 What is jitter?

The recipient may receive packets out of sequence and with delay, causing breaks and distortion in the media: this distortion is named "jitter". Figure 11.2 shows an example of how RTP packets may arrive out of sequence and be delayed. This is true of any Transmission Control Protocol over IP (TCP/IP) network where congestion is a factor. In RTP a sequence number is used to re-order out-of-sequence packets. The

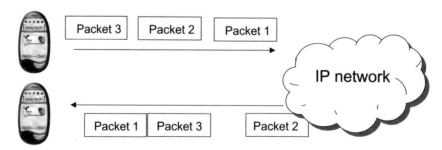

Figure 11.2 Packet jitter.

timestamp is used to allow the receiver to play packets at the right time and, therefore, minimize distortion.

11.2 RTCP

RTP Control Protocol (RTCP) packets are transmitted periodically to all participants in a session. There are four RTCP functions:

- To provide feedback on the QoS of real-time data distribution.

- To carry a persistent identifier of the RTP source (called a CNAME).

- To permit an adjustable RTCP packet distribution interval (the report interval).

- To convey session control information.

11.2.1 RTCP packet types

There are five types of RTCP packets:

- *SR*—a sender report, providing transmission and reception statistics, sent by active media senders.

- *RR*—a receiver report, providing reception statistics, sent by non-active senders.

- *SDES*—source description items, such as CNAME.

- *BYE*—indicates end of participation.

- *APP*—application-specific functions (defined by a profile).

11.2.2 RTCP report transmission interval

Every participant is required to send RTCP packets. If there are many participants (say, a conference), then there is a scalability problem as further users join in. To control the scalability problem, the rate at which RTCP packets are sent must be scaled down by dynamically calculating the interval between RTCP packet transmissions.

A certain percentage of session bandwidth should be dedicated to RTCP; this percentage and the interval between report transmissions are profile-defined.

11.3 RTP profile and payload format specifications

When conventional protocols are designed, they are generalized to accommodate the additional functionality of their applications. For RTP, this is achieved through modifications and additions to the existing headers as needed by an application. Therefore, RTP for a particular application, like audio, requires 1 or more companion documents: namely, a profile specification document and a payload format specification document.

11.3.1 Profile specification

In [RFC3550] the following fields have been identified as possible items to be defined by an application profile specification: RTP data header, payload types (codecs), RTP data header additions, RTP data header extensions, RTCP application-specific packet types, RTCP report interval (see Section 11.2.2), SR/RR extensions, SDES use, security services and algorithm, string (password) to encryption key mapping, congestion control behaviour, underlying protocol, RTP and RTCP mapping to transport-level addresses and encapsulation of multiple RTP packets in a lower layer packet.

11.3.2 Payload format specification

A payload format specification for an application of RTP defines how a particular payload is carried in RTP.

11.4 RTP profile and payload format specification for audio and video (RTP/AVP)

As defined in Section 11.3, an application using RTP must specify a profile and the payload formats used for it. In this section the profile and payload formats for audio and video applications are identified (the full specifications are defined in a combined document in [RFC3551]). Table 11.1 contains a summary of the profile specified for audio and video application and Table 11.2 contains a sample of the payload formats specified for audio and video.

11.4.1 Static and dynamic payload types

A static payload type is defined with a fixed identification number, a clock rate and a number of channels, when applicable. A dynamic payload type is defined without a

Table 11.1 RTP/AVP-specific profile.

Profile item	RTP/AVP-specific
RTP data header	Fixed RTP data header is used (1 marker bit)
Payload types	Defined below
RTP header additions	None
RTP header extensions	None
RTCP extension packet types	None
RTCP report interval	As recommended in [RFC3550]
SR/RR extensions	None
SDES use	As defined in [RFC3550]
Security	RTP default
String-to-key mapping	None
Congestion	[RFC1633] and [RFC2475] can be used, as can best effort services
Underlying protocol	UDP, TCP
Transport mapping	Standard mapping of RTP/RTCP to transport-level addresses
Encapsulation	Not specified

Table 11.2 Sample payload formats for audio and video.

PT	Encoding name	Media type	Clock rate (Hz)	Channels
0	PCMU	Audio	8,000	1
3	GSM	Audio	8,000	1
4	G723	Audio	8,000	1
Dynamic	L8	Audio	Variable	Variable
26	JPEG	Video	90,000	—
31	H261	Video	90,000	—
34	H263	Video	90,000	—
Dynamic	H263-1998	Video	90,000	—
96–127	Dynamic	A/V	?	—

number, but a number is dynamically allocated. Dynamic payload types are assigned numbers between 96 and 127.

In the SDP a static payload type does not need the rtpmap attribute. For example, the PCMU payload type can be represented as:

```
m=audio 49170 RTP/AVP 0
```

or

```
m=audio 49170 RTP/AVP 0
a=rtpmap:0 PCMU/8000
```

An example of a dynamic payload type is:

```
m=video 49172 RTP/AVP 96
a=rtpmap:96 H263-1998/90000
```

12

DNS

12.1 DNS resource records

On a large network, such as the Internet, it would be impractical to identify each system solely by its numerical Internet Protocol (IP) address. Names are more convenient; they help users and administrators locate network resources more easily. The Domain Name Service (DNS) is a distributed database holding the alphanumeric names and their corresponding IP addresses (and more) of every registered system on a Transmission Control Protocol (TCP)/IP network, such as the Internet or the IP Multimedia Subsystem (IMS). Each entry is referred to as a resource record (RR).

The alphanumeric names, better known as domain names, are hierarchical in nature where country, company, department and even a host (machine) name can be identified. Each step in the hierarchy is identified as a zone. The domain name below identifies a machine named pc27 in the engineering department of an Australian company called Foobar:

```
pc27.engineering.foobar.com.au
```

An entry in a DNS server that maps a domain name to an IP address is referred to as an address record, or A record. For IPv6 these records are referred to as AAAA records.

12.2 The naming authority pointer (NAPTR) DNS RR

NAPTR RRs are used to replace compact, regular expressions with a replacement field that may well be a pointer to another rule. [RFC2915] defines the NAPTR; however, it does not define why or how replacement fields are used. It is up to the applications using NAPTR to do this. The DNS-type code for NAPTR is 35.

The IMS. Miikka Poikselkä, Georg Mayer, Hisham Khartabil and Aki Niemi
Copyright 2004 by John Wiley & Sons, Ltd. ISBN 0-470-87113-X

Table 12.1 NAPTR RR fields.

Field	Description
Domain	The key for an entry (the domain name of the RR)
TTL	Time to live
Class	Class of record (NAPTR records are of class IN, or Internet)
Type	DNS-type code (for NAPTR it is 35)
Order	The order in which NAPTR need to be processed. This allows NAPTR records to specify a complete rule in an incremental fashion. Records are processed from lower order numbers to higher order numbers
Preference	Specifies the order in which NAPTR records with the same "order" values need to be processed. Records are processed from lower preference numbers to higher preference numbers
Flags	Indicate what happens next after this look-up. The "S" flag indicates that the next look-up should be an SRV look-up. The replacement field carries an SRV reference. The "U" flag indicates that the next step is not a DNS look-up. The regexp (see below) carries a URI. The "A" flag indicates that the next step is a DNS A, AAAA or an A6 look-up. The "P" flag indicates that the remainder of the look-ups are defined by the application that uses the NAPTR
Service	Specifies the services available in this domain. The replacement field is used to get to this service. It can also specify the protocol used to communicate with the server that offers this service. In SIP, three services are defined along with their resolution services (resolution services are defined after the "+" sign): "SIPS + D2T": Secure SIP, TLS over TCP "SIP + D2T": SIP over TCP "SIPS + D2S": Secure SIP, TLS over SCTP "SIP + D2S": SIP over SCTP "SIP + D2U": SIP over UDP
Regexp	Carries a substitution expression that is applied to the original domain name in order to construct a new domain name for the next look-up
Replacement	The next name used to query a DNS. This could be another NAPTR record, an SRV record or an address record. In SIP the replacement fields are SRV RRs and hence the flag field is set to "S"

The format of a NAPTR record is as follows (Table 12.1 describes each field in detail):

```
Domain TTL Class Type Order Preference Flag Service Regexp Replacement
```

12.2.1 NAPTR example

This example uses the Session Initiation Protocol (SIP) service. A NAPTR RR looks like:

Domain	TTL	Class	Type	Order	Preference	Flag	Service	Regexp	Replacement
example.com	7200	IN	35	50	50	"S"	"SIPS + D2T"	""	_sips._tcp.example.com
	7200	IN	35	90	50	"S"	"SIP + D2T"	""	_sip._tcp.example.com
	7200	IN	35	100	50	"S"	"SIPxD2U"	""	_sip._udp.example.com

In this example a client of the SIP service that does not support the Transport Layer Security (TLS) performs a NAPTR RR look-up for the "example.com." domain and selects the replacement:

```
_sip._tcp.example.com
```

12.3 ENUM—the E.164 to URI Dynamic Delegation Discovery System (DDDS) application

ENUM is described as the use of the DNS for storage of international public telecommunication numbers using international format (E.164) numbers [Draft-ietf-enum-rfc2916bis]. This is useful to look up what services are available for a specific domain.

Imagine that user *A* has a Public Switched Telephone Network (PSTN) phone and would like to call her friend user *B* who uses VoIP (Voice over IP). User *A* cannot enter a SIP address into her PSTN phone and, therefore, enters user *B*'s E.164 number. The call reaches a gateway that performs a DNS ENUM look-up and returns user *B*'s SIP address (see Figure 12.1). IP to IP calls are also feasible using E.164 numbers. The steps involved in this procedure are as follows:

1. User *A* dials the E.164 number, +135812345678.

2. The circuit-switched (CS) domain contacts a gateway.

3. The gateway formats the E.164 number into a fully qualified domain name (FQDN), 8.7.6.5.4.3.2.1.8.5.3.e164.arpa, and looks up the FQDN in DNS.

4. DNS returns the NAPTR record, sip:userB@example.com

5. The gateway contacts DNS for the IP address of the SIP server in the domain "example.com", following the procedures in Section 8.12.1.1.

6. DNS returns the IP address of the SIP server.

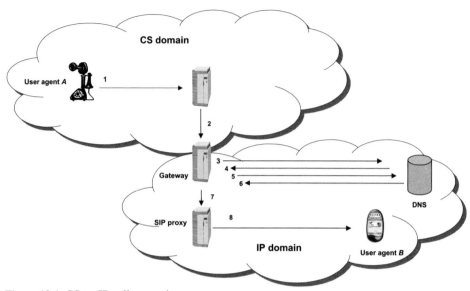

Figure 12.1 CS to IP cell example.

7. The gateway routes the call to the SIP server.

8. The SIP server contacts user *B* and the call is established.

The solution uses a NAPTR RR. First, the client turns the E.164 telephone number into a domain name. The domain "e164.arpa" is used to provide DNS with the infrastructure. The number is pre-pended to the domain by reversing the numbers, numbers are separated with a dot ".", each number becoming a zone. The NAPTR query returns a re-write rule using the regexp field. The NAPTR service parameter for ENUM is "E2U < enumservice > ".

12.3.1 ENUM service registration for SIP addresses-of-record (AORs)

[Draft-ietf-enum-sip] defines an enumservice for SIP that allows users to map a telephone number to a SIP AOR. The enumservice for SIP is "E2U + sip". An example of a NAPTR record looks like:

Domain	TTL	Class	Type	Order	Preference	Flag	Service	Regexp
8.7.6.5.4.3.2.1.8.5.3.e164.arpa	7200	IN	35	50	50	"U"	"E2U + sip"	"!^.*$!sip:userB@example.com!"

Table 12.2 SRV RR fields.

Field	Description
Service	Service identifier. An underscore "_" is pre-pended
Proto	Desired protocol. An underscore "_" is pre-pended
Name	The domain of the SRV record
TTL	Time to live
Class	Class of record (SRV records are of class IN, or Internet)
SRV	Identifies the entry as an SRV resource record
Priority	The priority of the target host in this record. Clients with lowest numbered priority are tried first. Weight (see below) is used if more than one record carries the same priority
Weight	Denotes a relative weight for entries with the same priority. Entries with larger weights typically have a higher probability of being selected
Port	The port number for the host identified in this record
Target	The domain name of the host targeted by this record

12.4 Service records (SRVs)

Before SRV RRs are introduced an entity must know the exact address of a server to contact for a service. With the introduction of SRV RRs in [RFC2782], DNS clients can query a domain for a service or protocol and get back the names of servers. Administrators can use SRV RRs to distribute the load by using several servers to service a single domain. The DNS-type code for SRVs is 33.

The format of an SRV record is as follows (Table 12.2 describes each field in detail):

```
_Service._Proto.Name TTL Class SRV Priority Weight Port Target
```

12.4.1 SRV example

Continuing with the example, the SRV RR looks like:

Service, Proto, Name	TTL	Class	Record type	Priority	Weight	Port	Target
_sip._tcp.example.com.	7200	IN	SRV	0	1	5060	server1.example.com
	7200	IN	SRV	0	3	5060	server2.example.com

In this example the client of the SIP service in the "examp.com" domain performs an SRV look-up of "_sip._tcp.example.com" followed by an A record look-up for "server2.example.com".

13

GPRS

13.1 Overview

The General Packet Radio Service (GPRS) is the packet-switched (PS) domain of the Global System for Mobile Communications (GSM) and the Universal Mobile Telecommunications System (UMTS) network. It provides Internet Protocol (IP) connectivity to attached user equipment (UE) via so-called Packet Data Protocol (PDP) contexts. As expressed in the name, it is a logical connection (context) that is related to a specific packet-based protocol.

The UE will be able to send IP packets over the air interface after it has established a PDP context.

This chapter concentrates only on those parts of GPRS that are necessary for a UE to access an IP Multimedia Subsystem (IMS) network. The detailed procedures within the Gateway GPRS Support Node (GGSN) or the Serving GPRS Support Node (SGSN), as well as the detailed message coding and flows are not discussed here. The aim of this chapter is to provide the reader with a short overview about the basic principles that are behind GPRS and its PDP contexts.

13.2 Packet Data Protocol (PDP)

A PDP context offers a packet data connection over which the UE and the network can exchange IP packets. The usage of these packet data connections is restricted to specific services. These services can be accessed via so-called access points.

13.2.1 Primary PDP context activation

This procedure is used to establish a logical connection with the quality of service (QoS) functionality through the network from the UE to the GGSN. PDP context activation is initiated by the UE and changes the session management state to active, creates the PDP context, receives the IP address and reserves radio resources. After a

The IMS. Miikka Poikselkä, Georg Mayer, Hisham Khartabil and Aki Niemi
Copyright 2004 by John Wiley & Sons, Ltd. ISBN 0-470-87113-X

PDP context activation the UE is able to send IP packets over the air interface. The UE can have up to 11 PDP contexts active concurrently.

13.2.2 Secondary PDP context activation

A secondary PDP context activation allows the subscriber to establish a second PDP context with the same IP address as the primary PDP context. The two contexts may have different QoS profiles, which makes the feature useful for applications that have different QoS requirements (e.g., IP multimedia). The access point name, though, will be the same for the primary and secondary PDP contexts.

13.2.3 PDP context modification

The UE, the SGSN or the GGSN initiate this procedure for updating the corresponding PDP context. Additionally, the radio access network is able to request a PDP context modification from the SGSN (e.g., when coverage to the UE has been lost). The procedures modify parameters that were negotiated during an activation procedure for one or several PDP contexts.

13.2.4 PDP context deactivation

This procedure is used to delete a particular logical connection between the UE and the GGSN. The initiative to deactivate a PDP context can come from the UE, the SGSN, the Home Location Register (HLR) or the GGSN.

13.3 Access points

Access points can be understood as IP routers that provide the connection between the UE and the selected service. Examples of such services are:

- Multimedia Messaging Service (MMS).

- Wireless Application Protocol (WAP).

- Direct internet access.

- IP Multimedia Subsystem (IMS).

Depending on the operator of the network, more than one of these services might be provided by the same access point. The UE needs to be aware of an access point name (APN)—the address of a GGSN—which gives access to the service-providing entity (e.g., an MMSC, the Internet or the P-CSCF). One GGSN may provide different services that can be accessed by different APNs.

When establishing a primary PDP context with an APN the UE receives an IP address or—in the case of IPv6—an IPv6 prefix that it has to use when communicating over that PDP context. This means that when a UE has established several connections to different APNs the UE will have different IP addresses for each of the provided services.

13.4 PDP context types

From the viewpoint of the IMS, it is important to distinguish between the PDP context types shown in Figure 13.1. A primary PDP context is used whenever the first connection with a specific APN is established. If further connections with the same APN are needed, secondary PDP contexts are established.

IMS signalling can be transported either in a dedicated signalling PDP context or a general purpose PDP context. If a dedicated signalling PDP context is established, all IMS-related media streams must be put into one or more separate secondary PDP contexts.

Over a general IMS PDP context the UE may send signalling and media streams together. Nevertheless, it still has the option to create separate PDP contexts for media in this case as well.

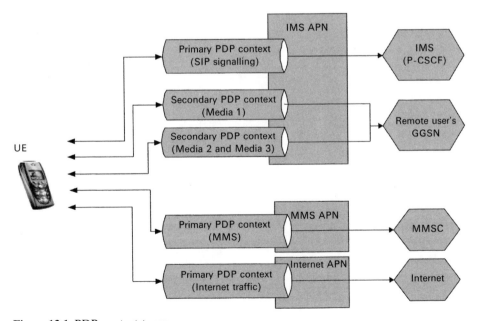

Figure 13.1 PDP context types.

14

TLS

This chapter makes an overview of the Transport Layer Security (TLS) [RFC2246].

14.1 Introduction

TLS provides transport layer security for Internet applications. It provides for confidentiality and data integrity over a connection between two end points. TLS operates on a reliable transport, such as TCP, and is itself layered into the TLS Record Protocol, and the TLS Handshake Protocol.

One advantage of TLS is that applications can use it transparently to securely communicate with each other. Another is that TLS is visible to applications, making them aware of the cipher suites and authentication certificates negotiated during the set-up phases of a TLS session; whereas, with the Internet Protocol Security (IPsec), security policies are usually not visible to each application individually, which makes it difficult to assess whether there is adequate security in place.

TLS allows for a variety of cipher suites to be negotiated, for the use of compression and for a TLS session to span multiple connections. This reduces the overhead of having to perform an expensive TLS handshake for each new parallel connection between the applications. It is also possible to resume a session: this means that the client and server can agree to use a previously negotiated session—if one exists in their session cache—instead of performing the full TLS handshake.

14.2 TLS Record Protocol

The TLS Record Protocol layers on top of a reliable connection-oriented transport, such as TCP. The Record Protocol provides data confidentiality using symmetric key cryptography and data integrity using a keyed message authentication checksum

(MAC). The keys are generated uniquely for each session based on the security parameters agreed during the TLS handshake. The Record Protocol is also used for encapsulating various upper layer protocols—most notably the TLS Handshake Protocol—in which case it can be used without encryption or message authentication. Other protocols encapsulated in the record protocol are the Alert Protocol and the Change Cipher Spec Protocol.

The basic operation of the TLS Record Protocol is as follows:

1. Read messages for transmit.

2. Fragment messages into manageable chunks of data.

3. Compress the data, if compression is required and enabled.

4. Calculate a MAC.

5. Encrypt the data.

6. Transmit the resulting data to the peer.

At the opposite end of the TLS connection, the basic operation of the sender is replicated, but in the reverse order:

1. Read received data from the peer.

2. Decrypt the data.

3. Verify the MAC.

4. Decompress the data, if compression is required and enabled.

5. Reassemble the message fragments.

6. Deliver the message to upper protocol layers.

14.3 TLS Handshake Protocol

The TLS Handshake Protocol is layered on top of the TLS Record Protocol. The Handshake Protocol is used to authenticate the client and the server, to exchange cryptographic keys and to negotiate the used encryption and data integrity algorithms before the applications start to communicate with each other. Figure 14.1 illustrates the actual handshake message flow. First, the client and server exchange Hello messages. The client sends a ClientHello message, which is followed by the server sending a ServerHello message. These two messages establish the TLS protocol version, the compression mechanism used, the cipher suite used, and possibly the

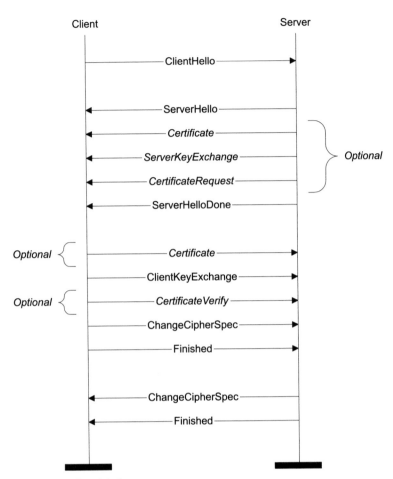

Figure 14.1 The TLS handshake.

TLS session ID. Additionally, both a random client nonce and a random server nonce are exchanged that are used in the handshake later on.

Then, the server may send any messages associated with the ServerHello. Depending on the selected cipher suite, it will send its certificate for authentication. The server may also send a key exchange message (e.g., if the server certificate is for signing only) and a certificate request message to the client, depending on the selected cipher suite. To mark the end of the ServerHello and the Hello message exchange, the server sends a ServerHelloDone message.

Next, if requested, the client will send its certificate to the server. In any case, the client will then send a key exchange message that sets the pre-master secret between the client and the server. The pre-master secret is used along with previously exchanged random nonce values to create a master secret for the session. Optionally, the client may also send a CertificateVerify message to explicitly verify the certificate that the server requested.

Then, both the client and the server send the ChangeCipherSpec messages and enable the newly negotiated cipher spec: in other words, the record layer is provided with the necessary security parameters resulting from the handshake. The first message passed in each direction using the new algorithms, keys and secrets is the Finished message, which includes a digest[1] of all the handshake messages. Each end inspects the Finished message to verify that the handshake was not tampered with.

As mentioned earlier, it is also possible to resume a TLS session. To do this, the client suggests in its ClientHello message a session ID used in the TLS session to be resumed. If the server is willing to resume the session and is able to find a matching session ID in its session cache, the handshake will be an abbreviated one, resulting in ChangeCipherSpec and Finished messages immediately after the ServerHello message.

14.4 Summary

The TLS protocol provides transport layer security for Internet applications and confidentiality using symmetric key cryptography and data integrity using a keyed MAC. It also includes functionality for client and server authentication using public key cryptography.

---

[1] Meaning the results of applying a one-way hash function to the handshake messages.

15

Diameter

This chapter makes an overview of the Diameter Protocol. It covers the basic components, protocol properties and protocol operation of Diameter. The services provided by Diameter are given a thorough inspection. This chapter also includes a short introduction to Diameter accounting.

15.1 Introduction

Diameter is an authentication, authorization and accounting (AAA) protocol developed by the Internet Engineering Task Force (IETF). Diameter is used to provide AAA services for a range of access technologies. Instead of building the protocol from scratch, Diameter is loosely based on the Remote Authentication Dial In User Service (RADIUS)[1] [RFC2865], which has previously been used to provide AAA services, at least for dial-up and terminal server access environments. As the basis for the Diameter work, the AAA Working Group first gathered requirements for AAA services as they apply to network access from different interest groups:

- IP Routing for Wireless/Mobile Hosts WG (MOBILEIP) [RFC2977].

- Network Access Server Requirements WG (NASREQ) [RFC3169].

- Roaming Operations WG (ROAMOPS) [RFC2477].

- Telecommunications Industry Association (TIA).

The final Diameter protocol is actually split into two parts: the Diameter base protocol and the Diameter applications. The base protocol is needed for delivering Diameter data units, negotiating capabilities, handling errors and providing for

---

[1] The name is derived from geometry, as Diameter $= 2 *$ Radius.

The IMS. Miikka Poikselkä, Georg Mayer, Hisham Khartabil and Aki Niemi
Copyright 2004 by John Wiley & Sons, Ltd. ISBN 0-470-87113-X

extensibility. A Diameter application defines application-specific functions and data units. Each Diameter application is specified separately. Currently, in addition to the base protocol [RFC3588], a few Diameter applications have been defined and some are in the process of being defined: namely, Mobile IP [Draft-ietf-aaa-diameter-mobileip], NASREQ [Draft-ietf-aaa-diameter-nasreq], Extensible Authentication Protocol (EAP) [Draft-ietf-aaa-eap], Diameter credit control [Draft-ietf-aaa-diameter-cc] and Diameter SIP application [Draft-ietf-aaa-diameter-sip-app]. In addition, the AAA transport profile [RFC3539] includes discussions and recommendations on the use of transports by AAA protocols. Third Generation Partnership Project (3GPP) Release 5 has also been allocated a specific set of Diameter command codes [RFC3589].

The Diameter base protocol uses both the Transmission Control Protocol (TCP) [RFC0793] and the Stream Control Transmission Protocol (SCTP) [RFC2960] as transport. However, SCTP is the preferred choice, mostly due to the connection-oriented relationship that exists between Diameter peers. It is beneficial to be able to categorize several independent streams to a single SCTP association, instead of keeping all streams open as independent TCP connections. Both Internet Protocol Security (IPsec) [RFC2401] and Transport Layer Security (TLS) [RFC2246] are used for securing the connections.

15.2 Protocol components

Diameter is a peer-to-peer protocol, since any Diameter node can initiate a request. Diameter has three different types of network nodes: clients, servers and agents. Clients are generally the edge devices of a network that perform access control. A Diameter agent provides either relay, proxy, redirect or translation services. A Diameter server handles the authentication, accounting and authorization requests for a particular domain, or realm. Diameter messages are routed according to the network access identifier (NAI) [RFC2486] of a particular user.

15.3 Message processing

Each Diameter node maintains a Diameter peer table, which contains a list of known peers and their corresponding properties. Each peer table entry is associated with an identity and can either be statically or dynamically assigned. Dynamically assigned peer entries have an expiration time associated with them, within which they are either refreshed or discarded. Each peer entry also includes a relative priority setting, which specifies the role of the peer as either primary, secondary or alternative. The

Table 15.1   Diameter local action entries.

Local action field	Description
Local	Messages resolving to a route entry with the local directive set can be processed locally and need not be routed to another Diameter peer
Relay	Messages resolving to a relay route entry will be routed to a next-hop peer without modification
Proxy	Messages that resolve to a proxy route entry will be routed to a next-hop peer, possibly after local policies are first applied
Redirect	Messages that resolve to a redirect route entry will result in a redirection of a downstream[a] Diameter peer to the correct next-hop peer

[a] "Downstream" is a term used to identify the direction of a message (e.g., toward the originator of a request).

status of the peer relates to a specific configuration of a finite state machine (FSM) of a peer connection, called the Diameter Peer State Machine. Each peer entry also specifies whether a peer supports TLS[2] and optionally includes other security information (e.g., cryptographic keys and certificates).

Diameter peer table entries are referenced by a Diameter realm-routing table, as part of the message-routing process. All realm-based routing look-ups are performed against a realm-routing table. The realm-routing table lists the supported realms (with each route entry containing certain routing information). Each route entry is either statically assigned or dynamically discovered. Dynamic entries are always associated with an expiry time. A route entry is also associated with an application identifier, which enables route entries to have a different destination depending on the Diameter application. The destination of a route entry corresponds to one or more peer entries in a Diameter peer table. As message processing can follow different procedures depending on the message, the local action field of a route entry can have one of four processing directives (see Table 15.1).

Each processing directive corresponds to a functional entity called a Diameter agent, as discussed in Section 15.5. All Diameter nodes maintain a transaction state (i.e., they keep track of request and answer pairs). A Diameter session is always bound to a particular service, defined by the corresponding Diameter application. A statefull agent maintains a session state (i.e., it stores information related to all of its authorized active sessions). Active sessions are either explicitly torn down or deactivate by expiration.

---

[2] Support for IPsec is mandatory for all nodes in Diameter, whereas support for TLS is optional for clients.

The local state can also be added to Diameter requests. This feature enables an otherwise stateless agent to encode its local state as part of the request, so that it gets reflected back in the corresponding answer message. Overall, maintaining the session state is necessary for some Diameter agents, such as translation agents and proxy agents.

15.4 Diameter clients and servers

Diameter clients originate AAA requests. The Diameter server[3] services the client's requests within a realm, according to the used Diameter application. Diameter agents provide value-added services for clients and servers.

Typically, the Diameter client is a network access server (NAS) that performs AAA services for a particular access technology. The NAS needs to authenticate the terminals that are attached to a network before allocating network resources for them: for example, a Wireless Local Area Network, or WLAN, access point may need to ascertain the user's identity before allowing access.[4]

15.5 Diameter agents

Diameter agents can be deployed in a network to perform load balancing, distribute system administration and maintenance, concentrate requests and perform additional message processing. Diameter relay agents are protocol-transparent (i.e., they support the Diameter base protocol), but are agnostic about Diameter applications. They simply accept requests and route messages based on the supported realms and known peers. Relay agents are usually used to reduce the configuration load on Diameter clients and servers: for example, NASs of a specific geographical area can be grouped behind a common relay agent, thus avoiding the need to configure each NAS with the information needed to communicate with other realms. Similarly, a Diameter server may have a relay agent that handles the configuration load caused by the addition and removal of individual NASs in the network.

Diameter redirect agents perform realm to server resolution. They use the Diameter Realm Routing Table to determine the next-hop Diameter peer to a particular request message. Instead of routing a request by itself, the redirect agent returns a special answer message that includes the identity of the next-hop peer. The originator of the request then contacts the next-hop peer directly. Since redirect agents operate

---

[3] Often referred to as the home server.
[4] Usually, there exists a prior (billing) relationship between the network provider and the user.

in a stateless fashion, they provide scalability and centralized message routing. However, centralized message routing is always a compromise between ease of configuration and fail-over resilience.

Diameter proxy agents carry out value-added message processing on the requests and answers. They are similar to relay agents in that message routing is based on a Diameter routing table, but different in that they also modify the messages by implementing policy enforcement (e.g., resource usage enforcement, admission control or provisioning). Since policy enforcement always requires an understanding of the offered service, proxy agents need to support specific Diameter applications as well as the Diameter base protocol.

Diameter translation agents perform protocol translation services between Diameter and another AAA protocol. Translation agents are used to allow legacy systems to communicate with the Diameter infrastructure. Sometimes, migration from an installed equipment base that is not easy to upgrade (e.g., embedded devices) takes a considerable amount of time.

Perhaps the most common legacy system in this case is RADIUS, to which Diameter has extensive support with the NASREQ application.

15.6 Message structure

Diameter messages consist of a Diameter header, followed by a certain number of Diameter attribute value pairs (AVPs). The Diameter header comprises binary data and as such is similar to an IP header [RFC0791] or a TCP header [RFC0793]. The format of the Diameter header is shown in Figure 15.1.

The command flags specify the type of Diameter message. A request message has the "R" bit (or the request bit) set, while an answer message has it cleared. The "P" bit (or the proxiable bit) indicates whether the message can be proxied or must instead be processed locally. The "E" bit (or the error bit) indicates that the answer message is an error message. The "T" bit indicates that the message is possibly a retransmission and is set by Diameter agents in fail-over situations to aid detection of duplicate messages. The "r" bits are unused flags that are reserved for future use. Together with the Command-Code, the command flags specify the semantics associated with the particular Diameter message.

The Command-Code field indicates the command associated with the Diameter message. The Command-Code field is 24 bits long and includes values from 0 to 255 that are reserved for RADIUS backward compatibility, as well as values 16777214 and 16777215 that are reserved for experimental use. The Command-Code namespace is maintained by the Internet Assigned Numbers Authority (IANA) and includes either codes that are used by the Diameter Base Protocol or specific codes used by the Diameter applications.

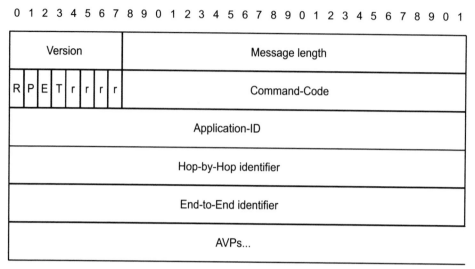

Figure 15.1 Diameter header.

The Application-ID field is used to identify the application for which the message is targeted. An Application-ID indicates either a standard application or a vendor-specific application. The Application-ID namespace is also maintained by the IANA.

The Hop-by-Hop identifier helps in matching requests and responses and is unique within a connection at any given time. The End-to-End identifier is used end-to-end to detect duplicate messages.

AVPs contain authentication, authorization and accounting information elements, as well as routing, security and configuration information elements that are relevant to the particular Diameter request or answer message. Each AVP contains an AVP header and some AVP-specific data. The AVP header is shown in Figure 15.2.

The AVP-Code field uniquely identifies the attribute together with the Vendor-ID field. The first 256 AVP numbers, or codes, are reserved for backward compatibility with RADIUS, whereas numbers above 256 belong to Diameter attributes. AVP-Code field values are maintained by the IANA.

AVP flags carry information on how the attribute should be handled by the receiving end. The "V" bit (or the vendor-specific bit) indicates that the AVP-Code belongs to a vendor-specific address space, denoted by the otherwise optional Vendor-ID field. The "M" bit (or mandatory bit) mandates the support for a particular AVP. Any message with an unrecognized AVP carrying the "M" bit is always rejected by the receiver. AVPs that do not have the M bit set are information-only

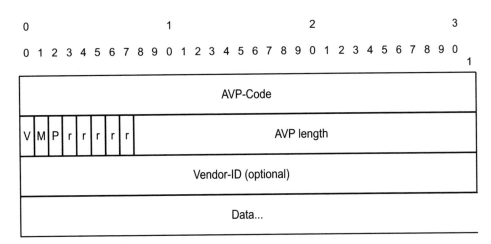

Figure 15.2 Diameter AVP header.

(i.e., they can be ignored by the receiver if they are not understood). When the "P" bit (or the protected bit) is set it indicates the need for encryption for end-to-end security. The "r" bits are unused flags reserved for future use.

15.7 Error handling

Diameter errors fall into two categories: protocol errors and application errors. Protocol errors occur at the base protocol level and are treated on a per-hop basis (e.g., errors in message routing may direct a downstream agent to take special action). Application errors are generated due to problems with specific Diameter applications.

When a request message generates a protocol error an answer message is sent back with the E bit set in the Diameter header, indicating a protocol error. The appropriate protocol error value is then placed in a Result-Code AVP in the answer message. On the other hand, when a Diameter entity generates an application error it sends back an answer message with the appropriate Result-Code AVP included. Since application errors do not require any Diameter agent involvement, the message is forwarded directly back to the originator of the request.

Table 15.2 describes the classes used in the Result-Code AVP data field: a 32-bit namespace for errors, maintained by the IANA. The Result-Code AVP can also be accompanied by other elements that give additional information about the error message (e.g., a human-readable error message must be included for all protocol errors in an Error-Message AVP).

Table 15.2 Diameter result codes.

Result code	Class	Description
1xxx	Information	Additional action required to complete the request
2xxx	Success	Request was successfully completed
3xxx	Protocol errors	Errors which are treated on a hop-by-hop basis
4xxx	Transient failures	Request failed but may succeed later
5xxx	Permanent failures	Request failed and should not be attempted again

15.8 Diameter services

Basically, the Diameter base protocol provides two types of services to Diameter applications: authentication and/or authorization, and accounting. An application can make use of authentication or authorization only or it can combine the two by requesting authentication with authorization. The accounting service can be invoked irrespective of whether authentication or authorization was requested first.

Each individual Diameter application defines its own authentication and authorization Command-Codes and AVPs. The accounting application described by the base protocol is shared among the applications—only the accounting AVPs are specific to an application.

15.8.1 Authentication and authorization

Authentication and authorization services are interlinked in Diameter. An auth request is issued by the client to invoke a service and, depending on the AVPs carried in the auth request, either authentication or authorization (or both) are performed on it.

Authentication clearly either succeeds or fails, whereas the Diameter base protocol provides authorization services in either stateless or statefull mode. In statefull authorization the server maintains a session state and the authorization session has a finite length. The total lifetime of a session consists of an authorization lifetime and a grace period, which together represent the maximum length of a session the server is willing to take responsibility for.[5] The authorization session can of course be terminated by the client or aborted by the server and, at the end of the authorization lifetime, it can also be re-authorized. These functions are provided by the Diameter base protocol.

The two authorization modes of operation correspond to a statefull authoriza-

---

[5] This usually means financial responsibility.

tion FSM and a stateless authorization FSM, with which all Diameter nodes that support authentication and authorization services have to comply.

15.8.2 Accounting

The Diameter base protocol provides accounting services to Diameter applications. When an accounting session is not active, there are no resources reserved for it in either the Diameter client or the Diameter server.[6] A successful accounting request (ACR) activates an accounting session, in which the accounting records exchanged fall into two categories, based on the accounting service type:

- Measurable length services have clearly defined beginnings and ends. An accounting record is created when the service begins and another is sent when the service ends. Optionally, interim accounting records can be produced at certain intervals within the measurable length session.

- One-time events are services without a measurable service length. In a one-time event accounting record the beginning of the service and the end of the service actually coincide. Therefore, a one-time event only produces a single accounting record.

The accounting strategy of a session is dictated by the accounting server or the server authorizing a user session: this is called the server-directed model for accounting. The accounting server directs the client to use either measurable length service accounting or one-time event accounting. It also optionally specifies the time interval to use when generating interim accounting records.

The Diameter accounting protocol has built-in fault resilience to overcome small message loss and temporary network faults, as well as real-time delivery of accounting information. Accounting records are correlated with the Session-Id AVP, which is a globally unique identifier and present in all AAA messages. Alternatively, if a service consists of several different sessions each with a unique Session-Id AVP, then an Accounting-Multi-Session-Id AVP is used for correlation of accounting records.

15.9 Specific Diameter applications used in 3GPP

This section lists the two Diameter applications used in the 3GPP IMS. First, is the Diameter SIP application [Draft-ietf-aaa-diameter-sip-app], which is used in the Cx, Dx, Sh and Dh interfaces (described in Section 2.3). The second is the Diameter credit control application [Draft-ietf-aaa-diameter-cc], which is utilized in the Ro interface for online charging functionality (see Section 3.10.1.2).

---

[6] In this case the accounting server.

15.10 Diameter SIP application

The Diameter SIP application [Draft-ietf-aaa-diameter-sip-app] defines a Diameter application that can be used by a SIP server to authenticate users and to authorize usage of different SIP resources. The Diameter SIP application is close to the 3GPP IMS Cx interface in functionality, but it is designed to be generic enough for other SIP deployment scenarios to be able to benefit from it. Table 15.3 illustrates a mapping between Cx interface parameters and Diameter SIP application AVPs.

The general architecture used as the basis for the Diameter SIP application is illustrated in Figure 15.3 (see p. 342). Proxy *A* and proxy *B* have different roles: proxy *A* is configured as an edge proxy to a domain and proxy *B* as the home proxy of a domain. Proxy *A* closely resembles the Interrogating Call Session Control Function (I-CSCF) function and proxy *B* the Serving-CSCF (S-CSCF) function in the IMS. For redundancy and fault tolerance, there may be additional Diameter subscriber locator (SL) nodes that either implement the Diameter relay or Diameter redirect agent functionality.

The Diameter SIP application defines a set of Command-Codes that are an extension to the Diameter base protocol (see Table 15.4). These Command-Codes perform the following functions:

- UAR/UAA—determines whether a user is authorized to receive a certain service and, if so, indicates the local server capable of providing that service.

- SAR/SAA—assigns a specific SIP server for a particular user and delivers the user profile to it in a synchronous way.

- LIR/LIA—determines the next-hop SIP entity at an edge proxy.

- MAR/MAA—authenticates and authorizes a user for a specific SIP service (e.g., SIP registration). Authentication can either be performed in the Diameter server or delegated to the SIP server.

- RTR/RTA—de-registration initiated by the Diameter server.

- PPR/PPA—asynchronously delivers the user profile to the SIP server from the Diameter server.

15.11 Diameter credit control application

The Diameter credit control application [Draft-ietf-aaa-diameter-cc] is used for real-time credit and cost control for a variety of different services. As opposed to non-real-time accounting, which is post-paid, this is the pre-paid mode of operation that has proved a great success in mobile phone markets: instead of, say, a monthly

Table 15.3 Mapping Cx parameters to the Diameter SIP application.

Cx parameter	Diameter SIP application AVP Name[a]
Visited network identifier	SIP-Visited-Network-Identifier
Public user ID	SIP-AOR
Private user ID	User-Name
S-CSCF name	SIP-Server-URI
S-CSCF capabilities	SIP-Server-Capabilities SIP-Mandatory-Capability SIP-Optional-Capability
Result	Result-Code SIP-Reason-Code
User profile	SIP-User-Data SIP-User-Data-Request-Type SIP-User-Data-Already-Available
Server assignment type	SIP-Server-Assignment-Type
Authentication data	SIP-Auth-Data-Item
Item number	SIP-Item-Number
Authentication scheme	SIP-Authentication-Scheme
Authentication information	SIP-Method SIP-Authenticate SIP-Authentication-Context
Authorization information	SIP-Authorization SIP-Authentication-Info
Confidentiality key	SIP-Confidentiality-Key
Integrity key	SIP-Integrity-Key
Number authentication items	SIP-Number-Auth-Items
Reason for de-registration	SIP-Deregistration-Reason SIP-Reason-Info
Charging information	SIP-Accounting-Information SIP-Accounting-Server-URI SIP-Credit-Control-Server-URI
Route information	Destination-Host
Type of authorization	SIP-User-Authorization-Type

[a] Note that those AVP names without the SIP-prefix are Diameter Base Protocol AVPs.

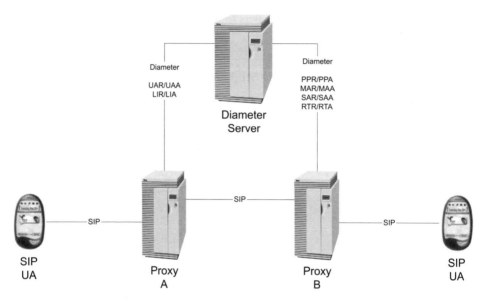

Figure 15.3 Diameter SIP application architecture.

Table 15.4 Diameter SIP application Command-Codes.

Command name	Abbreviation
User-Authorization-Request	UAR
User-Authorization-Answer	UAA
Server-Assignment-Request	SAR
Server-Assignment-Answer	SAA
Location-Info-Request	LIR
Location-Info-Answer	LIA
Multimedia-Auth-Request	MAR
Multimedia-Auth-Answer	MAA
Registration-Termination-Request	RTR
Registration-Termination-Answer	RTA
Push-Profile-Request	PPR
Push-Profile-Answer	PPA

invoice, users buy credit up front which is then debited in real time as services are used. The credit control application specifies a pair of new Diameter Command-Codes for credit control requests (see Table 15.5). The application has two main parts:

Table 15.5 Diameter credit control application Command-Codes.

Command Name	Abbreviation
Credit-Control-Request	CCR
Credit-Control-Answer	CCA

- Cost and credit control—services need to be rated in real time, which means that the cost of a given service needs to be assessed in real time. The user's account also needs to be debited or credited in real time, based on the accrued service cost.

- Credit authorization—the user's account may also need to be inspected to see that her balance is sufficient to cover the service. Credit needs to be reserved and unused reservations need to be refunded. If the user's balance is insufficient to cover the service charges, the service element needs to be able to deny the service altogether.

As in Diameter base protocol accounting, the credit control sessions have two distinct models: credit authorization with credit reservation and credit authorization with direct debiting. The first is similar to accounting measurable length services and consists of an initial interrogation where the credit control server rates the service request and reserves a sufficient amount of credit from the user's account. This may optionally be followed by any number of intermediary interrogations, in which the server deducts used credit from the user's account, and may involve the server performing service rating and credit reservation to accommodate a service that is continuing. A final interrogation reports back to the credit control server that the service has been terminated, whereupon the user's account is debited for the used amount.

The second model is similar to accounting one-time events. On receiving a credit authorization request, the server deducts a suitable amount of credit from the user's account. This model can also be seen as a credit control session, except that the start and end of the session happen to coincide. The stateless mode also supports operations, such as service price queries, user account balance checks and credit refunds.

15.12 Summary

The Diameter protocol is used to provide AAA services for a range of access technologies. Diameter is loosely based on an existing AAA protocol called RADIUS, which has been used widely for dial-up and terminal server access. The Diameter protocol uses a binary header format and is capable of transporting a range of data units called AVPs. The Diameter base protocol specifies the delivery mechanisms, capability negotiation, error handling and extensibility of the protocol, whereas individual Diameter applications specify service-specific functions and AVPs. In addition to the base protocol which includes accounting, the 3GPP's IP Multimedia Subsystem (IMS) currently makes use of two other Diameter applications: namely, the Diameter SIP application and the Diameter credit control application.

16

MEGACO

This chapter makes an overview of the Media Gateway Control Protocol (MEGACO) Version 1 [RFC3525], also known as H.248 [ITU.H248.1]. This protocol is used in the Third Generation Partnership Project's (3GPP) Internet Protocol Multimedia Subsystem (IMS) Mp reference point, as explained in Section 2.3.16.

16.1 Introduction

MEGACO is used between a media gateway and the media gateway controller to handle signalling and session management during a multimedia conference. The media gateway controller and the media gateway share a master/slave relationship.

16.2 Connection model

The connection model for the protocol describes the main objects within a media gateway as terminations and contexts that can be controlled by the media gateway controller. A termination sources or sinks[1] one or more streams, and each termination holds information about the actual media streams.

Different terminations are linked together by a context. The set of terminations that are not associated with other terminations are defined as being represented by a special type of context (namely, the null context). A context describes the topology of terminations associated with it: for example, it includes parameters about mixing in case the context contains more than two terminations.

---

[1] In layman terms: either originates or terminates.

The IMS. Miikka Poikselkä, Georg Mayer, Hisham Khartabil and Aki Niemi

16.3 Protocol operation

The protocol contains a set of commands that are used to manipulate the logical entities described in the connection model (namely terminations and contexts). Specifically, the set of commands offered by MEGACO are:

- Add—adds a termination to a context. It is also used to implicitly create a context (a context is created as soon as the first termination is added to it).

- Modify—modifies the state properties of a termination and properties specific to media streams.

- Subtract—removes a termination from a context. It is also used to implicitly delete a context. Similar to creating one, subtracting the last termination of a context deletes the context.

- Move—moves a specific termination from one context to another.

- AuditValue—returns the current termination properties as well as the events,[2] signals[3] and statistics of a termination.

- AuditCapabilities—returns all possible values for the termination properties as well as the signals and events allowed by a particular media gateway.

- Notify—used by the media gateway to report to the media gateway controller when certain events occur.

- ServiceChange—used by the media gateway to inform the media gateway controller of specific changes in service (e.g., when a particular termination has been taken out of service or is returned to service).

Each command can carry a number of parameters, called descriptors. A command may also return descriptors as output. Descriptors consist of a name and a list of items, some of which may have values. The list of descriptors in MEGACO is shown in Table 16.1.

---

[2] "Events" describe situations in a stream which the media gateway can detect and report (e.g., on-hook, off-hook).

[3] "Signals" are media generated by the media gateway (e.g., announcements and tones).

Table 16.1   MEGACO descriptors.

Name	Description
Audit	Indicates the information requested in Audit commands
DigitMap	Describes patterns for a sequence of events that when matched should be reported
Error	Indicates an error code and, optionally, text describing the protocol error
EventBuffer	Describes events for the media gateway to detect while event buffering is used
Events	Describes events for the media gateway to detect and instructions about what to do when that happens
Local	Describes the media flow properties of flows received from a remote end
LocalControl	Describes properties that are of interest to the media gateway and the media gateway controller
Media	Specifies the parameters for all media streams
Modem	Identifies the modem type and other properties for a termination
Mux	Describes the multiplex type and identifies the terminations forming the input mux
ObservedEvents	Reports observed events
Packages	Identifies the properties, events, signals and statistics of a termination, which are grouped into packages
Remote	Describes the media flow properties of flows sent to a remote end
ServiceChange	Describes the details of a service change
Signals	Describes signals applied to terminations
Statistics	Describes the statistics kept about a termination
Stream	A list of descriptors for a single stream, consisting of Local, Remote and LocalControl descriptors
TerminationState	Describes the properties of a termination
Topology	Describes the topology of a context (i.e., flow directions between terminations)

17

COPS

This chapter makes an overview of the Common Open Policy Service (COPS) [RFC2748] protocol and its use in supporting policy provisioning (COPS-PR) [RFC3084]. It also describes the specific extensions to COPS-PR used in the Go interface, as explained in Section 2.3.17.

17.1 Introduction

The COPS protocol is an Internet Engineering Task Force (IETF) protocol used for the general administration, configuration and enforcement of policies. It defines a simple query and response protocol for exchanging policy information between a policy server and its clients. The clients are denoted as policy enforcement points (PEPs) and the server as a policy decision point (PDP),[1] respectively. The protocol employs a client/server model in which a PEP sends requests, updates and deletes to the PDP, which in turn returns policy decisions back to the PEP. A special type of PDP is the local policy decision point (LPDP), which is used by PEPs to request local policy decisions when there is no available PDP to communicate with. Figure 17.1 illustrates the model.

There are two main models for COPS policy control:

- Outsourcing—the PEP assigns (outsources) responsibility of authorizing certain events at the PEP to an external entity (PDP). This model assumes a one-to-one correlation between events at a PEP and decisions from a PDP.

- Configuration—unlike the previous model, there exists no direct mapping between events at the PEP and decisions from the PDP. The PDP may proactively configure the PEP based on any external events as well as events

---

[1] In this chapter PDP stands for policy decision point and should not be confused with Packet Data Protocol.

The IMS. Miikka Poikselkä, Georg Mayer, Hisham Khartabil and Aki Niemi
Copyright 2004 by John Wiley & Sons, Ltd. ISBN 0-470-87113-X

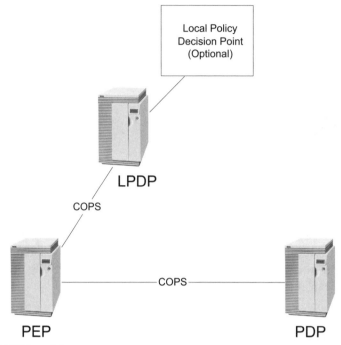

Figure 17.1 COPS model.

originating at the PEP. This may be performed by the PDP in bulk or in portions, but the overall timing is more flexible than the outsourcing model's.

The COPS protocol uses persistent Transmission Control Protocol (TCP) connections between the PEPs and the PDP for reliable transport of protocol messages. The COPS protocol is statefull since the request/decision state is shared between the client and the server. Also, unlike many other client/server protocols, the state of a given request/decision pair can be associated with another request/decision pair. The server is also able to push state to the client and remove it once it is no longer valid. COPS has a built-in extensibility: policy objects are self-identifying and support vendor-specific objects. Extensions may describe the message formats and objects that carry the policy data without requiring any changes to the protocol itself.

17.2 Message structure

COPS is a binary protocol whose messages consist of the COPS header followed by typed objects. The COPS header (Figure 17.2) consists of the following fields:

- Version—indicates the COPS version number (its default is 1).

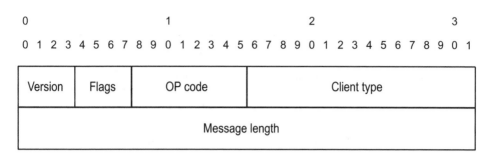

Figure 17.2 COPS header.

- Flags—denotes certain message flags. The only one currently defined is 0x1, which indicates whether the message was solicited by another COPS message.

- OP code—denotes the COPS operation. Table 17.1 lists the defined COPS operation codes.

- Client type—identifies the policy client and determines the interpretation of the accompanying typed objects, with the exception of a Keep-Alive message that always has client type set to zero.

- Message length—denotes the octet length of the message.

The format of COPS-specific objects is illustrated in Figure 17.3, and consists of:

- Length—object length in octets.

- C-Num—typically denotes the class of the information contained in the object.

- C-Type—typically identifies the subtype or version of the information contained in the object.

The remainder of the object encapsulates actual object information, which varies from object to object. COPS-specific objects are described in Table 17.2. Objects related to the use of COPS for provisioning are described in the following sections.

17.3 COPS usage for policy provisioning (COPS-PR)

COPS-PR defines COPS usage for policy provisioning and is independent of the provisioned policy. The PEP describes its configurable parameters in policy requests, denoted as configuration requests. If there is a change in these parameters, an update request is sent. The requests do not necessarily map directly to any decisions, which are typically issued by the PDP only when the PDP responds to external events: for

Table 17.1 COPS operation codes.

Value	Short	Where	Name	Description
1	REQ	PEP → PDP	Request	Requests a decision from a PDP and establishes a client handle that identifies the specific state associated with this PEP
2	DEC	PDP → PEP	Decision	Returning one or more decisions as a response to a Request
3	RPT	PEP → PDP	Report state	Reports back to the PDP the PEP's success or failure in carrying out the PDP's decision or reports a change in state
4	DRQ	PEP → PDP	Delete request state	Indicates to the PDP that the state associated with the client handle is no longer relevant
5	SSQ	PDP → PEP	Synchronize state Request	Indicates that the remote PDP wishes the PEP to re-send its state
6	OPN	PEP → PDP	Client-Open	Specifies the PEP's supported client types, the last PDP connection for a given client type and client-specific feature negotiation
7	CAT	PDP → PEP	Client-Accept	Responds positively to the Client-Open message and returns a timer object for the Keep-Alive timer
8	CC	PEP → PDP, PDP → PEP	Client-Close	Indicates that a particular client type is no longer supported
9	KA	PEP → PDP, PDP → PEP	Keep-Alive	Resets the Keep-Alive timers
10	SSC	PEP → PDP	Synchronize complete	Reports successful completion of a state synchronization

example, the PDP would issue a configuration decision only when a service-level agreement (SLA) is changed. Hence, both policy requests and decisions may be fairly infrequent.

The data model in COPS-PR is based on something called a policy information base (PIB). Policy-provisioning data are identified by a PIB, and each area of policy provisioning may have one or more PIBs defined. In other words, a given PEP–PDP connection may see multiple PIBs in transit, but each PIB would be associated with a specific client type.

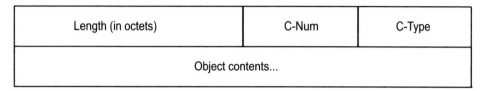

Figure 17.3 COPS-specific objects.

Table 17.2 COPS-specific object description.

C-Num	Name	Where	Description
1	Handle	Most	Unique value to identify an installed state
2	Context	REQ, DEC	Type of event that triggered a particular query
3	In interface	REQ	The address and interface identifier for the incoming interface on the PEP
4	Out interface	REQ	The address and interface identifier for the outgoing interface on the PEP
5	Reason code	DRQ	The reason for deleting the request state
6	Decision	DEC	Decision made by the PDP
7	LPDP decision	DEC	Decision made by the LPDP
8	Error	CC	Identifies a particular protocol error
9	Client-specific info	REQ, DEC, RPT, OPN	Client-specific information
10	Keep-Alive timer	CAT	Keep-Alive time value
11	PEP identification	OPN	Identifies the PEP client to the PDP
12	Report type	RPT	The type of report on the request state, which is associated with a specific handle
13	PDP redirect address	CC	Address to which the PDP may optionally redirect the PEP
14	Last PDP address	OPN	Address of the PDP the PEP last connected
15	Accounting timer	CAT	Optionally indicates the timer value between periodic reports of accounting type
16	Message integrity	Any	Sequence number and a message authentication checksum used for integrity protection of the requests

As the name suggests, it is quite similar to the management information base (MIB) [RFC3418] used for the Simple Network Management Protocol (SNMP) [RFC3411]. In fact, they share a common data structure, as both use Structure for Management Information Version 2 (SMIv2) definitions. In COPS-PR this is used for Encoded Provisioning Instance Data (EPID) encoding rules.

The PIB is conceptually structured as a tree, with the PIB being the root. As such a PIB represents a virtual database for the contents of the tree. Each sub-tree represents a provisioning class (PRC), whereas the leaves represent instantiations of this class (namely, provisioning instances, or PRIs). There may be multiple PRIs per PRC and there may be multiple PRCs associated with a PIB. Each PRI has an identifier, a provisioning instance identifier (PRID), which is the name carried by the COPS object that identifies that particular instance of a provisioning class.

17.4 The PIB for the Go interface

As described in Section 2.3.17, the Go interface uses the COPS protocol to perform media authorization and charging correlation. It also employs COPS-PR extensions. To transmit Go-specific data, the Third Generation Partnership Project (3GPP) has defined a PIB for the Go interface. The details of the 3GPP Go PIB can be found in [3GPP TS 29.207].

17.5 Summary

The COPS protocol is an IETF protocol used for general administration, configuration and enforcement of policies. COPS-PR extensions define COPS usage for policy provisioning as well as the concept of a PIB. The 3GPP Go defines such a PIB for representing information exchanged over the Go interface.

18

IPsec

18.1 Introduction

The Internet Protocol Security (IPsec) provides for various security services on the IP layer, in IPv4 as well as IPv6, thus offering protection for protocols in the upper layers. IPsec is typically used to secure communications between hosts and security gateways. The set of security services that IPsec provides includes:

- access control;

- data integrity protection;

- data origin authentication;

- anti-replay protection;

- confidentiality;

- limited traffic flow confidentiality.

IPsec can operate in two modes: tunnel mode and transport mode. Transport mode is mainly used to provide security services for upper layer protocols. Tunnel mode is typically used to tunnel IP traffic between two security gateways. The difference is that in transport mode IPsec offers limited protection to IP headers, whereas in tunnel mode the full IP datagram is protected. More specifically, in the case of an Encapsulated Security Payload (ESP) there are no security services provided for the IP headers that precede the ESP header, whereas an Authentication Header (AH) does extend protection to some parts of the IP header.

The components of the IPsec security architecture [RFC2401] are:

- Security protocols—the AH [RFC2402] and the ESP [RFC2406].

- Security associations—definition of the Security Policy Database (SPD) and the Security Association Database (SAD), as well as the management and usage of security associations.

The IMS. Miikka Poikselkä, Georg Mayer, Hisham Khartabil and Aki Niemi
Copyright 2004 by John Wiley & Sons, Ltd. ISBN 0-470-87113-X

- Key management—the distribution of cryptographic keys for use with the security protocols (namely, the Internet Key Exchange, or IKE [RFC2409]).

- Algorithms used for encryption and authentication.

In the following sections we will make an overview of each of these components.

18.2 Security associations

The concept of a security association is germane to IPsec. Security protocols make use of security associations (SAs) as they provide security services (the main responsibility of key management is to establish and manage SAs). An SA is a relationship between two entities that defines how they are going to use security services to secure their communications. It includes information on authentication and/or encryption algorithms, cryptographic keys and key lengths as well as the initialization vectors (IV) that are shared between the entities. An SA is unidirectional; so, typically two SAs are needed for a bidirectional flow of traffic—one for inbound (read) traffic and one for outbound (write) traffic. An SA is uniquely identified by the following three items:

- security parameter index (SPI);

- destination IP address;

- security protocol (either AH or ESP).

The management of SAs involves two databases: the SPD and the SAD. The SPD contains the policies by which all inbound and outbound traffic is categorized on a host or a security gateway. The SAD is a container for all active SAs, and related parameters. A set of selectors—IP layer and upper layer (e.g., TCP and UDP) protocol field values—is used by the SPD to map traffic to a specific SA.

18.3 Internet Security Association and Key Management Protocol (ISAKMP)

ISAKMP is used for negotiating, establishing, modification, and deletion of SAs and related parameters. It defines the procedures and packet formats for peer authentication creation and management of SAs and techniques for key generation. It also includes mechanisms that mitigate certain threats (e.g., denial-of-service, or DOS, and anti-replay protection).

In ISAKMP, SA and key management is separate from any key exchange protocols; so, in a sense ISAKMP is an "abstract" protocol—it provides a framework for authentication and key management, and supports many actual key exchange protocols (e.g., IKE). ISAKMP defines header and payload formats, but needs an instantiation to a specific set of protocols. Such an instantiation is denoted as the ISAKMP Domain of Interpretation (DOI): an example of this for the IPsec/IKE is the IPsec DOI [RFC2407].

ISAKMP operates in two phases. During phase 1 the peers establish an ISAKMP SA (namely, they authenticate and agree on the used mechanisms to secure further communications). In phase 2 this ISAKMP SA is used to negotiate further protocol SAs (e.g., an IPsec/ESP SA). After the initial establishment of an ISAKMP SA, multiple protocol SAs can be established.

18.4 Internet Key Exchange (IKE)

The IKE is a key exchange protocol which, in conjunction with ISAKMP, negotiates authenticated keying material for SAs. IKE can use two modes to establish a phase 1 ISAKMP SA: main mode and aggressive mode. Both modes use the Ephemeral Diffie–Hellman key exchange algorithm[1] to generate keying material for the ISAKMP SA. The difference between these modes is that in the main mode, while consuming more message round trips, the identities of the negotiating entities are protected, whereas in aggressive mode they are revealed to the outside world. After establishing the ISAKMP SA in phase 1, protocol SAs can be negotiated, while negotiation is secured using the ISAKMP SA.

18.5 Encapsulated Security Payload (ESP)

The ESP is used to provide security services in IPv4 and IPv6. It can be used alone or in unison with an AH. It can provide either confidentiality (i.e., encryption) or integrity protection (i.e., authentication), or both. As mentioned previously, the ESP can operate in transport mode and in tunnel mode.

The ESP header is inserted into the IP datagram after the IP header and before any upper layer protocol headers in transport mode, or before an encapsulated IP datagram in tunnel mode. Figure 18.1 illustrates the ESP packet format.

The fields in the ESP header are as follows:

- The SPI—a unique and random 32-bit value that, together with the destination IP address and security protocol, uniquely identifies the SA for the packet.

---

[1] In the Ephemeral Diffie-Hellman algorithm the public key (out of the temporary D–H key pair) is tasked with guarding against man-in-the-middle (MITM) attacks.

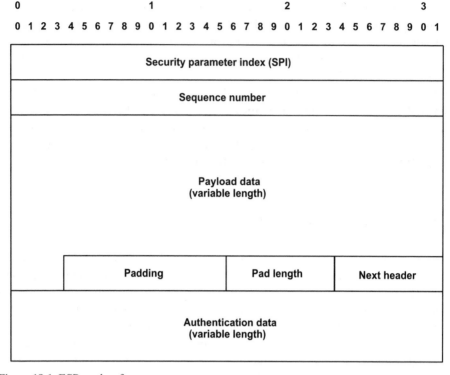

Figure 18.1 ESP packet format.

- The sequence number—a monotonically increasing 32-bit counter used to protect against replay attacks. When an SA is established the sequence number is reset to zero.

- Payload data—a variable length field that typically contains the data payload, whose type is denoted by the next header field. It may also contain cryptographic synchronization data, such as an IV.

- Padding—used to fill the payload data to a specific block size multiple required by a particular encryption algorithm, or to randomize the length of the payload in order to protect against traffic flow analysis.

- Pad length—an 8-bit field whose value indicates in bytes the length of the padding field.

- Next header—an 8-bit field whose value indicates the type of data contained in the payload datafield.

- Authentication data—a variable length field containing an integrity check value (ICV), which is computed (using an authentication algorithm) from the rest of the ESP packet, to provide data integrity protection.

To process outbound traffic, a host or security gateway first uses a set of selectors in the SPD to determine the outbound SA used. It then follows a set of steps to process the outbound packet:

1. Either the entire original outbound IP datagram is encapsulated in an ESP payload field (tunnel mode) or just the original upper layer protocol information from the outbound IP datagram is encapsulated (transport mode).

2. Appropriate padding is added to the payload data.

3. The results are encrypted using an encryption key and an algorithm.

4. The sequence number is incremented as appropriate.[2]

5. If authentication is enabled, then the ICV is calculated.

6. Possible fragmentation of the IP datagram is performed.

On receiving an IP datagram the recipient follows the following steps to process the packet:

1. Possible reassembly of the IP datagram is performed.

2. Using the SPI, security protocol and destination IP address, an appropriate SA is looked up from the SAD.

3. If anti-replay protection is enabled, the sequence number is inspected.

4. If authentication is enabled, then the ICV is verified.

5. The packet is decrypted, padding is removed and the original IP datagram is reconstructed.

18.6 Summary

IPsec provides security services in the IP layer, in both IPv4 and IPv6, offering protection for protocols at higher layers. The IKE is used for key exchange—creating and managing SAs and related security parameters—and the ESP is used for confidentiality and integrity protection.

---

[2] In case this was the first packet to be protected using a new SA, the sequence number would start from zero.

19

Signalling Compression

Signalling compression (SigComp) is a mechanism that application protocols use to compress messages before sending to the network. It is presented to applications as a layer between application protocols and transport protocols. SigComp uses a universal decompressor virtual machine (UDVM) to decompress messages. Figure 8.1 demonstrates SigComp's location in the protocol stack. SigComp is defined in [RFC3320].

A message that is compressed using SigComp is referred to as a SigComp message.

19.1 SigComp architecture

SigComp's architecture can be broken down into five entities (see Figure 19.1). The entities are described as follows:

- *Compressor dispatcher*—this is the interface between the application and the SigComp system. It invokes a compressor, indicated by the application using a compartment identifier. The compressor dispatcher forwards the returned compressed message to its destination.

- *Decompressor dispatcher*—this is the interface between the SigComp system and the application. It invokes a UDVM that decompresses the message. The decompressor dispatcher then passes the decompressed message to the application. If the application wishes the decompressor to retain the state of the message, it returns what is called a compartment identifier.

- *Compressors*—this entity compresses the application message. It uses a compartment that is identified using the compartment identifier. The compressed message is passed to the compressor dispatcher. DEFLATE is an example of a compression algorithm.

The IMS. Miikka Poikselkä, Georg Mayer, Hisham Khartabil and Aki Niemi

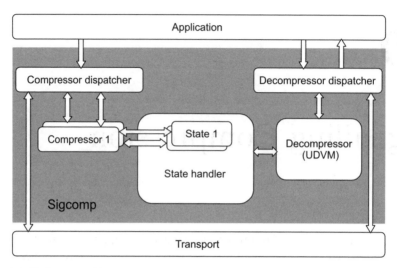

Figure 19.1 SigComp architecture.

- *UDVM*—This entity decompresses a compressed message. A new instance is invoked for every new SigComp message. UDVM uses the state handler to create a state for a new message or make use of an existing state.

- *State handler*—this holds information that is stored between SigComp messages (referred to as the state of the messages). It can store and recover the state.

19.2 Compartments

An application like a Session Initiation Protocol (SIP) can group messages that are related. Depending on the signalling protocol, this grouping is done according to how the protocol relates messages: for instance, in SIP, messages are grouped using dialogs or, in cases where there is a configured next-hop address, messages are grouped by the same next-hop address. The application allocates a compressor per compartment and state memory in which to store state information. It also determines when a compartment should be created or destroyed. A compartment is identified uniquely using a compartment identifier.

It is the application's responsibility to determine the compartment identifier for a decompressed message. When an application receives a decompressed message, it determines or creates the compartment for the message and supplies the compartment identifier to the SigComp system.

19.3 Compressing a SIP message in IMS

19.3.1 Initialization of SIP compression

During the registration phase the user equipment (UE) and the Proxy Call Session Control Function (P-CSCF) announce their willingness to perform compression by providing details about their compression capabilities, such as memory size and processing power, upload states and compression instructions (see Section 8.13.15). Due to the strong security requirements in the Third Generation Partnership Project's (3GPP) Internet Protocol Multimedia Substem (IMS), announcements and state creations are only allowed after the security association has been established. Otherwise, a malicious user could upload false states that would make compression vulnerable.

19.3.2 Compressing a SIP message

When the UE or P-CSCF wants to send a compressed Session Initiation Protocol (SIP) message, it follows the framework described in [RFC3320], which states that a SIP application in the UE should pass a message to a compressor dispatcher. The compressor dispatcher invokes a compressor, fetches the necessary compression states, using a compartment identified by its ID and supplied by the application, and uses a certain compression algorithm to encode the message. Finally, the compressor dispatcher relays the compressed message to the transport layer to be delivered to the remote end (the P-CSCF).

The compressor is responsible for ensuring that the remote end can decompress the generated message. It is possible to include all the needed information in every SigComp message (i.e., every bytecode) to decompress the message. However, this would reduce the archived compression ratio significantly; so, it is better to ask the other end to create states. The information saved in these state items can then be accessed for future SigComp message decompression for messages that arrive from the same source and are related, avoiding the need to upload the data on a per-message basis.

19.3.3 Decompressing a compressed SIP message

When a decompressor dispatcher receives a message it inspects the prefix of the incoming message. As all SigComp messages contain a prefix (the five most significant bits of the first byte are set to one) the decompression dispatcher is able to identify that the message is compressed. This prefix does not occur in

UTF-8-encoded text messages. The decompressor dispatcher forwards the message to the Universal Decompressor Virtual Machine (UDVM). The UDVM requests previously created states from the state handler and uses the provided states (or provided bytecode in the message if no states exist) to decompress the message. After decompression the UDVM returns the uncompressed message to the decompression dispatcher which further passes the message to an application.

20

DHCPv6

The Dynamic Host Configuration Protocol for Internet Protocol Version 6 (DHCPv6), as defined in [RFC3315], is a client–server protocol that allows devices to be configured with management configuration information. DHCP specifically provides clients with dynamically assigned IP addresses using a DHCP server. It also provides other configuration information carried as options: options are extensions to DHCP. "Extending DHCP" means defining a new option.

DHCP messages are exchanged between clients and servers using User Datagram Protocol (UDP) as the transport protocol. Clients listen on port 546 for messages, while servers listen on port 547. A client uses a link-local address for transmitting and receiving DHCP messages. A client sends DHCP messages to servers using the link-scoped multicast address "FF02::1:2", also known as All_DHCP_Relay_Agents_and_Servers. DHCP relay agents on the client's link allow a DHCP client to send a message to a DHCP server that is not attached to the same link.

A client needs, first, to locate a DHCP server before requesting an IP address and other configuration information. It does so by sending a DHCP solicit message to the multicast address identified earlier. A server willing to answer the request answers with a DHCP advertise message. The client then chooses one of the servers and a DHCP request message. The server responds with a relay message confirming the assigned IP addresses and other configuration information. Because assigned IP addresses expire, the client needs to send a DHCP renew message to the server in order to extend the assigned IP address lifetime. Figure 20.1 presents the format of a client–server DHCP message. The fields are described as follows:

- Message type—this is the DHCP message type, such as solicit, advertise, request, confirm, renew, bind, reply, release, decline, reconfigure, information-request, relay-forw and relay-repl.

The IMS. Miikka Poikselkä, Georg Mayer, Hisham Khartabil and Aki Niemi

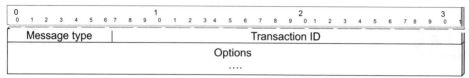

Figure 20.1 Client–Server DHCP message format.

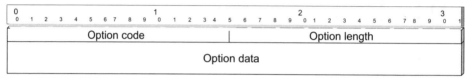

Figure 20.2 DHCP options format.

- Transaction ID—an ID for the message exchange.

- Options—zero or more options.

20.1 DHCP options

As described earlier, options are used to hold additional information. The format of DHCP options is depicted in Figure 20.2. The fields are described as follows:

- Option code—the specific option type carried in this option.

- Option length—the length, in octets, of the option data.

- Option data—the data for the option.

20.2 DHCP options for SIP servers

[RFC3319] specifies two DHCPv6 options that allow Session Initiation Protocol (SIP) clients to locate their outbound SIP proxy server. One option carries a list of SIP server domain names and the other carries a list of 128-bit IPv6 addresses.

The "SIP server domain name list" option can carry multiple domain names. Each domain name has a naming authority pointer (NAPTR) record that differs from any other. The client tries the records in the order listed, as specified in Section 8.12.1.1, trying the next record if the previous one fails.

The option is populated with the following:

- Option code—OPTION_SIP_SERVER_D (21).

- Option length—variable.

- Option data—SIP server domain name list.

The "SIP server IPv6 address list option" can carry multiple IP addresses. Servers are listed in order of preference.

The option is populated with the following:

- Option code—OPTION_SIP_SERVER_A (22).

- Option length—variable.

- Option data—IPv6 addresses of SIP servers.

21

XCAP

In many services today the service provider needs—in order to carry out a request—to have access to information that can only be set by users. Such services include presence, messaging and conferencing.

In XCAP a user is able to upload information to an XCAP server, which provides this uploaded information to application servers that use this information to satisfy a request demanded by the user. With XCAP, the user is also allowed to manipulate, add and delete such data. An example of the data that a user can upload is the user's resource list for presence (defined in Section 23.3).

XCAP uses the Hyper Text Transfer Protocol (HTTP) to upload and read the information set by users and the Information is represented using XML (Extensible Markup Language).

21.1 XCAP application usage

Applications, like presence, need to define an XCAP application usage, which defines the way that a unique application can make use of XCAP. It defines the XML document for the application, along with the following:

- Application usage ID (AUID)—this ID uniquely identifies an application usage.

- Additional constraints—these cover data constraints that are not possible to represent using an XML schema.

- Data semantics—the semantics for each element and attribute in an application usage XML document.

- Naming convention—defines how an application constructs the URI (uniform resource identifier) representing the document that is to be read or written by the application so that it can carry out its tasks.

The IMS. Miikka Poikselkä, Georg Mayer, Hisham Khartabil and Aki Niemi
Copyright 2004 by John Wiley & Sons, Ltd. ISBN 0-470-87113-X

- Resource interdependencies—in some XCAP application usages, many entities need to contribute to populating the XML document element and attribute values to make the XML document ready. This interdependency is specified in a section of the XCAP application usage document.

- Authorization policy—By default, the creator of the XML document is the sole resource with access to the document. Application usages can define a different authorization policy to override the default.

Examples of application usages can be seen in Sections 23.4 and 23.5.

22

CPCP

The Conference Policy Control Protocol (CPCP) is a client–server protocol that can be used by users to manipulate the rules associated with the conference. These rules include directives on the lifespan of the conference, who can and cannot join the conference, definitions of roles available in the conference and the responsibilities associated with those roles, and policies on who is allowed to request which roles.

The conference policy is represented by a URI (uniform resource identifier). There is a unique conference policy URI for each conference, which points to a conference policy server where a user can manipulate that conference policy. Note that CPCP is not the only mechanism to manipulate conference policy—other mechanisms exist as well, such as the Web interface. CPCP allows clients to manipulate the conference policy at the conference policy server (CPS), which is able to inform the focus about changes in conference policy, if necessary: for example, if new users are added to the dial-out list, then the conference policy server informs the focus, which makes the invitations as requested.

CPCP can specifically be used to create conferences, terminate conferences, define start and stop times, define the dial-out and dial-in lists, bring about conference security and give out general conference information. A conference policy also includes the conference media policy and floor control.

CPCP allows the creator of a conference to:

- determine the lifespan of the conference;

- decide who can and who cannot join the conference;

- define the dial-out and dial-in lists;

- define the roles available in the conference and the responsibilities associated with those roles;

The IMS. Miikka Poikselkä, Georg Mayer, Hisham Khartabil and Aki Niemi
Copyright 2004 by John Wiley & Sons, Ltd. ISBN 0-470-87113-X

- give out general conference information;

- set up media policy;

- set up conference security.

XCAP can be used as a protocol for CPCP as defined in [Draft-ietf-xcon-cpcp]. Chapter 25 defines conferencing in more detail.

Part IV

Services

23

Presence

Presence is in essence two things: it involves making my status available to others and the statuses of others available to me. Presence information may include:

- person and terminal availability;

- communication preferences;

- terminal capabilities;

- current activity;

- location.

It is envisioned that presence will facilitate all mobile communication, not only instant messaging, which has been the main driver for presence. Instant messaging has been the main interactive, almost real time communication service in the Internet and presence is a great compliment in that you know if a friend is online before you begin a chat session with her. However, in the mobile environment, it is envisioned that presence information will not only support instant messaging, it will also be used as an indicator of the ability to engage in any session, including voice calls, video and gaming: all mobile communications will be presence-based.

Presence-specific and presence-enhanced applications and services will be available in the near future. A typical example of a presence-specific application will be a phonebook with embedded presence information, making it dynamic. Dynamic presence (Figure 23.1) will be the initial information the user sees before establishing communication. This information affects the choice of communication method and timing.

The IMS. Miikka Poikselkä, Georg Mayer, Hisham Khartabil and Aki Niemi
Copyright 2004 by John Wiley & Sons, Ltd. ISBN 0-470-87113-X

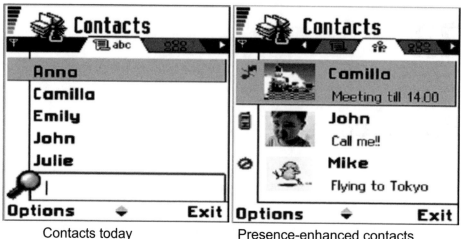

Contacts today   Presence-enhanced contacts

Figure 23.1 Dynamic presence.

23.1 SIP for presence

The Session Initiation Protocol (SIP) has been extended for presence by the creation of an event package called "presence". Event packages are described in Section 8.13.1. When subscribing to such an event, a subscriber places the "presence" token in the Event header.

Some definitions have been created to describe the subscriber and the notifier for the purpose of presence:

- *Presentity*—the presence entity, a resource that provides presence information to a presence service.

- *Watchers*—entities that request information (presence), about resources (presentities).

Two SIP entities are defined for presence in [Draft-ietf-simple-presence]:

- *Presence agent (PA)*—capable of storing subscriptions and generating notifications.

- *Presence user agent (PUA)*—manipulates presence information for a presentity and publishes such presence information.

As mentioned earlier, the NOTIFY request body carries the state information. In this case, the state information is the presence state of a presentity. The Multipurpose Internet Mail Extension (MIME) type of such content is "application/ pidf + xml" as defined in [Draft-ietf-impp-cpim-pidf]. This presence XML (Extensible Markup Language) document can be extended to carry more information than defined in [Draft-ietf-impp-cpim-pidf].

The presentity uploads presence information using the PUBLISH method. The PUBLISH method is described in Section 8.13.2.

23.2 Presence service architecture in IMS

A user's presence information can be obtained from a multiplicity of entities in the Internet Protocol Multimedia Subsystem (IMS) network: it could be a PUA located in a foreign network, a PUA at the terminal or a PUA located as an entity in the network. The presence server is an example of an IMS application server. Watchers may be in the same home domain as the presentity or in a foreign domain.

Figure 23.2 represents a reference architecture to support a presence service in the IMS. The entities are defined as follows:

- *Presence server*—manages presence information uploaded by PUAs and handles presence subscription requests.

- *Watcher presence proxy*—identifies the target network for a presentity and resolves its address.

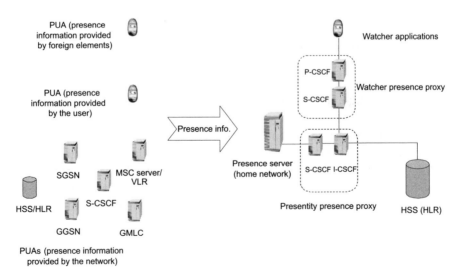

Figure 23.2 Reference architecture to support a presence service in the IMS.

- *Presentity presence proxy*—Identifies the presence server assigned to a certain presentity.

- *Presence user agents*—assemble and provide presence information to the server.

23.3 Resource (presentity) list

It is envisioned that users (watchers) will have many presentities (friends) whose presence information is of interest to them. For congestion control and bandwidth limitations, it is uneconomical to have the user's terminal send a multiplicity of SUBSCRIBE requests, one for each presentity.

To resolve this problem, [Draft-ietf-simple-event-list] describes an event notification extension that allows users to subscribe to a list of resources with a single SUBSCRIBE request. The list is identified by a uniform resource identifier (URI) and contains zero or more URIs pointing to atomic resources or to other lists. In a presence system these resources are presentities. The entity that processes the list SUBSCRIBE request is referred to as the resource list server (RLS). The RLS can generate individual subscriptions for each resource in the list. These subscriptions may or may not be SIP SUBSCRIBE requests. A client sending SUBSCRIBE requests to a list includes a Supported header with an option tag of value "eventlist". If this option tag is not included and the URI in the request-URI represents a list, the RLS returns a "421 Extension Required" error response.

If the subscription is accepted, then the RLS generates NOTIFY requests carrying state information about the list. The NOTIFY request contains a Require header with an option tag of value "eventlist". It also contains a body of MIME type "multipart/related", which internally carries a MIME type of "application/rlmi + xml" that holds the resource list meta-information.

23.4 XCAP usage for resource (presentity) lists

Section 23.3 discussed resource (presentity) lists and how to gain state information about these lists. This section discusses how these lists are created and maintained.

The list creation solution takes advantage of the XML Configuration Access Protocol (XCAP) (see Chapter 21) as an application usage. [Draft-ietf-simple-xcap-list-usage] defines the XML schema along with its semantics. It also defines the following:

- Application usage ID (AUID)—"resource-lists".

- Additional constraints—none.

- Naming convention—none.

- Resource interdependencies—the list is represented by a URI. If the client does not populate the XML element carrying the URI value, the XCAP server needs to do so.

- Authorization policy—default.

The Ut interface in the IMS architecture is used to manipulate resource lists.

23.5 Setting presence authorization

Presence information can be available at different levels and different scopes to different watchers. This means that different watchers may be authorized to view different parts of the presence information of a presentity. The choice of who sees what belongs to the presentity. The presentity can set such authorization levels using an XCAP-defined solution in the form of permission statements.

[Draft-ietf-geopriv-commom-policy], [Draft-rosenberg-simple-rules], [Draft-rosenberg-simple-commom-policy-caps] and [Draft-rosenberg-simple-pres-policy-caps] define the XML schema along with its semantics as well as those sections that are mandatorily defined for XCAP usage.

23.6 Publishing presence

Publishing presence information can be achieved using the SIP extension defined in Section 8.13.2. The Event header of a PUBLISH request carries the "presence" token. The default expiration time of a publication is 3,600 seconds. The body of a PUBLISH request carrying presence information is of MIME type "application/pidf + xml".

23.7 Watcher information event template package

Users subscribe to receive information about the state of a particular resource using an event package (referred to as the main package in this section). These subscribers can be referred to as watchers. A watcher information template package allows a user to gain knowledge about watchers and the state of their subscriptions to the main package. A watcher information template package is identified with the token "winfo".

The primary use of this template package is for presence. Users subscribe to this template package to see who is subscribing to their presence information and the state of that subscription.

The information carried to the watcher information subscriber contains two important items: the status of each subscription made by the watchers of the main package and the event that caused the transition from the previous status to the current one.

The states of the watcher information package are described in [Draft-ietf-simple-winfo-package] as follows:

- Init—no state is allocated for a subscription.

- Terminated—a policy exists that forbids a watcher from subscribing to the main event package.

- Active—a policy exists that authorizes a watcher to subscribe to the main package.

- Pending—no policy exists for that watcher.

- Waiting—similar to pending, but tells the template package subscriber that a user has attempted to subscribe to the main package and that the subscription expired before a policy was created.

23.8 Example signalling flows of presence service operation

23.8.1 Successful subscription to presence

Figure 23.3 shows an example flow of a watcher who has successfully subscribed to the presence information of a presentity residing in a different network while the

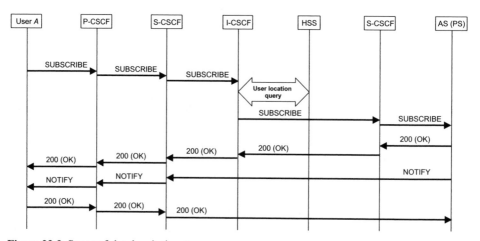

Figure 23.3 Successful subscription to presence.

watcher resides in her home network. The flow shows an initial NOTIFY request to the SUBSCRIBE request.

23.8.2 Successful publication of presence information

Figure 23.4 shows an example flow of a presentity that has successfully published presence information. In this scenario the user equipment (UE) is behaving as a PUA. This typically results in the packet-switched (PS) domain generating notifications to the watchers.

23.8.3 Subscribing to a resource list

Figure 23.5 shows an example flow of a watcher who is subscribing to the presence information of a resource list that was created earlier (possibly using XCAP). The list of presentities (resources) is referenced using a SIP URI. The flow shows the immediate NOTIFY request that is sent after receiving the SUBSCRIBE request.

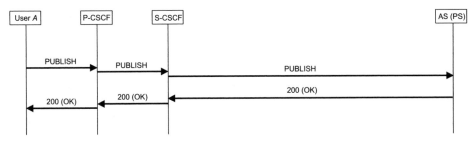

Figure 23.4 Successful publication.

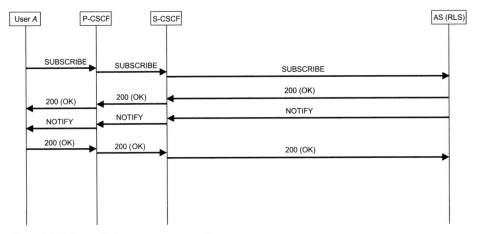

Figure 23.5 Subscription to a resource list.

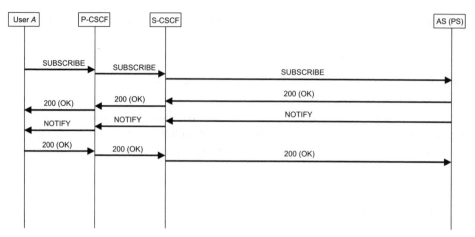

Figure 23.6 Subscription to watcher information.

If the RLS does not hold any presence information about the resources in the list, then the NOTIFY request may carry an empty body.

23.8.4 Subscribing to watcher information

Figure 23.6 depicts how the message actually flows between the PUA and the PS domain when a user subscribes to receive notifications about the changes of state of the watchers. The path of the messages is the same as that of the PUBLISH request. The flow shows the immediate notification following a successful subscription request.

24

Messaging

There are currently many forms of messaging services available. In general, messaging entails sending a message from one entity to another. Messages can take many forms, include many types of data and be delivered in various ways. It is usual to have messages carry multimedia as well as text and be delivered either in near-real time as in many instant messaging systems or into a mailbox as in email today. In this chapter we give some details about messaging in the Internet Protocol Multimedia Subsystem (IMS) context.

24.1 Overview of IMS messaging

IMS messaging takes three forms:

- immediate messaging;
- session-based messaging;
- deferred delivery messaging.

Each form of IMS messaging has its own characteristics; so, even though messaging in its simplest form can be thought of as a single service—after all, all forms of messaging are really about sending a message from A to B—the fact that these characteristics differ makes them each a service on their own. However, the way in which applications are built on top of these services may well hide the fact that these are different forms of messaging. In fact, one of the key requirements for IMS messaging is easy interworking between different messaging types.

The IMS. Miikka Poikselkä, Georg Mayer, Hisham Khartabil and Aki Niemi
Copyright 2004 by John Wiley & Sons, Ltd. ISBN 0-470-87113-X

24.2 IMS messaging architecture

Of the three IMS messaging types, immediate messaging and session-based messaging utilize the IMS architecture directly. Deferred delivery messaging utilizes the packet-switched (PS) domain as well, even though it is a separate infrastructure to the IMS.

24.3 Immediate messaging

Immediate messaging is the familiar instant messaging paradigm adopted in the IMS framework. It uses the Session Initiation Protocol (SIP) MESSAGE method (see Section 8.13.3 for details) to send messages between peers in near-real time. Figure 24.1 illustrates a typical message flow.

In immediate messaging, the user equipment (UE) simply generates a MESSAGE request, fills in the desired content—which typically consists of text, but can also contain snippets of multimedia such as sounds and images—and populates the request-URI (Uniform Resource Identifier) with the address of the recipient. The request is then routed via the IMS infrastructure similar to the manner used for an INVITE, until the immediate message finds its way to the UE of the recipient user.

There might of course be a reply to this message; in fact, a full dialogue of immediate messages back and forth between the two users is likely. However, in contrast to session-based messaging, the context of this session only exists in the minds of the participating users. There is no protocol session involved: each immediate message is an independent transaction and is not related to any previous requests.

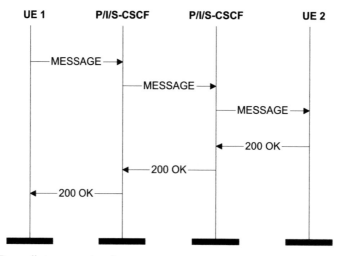

Figure 24.1 Immediate messaging flow.

24.4 Session-based messaging

Session-based messaging relates to a familiar paradigm of messaging already in use in the Internet: Internet Relay Chat (IRC) [RFC2810]. In this mode of messaging the user takes part in a session in which the main media component often consists of short textual messages. As in any other session a message session has a well-defined lifetime: a message session starts when the participants begin the session and stops when the participants close the session. After the session is set up between the participants, media then flows directly from peer to peer. Figure 24.2 illustrates the typical message flow of a message session.

Session-based messaging can be peer to peer, in which case the experience closely mimics that of a normal voice call. An ordinary invitation to a session is received, the only difference being that the main media component is a session of messages. However, this is not an actual limitation to session-based messaging, since it is of course possible to combine other media sessions with message sessions. In fact, many useful and exciting applications are enabled by this functionality: for example, video calls with a text side channel might be a valuable application for hearing-impaired people.

Session-based messaging forms a natural unison with conferencing as well. Using the conferencing functionality, session-based messaging can turn into a multi-party chat conference. In this mode of operation, session-based messaging can enable applications similar to modern day voice conferences. A chat conference can also be comparable with a channel in the IRC[1] or a chat group in a typical Internet chat service. A service provider will typically offer the possibility for users to have both private chats, where the set of participants is restricted, and public chats, some of which are maintained by the service provider.

24.5 Deferred delivery messaging

Deferred delivery messaging is in fact the well-known Multimedia Messaging Service (MMS). It was decided that the requirements put forth in the Third Generation Partnership Project's (3GPP) IMS Messaging Stage 1 should coincide with those of the MMS.

In Release 6, work is ongoing to define tighter integration of the MMS with the IMS, especially with regards to addressing and using SIP as a way to notify the UE of received MMs.

---

[1] In this respect a channel such as #Helsinki on an IRC server could be simply represented by a SIP URI: sip:helsinki@some.chat.net.

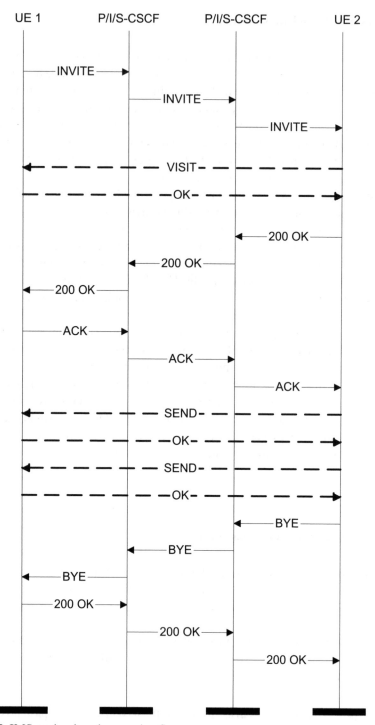

Figure 24.2 IMS session-based messaging flow.

25

Conferencing

A conference is a conversation between multiple participants. There are many different types of conferences, including loosely coupled conferences, fully distributed multiparty conferences, tightly coupled conferences. In this chapter only the latter is described since it is the only one that is of interest to the Internet Protocol Multimedia Subsystem (IMS).

Conferencing is not just limited to audio; the popularity of video and text conferencing, better known as chatting, has been growing rapidly over the past few years. This popularity is due to conferencing's ability to simulate a face-to-face meeting in so many realistic ways: for example by enabling file and whiteboard sharing and conveying emotions using video, all in real time.

25.1 Conferencing architecture

In tightly coupled conferences, there is always a central point of control where each conference participant has a connection. This central point provides a variety of conference services including media mixing, transcoding and participant list notifications.

The central point is referred to as a "focus" a Session Initiation Protocol user agent (SIP UA) that is addressed by a conference uniform resource identifier (a SIP URI). The conference URI identifies a conference and the focus sets up a signalling dialog between each participant.

A conference policy is a set of rules associated with a certain conference. These rules include directives on the lifespan of the conference, who can and who cannot join the conference (membership policy), definitions of roles available in the conference and the responsibilities associated with those roles, and policies on who is allowed to request which roles. The conference policy also includes the media policy: the mixing characteristics of a conference.

There are many ways to create a conference: one method uses SIP, which creates *ad hoc* conferences. *Ad hoc* conferences are ones that are created without scheduling

The IMS. Miikka Poikselkä, Georg Mayer, Hisham Khartabil and Aki Niemi
Copyright 2004 by John Wiley & Sons, Ltd. ISBN 0-470-87113-X

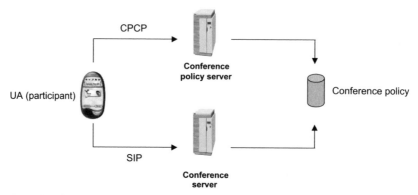

Figure 25.1 Conferencing architecture.

and are short-lived. Scheduled conferences are created using the Conference Policy Control Protocol (CPCP) (see Chapter 22) and give the creator more control over the policy.

A conference factory URI (a conference factory is a conference server that creates a new conference instance and assigns it a URI when it receives an INVITE request addressed to it using the conference factory URI) can be allocated and globally published in order to automatically create an *ad hoc* conference, using SIP call control means (INVITE request): a conference factory URI is globally routable.

CPCP is a client–server protocol that can be used by users to manipulate the rules associated with the conference. The conference policy is represented by a URI as is unique to each conference. The conference policy URI points to a conference policy server that can be used to manipulate that conference policy.

Figure 25.1 presents an architecture for conferencing. It identifies the interfaces between entities and the protocols used between them. The UA in the figure identifies a participant, who creates an *ad hoc* conference or joins a conference created using CPCP by sending a SIP INVITE request. In either case, the conference server spawns a focus.

25.2 SIP event package for conference state

This is yet another event notification package (see Section 8.13.1). The conference-state event package is used to learn about changes in the conference participants: in other words, a user can learn, through notifications, who has joined or left a conference. This event package also allows participants to learn the status of a user's participation in a conference and the sidebars in the conference.

Users can subscribe to a conference state by sending a SIP SUBSCRIBE request to the conference URI that identifies the focus. The focus acts as a notifier for this event package.

The name of this event package is "conference". This token appears in the Event header of the SUBSCRIBE request. The body of a notification carries the conference-state information document in the Multipurpose Internet Mail Extension (MIME) type "application/conference-info + xml" as defined in [Draft-ietf-sipping-conference-package].

Two pieces of information are provided about user status: the user's current level of participation in a conference (labelled activity-status) and the method by which the participant entered or left the conference (labelled history-status). The activity-status carries one of the following statuses: connected, disconnected or on-hold. The history-status carries one of the following statuses: dialled-in, dialled-out, departed, booted or failed.

Conference status information also carries information about the media streams that each participant is connected to and about sidebars in a conference, by indicating the URIs of those sidebars and the participants in each sidebar.

25.3 Example signalling flows of conferencing service operation

25.3.1 Creating a conference with a conference factory URI

When creating a conference with a conference factory URI the conference participant generates an initial INVITE request in accordance with Section 8.10 and sets the request URI of the INVITE request to the conference factory URI.

The conference server creates a focus for the newly created conference, assigns it a conference URI and returns the conference URI in the Contact header of the 200 (OK) response. The URI contains an "isfocus" parameter indicating that this is a focus URI. On receiving a 200 (OK) response to the INVITE request with the "isfocus" parameter indicated in the Contact header, the conference participant stores the content of the received Contact header as the conference URI. This conference URI can be used to refer other users to the conference as well as to subscribe to the conference-state event package. Figure 25.2 shows an example call flow of creating a conference using a conference factory URI.

25.3.2 Referring a user to a conference using the REFER request

When generating a REFER request that is destined to a user in order to invite that user to a specific conference, the Refer-To header of the REFER request is set to the conference URI, including the "isfocus" parameter. The referrer treats the incoming NOTIFY requests that are related to the previously sent REFER request in accordance with [RFC3515]. Figure 25.3 shows an example flow.

Another way of inviting a user to a conference is by sending the REFER request to the focus itself and putting the URI of the participant being invited into the

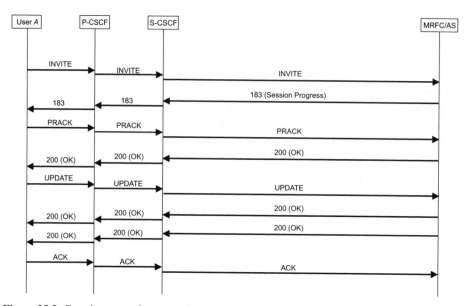

Figure 25.2 Creating a conference using a conference factory URI.

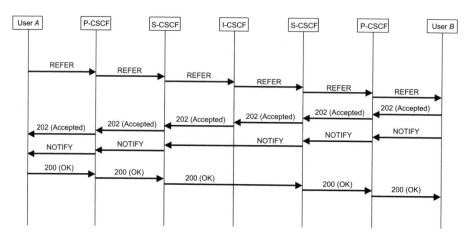

Figure 25.3 Referring a user to a conference using the REFER request.

Refer-To header. This causes the focus to generate an INVITE request inviting the user to the conference.

25.3.3 Subscribing to a conference state

Figure 25.4 shows an example flow of a user subscribing to a conference state. The flow shows the immediate notification carrying the conference state at that moment.

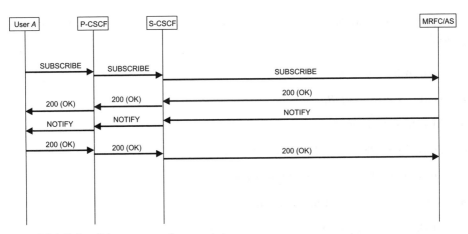

Figure 25.4 Subscribing to a conference state.

More notifications can then follow depending on the change of state of the conference.

25.3.4 Conference creation using CPCP

Using the Ut interface a CPCP message can be sent to the conference server to set up a conference. Figure 25.5 shows an example of a conference being created

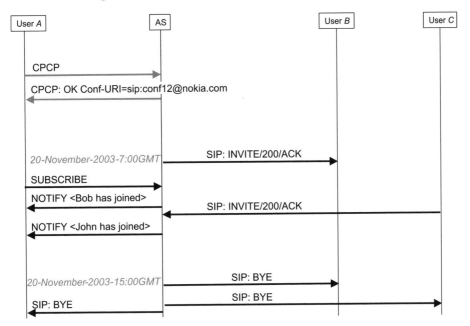

Figure 25.5 Conference creation using CPCP.

and participants added. User *A* creates the conference using the following procedure:

- Dial-in list—user *C*.

- Dial-out list—user *B*.

- Moderator—user *A*.

- Start time—28-November-2003-7:00GMT.

- Stop time—28-November-2003-15:00GMT.

- Subject—organizing tonight's activities.

- Media—audio/adaptive multi-rate (AMR).

At 7:00 GMT on 28 November 2003 the conference server creates a focus. The focus then reads the dial-out list and discovers that user *B* is on it. It then sends user *B* an invitation to join the conference. User *A* subscribes to the conference-state event package and notices that user *B* has joined. After a while, user *C* joins and user *A* is informed about it. At 15:00 GMT, the focus sends a message terminating the conference.

For brevity the flow does not show all the messages exchanged, just the significant ones.

References

Third Generation Partnership Project (3GPP)

3GPP TS 21.111	"USIM and IC card requirements".
3GPP TS 22.101	"Service principles".
3GPP TS 22.228	"Service requirements for the IP multimedia core network subsystem"
3GPP TS 22.340	"IP Multimedia Subsystem (IMS) messaging; Stage 1", March 2003.
3GPP TS 23.002	"Network architecture".
3GPP TS 23.003	"Technical Specification Group Core Network"; Numbering, addressing and identification".
3GPP TS 23.060	"General Packet Radio Service (GPRS); Service description; Stage 2".
3GPP TS 23.107	"Quality of Service (QoS) concept and architecture".
3GPP TS 23.207	"End-to-End QoS Concept and Architecture".
3GPP TS 23.218	"IP Multimedia (IM) session handling; IM call model; Stage 2".
3GPP TS 23.221	"Architectural requirements".
3GPP TS 23.228	"IP Multimedia (IM) Subsystem; Stage 2".
3GPP TR 23.815	"Charging implications of IMS architecture".
3GPP TS 24.008	"Mobile Radio Interface Layer 3 specification; Core Network Protocols; Stage 3".
3GPP TS 24.141	"Presence service using the IP Multimedia (IM) Core Network (CN) subsystem; Stage 3".
3GPP TS 24.147	"Conferencing using the IP Multimedia (IM) Core Network (CN) subsystem; Stage 3".
3GPP TS 24.247	"Messaging service using the IP Multimedia (IM) Core Network (CN) subsystem; Stage 3".
3GPP TS 24.228	"Signalling flows for the IP multimedia call control based on SIP and SDP; Stage 3".
3GPP TS 24.229	"IP Multimedia Call Control based on SIP and SDP; Stage 3".
3GPP TS 24.247	"Messaging using the IP Multimedia (IM) Core Network (CN) subsystem; Stage 3", October 2003.

3GPP TS 26.234	"Transparent end-to-end packet-switched streaming service (PSS); Protocols and codecs".
3GPP TS 26.235	"Packet-switched multimedia applications; Default codecs".
3GPP TS 26.236	"Packet-switched conversational multimedia applications; Transport protocols".
3GPP TS 29.198	"Open Service Access (OSA); Application Programming Interface (API), Multiple parts".
3GPP TS 29.207	"Policy control over Go interface", June 2003.
3GPP TS 29.208	"End-to-end Quality of Service (QoS) Signaling flows".
3GPP TS 29.228	"IP Multimedia (IM) Subsystem Cx and Dx Interfaces; Signaling flows and message contents".
3GPP TS 29.229	"Cx and Dx interfaces based on the Diameter protocol; Protocol details", June 2003.
3GPP TS 29.329	"Sh interface based on the Diameter protocol", June 2003.
3GPP TS 31.102	"Characteristics of the USIM Application".
3GPP TS 31.103	"Characteristics of the IP Multimedia Services Identity Module (ISIM) application".
3GPP TS 32.200	"Telecommunication management; Charging management; Charging principles", June 2003.
3GPP TS 32.225	"Telecommunication management; Charging management; Charging data description for the IP Multimedia Subsystem (IMS)", June 2003.
3GPP TS 32.235	"Charging data description for the IP Multimedia Subsystem (IMS)".
3GPP TS 33.102	"3G security; Security architecture", June 2003.
3GPP TS 33.120	"Security Objectives and Principles", April 2001.
3GPP TS 33.203	"3G security; Access security for IP-based services", June 2003.
3GPP TS 33.210	"3G security; Network Domain Security (NDS); IP network layer security", June 2003.
3GPP TS 33.220	"3G security; Generic Authentication Architecture (GAA); Generic Bootstrapping Architecture", 3GPP, December 2003.
3GPP TS 33.222	"Generic Authentication Architecture (GAA); Access to Network Application Functions using Secure Hypertext Transfer Protocol (HTTPS)", February 2004.
3GPP TS 33.310	"Network Domain Security; Authentication Framework (NDS/AF)", December 2003.

Internet Engineering Task Force (IETF)

RFC0791	Postel, J., "Internet Protocol", STD 5, September 1981.
RFC0793	Postel, J., "Transmission Control Protocol", STD 7, September 1981.

RFC1305   Mills, D., "Network Time Protocol (Version 3) Specification,
                 Implementation", March 1992.

RFC1321   Rivest, R., "The MD5 Message-Digest Algorithm", April 1992.

RFC1633   Braden, R., Clark, D. and S. Shenker, "Integrated Services in the
                 Internet Architecture: An Overview", June 1994.

RFC1851   Karn, P., Metzger, P. and W. Simpson, "The ESP Triple DES
                 Transform", September 1995.

RFC2104   Krawczyk, H., Bellare, M. and R. Canetti, "HMAC: Keyed-Hashing
                 for Message Authentication", February 1997.

RFC2246   Dierks, T., Allen, C., Treese, W., Karlton, P., Freier, A. and P.
                 Kocher, "The TLS Protocol Version 1.0", January 1999.

RFC2327   Handley, M. and V. Jacobson, "SDP: Session Description Protocol",
                 April 1998.

RFC2396   Berners-Lee, T., Fielding, R., Irvine U.C. and L. Masinter, "Uniform
                 Resource Identifiers (URI): Generic Syntax", August 1998.

RFC2401   Kent, S. and R. Atkinson, "Security Architecture for the Internet
                 Protocol", November 1998.

RFC2402   Kent, S. and R. Atkinson, "IP Authentication Header", November
                 1998.

RFC2403   Madson, C. and R. Glenn, "The Use of HMAC-MD5-96 within ESP
                 and AH", November 1998.

RFC2404   Madson, C. and R. Glenn, "The Use of HMAC-SHA-1-96 within ESP
                 and AH", November 1998.

RFC2406   Kent, S. and R. Atkinson, "IP Encapsulating Security Payload
                 (ESP)", November 1998.

RFC2407   Piper, D., "The Internet IP Security Domain of Interpretation for
                 ISAKMP", November 1998.

RFC2408   Maughan, D., Schneider, M. and M. Schertler, "Internet Security
                 Association and Key Management Protocol (ISAKMP)", November
                 1998.

RFC2409   Harkins, D. and D. Carrel, "The Internet Key Exchange (IKE)",
                 November 1998.

RFC2412   Orman, H., "The OAKLEY Key Determination Protocol",
                 November 1998.

RFC2451   Pereira, R. and R. Adams, "The ESP CBC-Mode Cipher Algorithms",
                 November 1998.

RFC2475   Blake, S., Black, D., Carlson, M., Davies, E., Wang, Z. and W. Weiss,
                 "An Architecture for Differentiated Service", December 1998.

RFC2477   Aboba, B. and G. Zorn, "Criteria for Evaluating Roaming Protocols",
                 January 1999.

RFC2486	Aboba, B. and M. Beadles, "The Network Access Identifier", January 1999.
RFC2543	Handley, M., Schulzrinne, H., Schooler, E. and J. Rosenberg, "SIP: Session Initiation Protocol", March 1999.
RFC2578	McCloghrie, K., Perkins, D., Schoenwaelder, J., Case, J., McCloghrie, K., Rose, M. and S. Waldbusser, "Structure of Management Information Version 2 (SMIv2)", STD 58, April 1999.
RFC2616	Fielding, R., Gettys, J., Mogul, J., Nielsen, H., Masinter, L., Leach, P. and T. Berners-Lee, "Hypertext Transfer Protocol—HTTP/1.1", June 1999.
RFC2617	Franks, J., Hallam-Baker, P., Hostetler, J., Lawrence, S., Leach, P., Luotonen, A. and L. Stewart, "HTTP Authentication: Basic and Digest Access Authentication", June 1999.
RFC2633	Ramsdell B., "S/MIME Version 3 Message Specification", June 1999.
RFC2748	Durham, D., Boyle, J., Cohen, R., Herzog, S., Rajan, R. and A. Sastry, "The COPS (Common Open Policy Service) Protocol", January 2000.
RFC2753	Yavatkar, R., Pendarakis, D. and R. Guerin, "A Framework for Policy-based Admission Control", January 2000.
RFC2782	Gulbrandsen, A., Vixie, P. and L. Esibov, "A DNS RR for specifying the Location of Services (DNS SRV)", February 2000.
RFC2806	Vaha-Sipila, A., "URLs for Telephone Calls", April 2000.
RFC2810	Kalt, C., "Internet Relay Chat: Architecture", April 2000.
RFC2833	Schulzrinne, H. and S. Petrack, "RTP Payload for DTMF Digits, Telephony Tones and Telephony Signals", May 2000.
RFC2865	Rigney, C., Willens, S., Rubens, A. and W. Simpson, "Remote Authentication Dial In User Service (RADIUS)", June 2000.
RFC2915	Mealling, M. and R. Daniel, "The Naming Authority Pointer (NAPTR) DNS Resource Record", September 2000.
RFC2960	Stewart, R., Xie, Q., Morneault, K., Sharp, C., Schwarzbauer, H., Taylor, T., Rytina, I., Kalla, M., Zhang, L. and V. Paxson, "Stream Control Transmission Protocol", October 2000.
RFC2977	Glass, S., Hiller, T., Jacobs, S. and C. Perkins, "Mobile IP Authentication, Authorization, and Accounting Requirements", October 2000.
RFC3084	Chan, K., Seligson, J., Durham, D., Gai, S., McCloghrie, K., Herzog, S., Reichmeyer, F., Yavatkar, R. and A. Smith, "COPS Usage for Policy Provisioning (COPS-PR)", March 2001.
RFC3169	Beadles, M. and D. Mitton, "Criteria for Evaluating Network Access Server Protocols", September 2001.

RFC3261   Rosenberg, J., Schulzrinne, H., Camarillo, G., Johnston, A., Peterson, J., Sparks, R., Handley, M. and E. Schooler, "SIP: Session Initiation Protocol", June 2002.

RFC3262   Rosenberg, J. and H. Schulzrinne, "Reliability of Provisional Responses in the Session Initiation Protocol (SIP)", June 2002.

RFC3263   Rosenberg, J. and H. Schulzrinne, "SIP: Locating SIP Servers", June 2002.

RFC3264   Rosenberg, J. and H. Schulzrinne, "An Offer/Answer Model with SDP", June 2002.

RFC3265   A. B. Roach, "Session Initiation Protocol (SIP)-specific Event Notification", June 2002.

RFC3310   Niemi, A., Arkko, J. and V. Torvinen, "Hypertext Transfer Protocol (HTTP) Digest Authentication Using Authentication and Key Agreement (AKA)", September 2002.

RFC3311   Rosenberg, J., "The Session Initiation Protocol (SIP) UPDATE Method", October 2002.

RFC3312   Camarillo, G., Marshall, W., Rosenberg, J., "Integration of Resource Management and Session Initiation Protocol", October 2002.

RFC3313   Marshal, W., "Private Session Initiation Protocol (SIP) Extensions for Media Authorization", January 2003.

RFC3315   Droms, R., Bounds, J., Volz, B., Lemon, T., Perkins, C. and M. Carney, "Dynamic Host Configuration Protocol for IPv6 (DHCPv6)", July 2003.

RFC3319   Schulzrinne, H. and B. Volt, "Dynamic Host Configuration Protocol (DHCPv6) Options for Session Initiation Protocol (SIP) Servers", July 2003.

RFC3320   Price, R., Bormann, C., Christoffersson, J., Hannu, H., Liu, Z. and J. Rosenberg, "Signaling Compression (SigComp)", January 2003.

RFC3323   Peterson, J., "A Privacy Mechanism for the Session Initiation Protocol (SIP)", November 2002.

RFC3325   Jenning, C., Peterson, J. and M. Watson, "Private Extensions to the Session Initiation Protocol (SIP) for Asserted Identity within Trusted Networks", November 2002.

RFC3327   Willis, D. and B. Hoeneisen, "Session Initiation Protocol (SIP) Extension Header Field for Registering Non-Adjacent Contacts", December 2002.

RFC3329   Arkko, J., Torvinen, V., Camarillo, G., Niemi, A. and T. Haukka, "Security Mechanism Agreement for the Session Initiation Protocol (SIP)", January 2003.

RFC3388	Camarillo, G., Eriksson, G., Holler, J. and H. Schulzrinne, "Grouping of Media Lines in the Session Description Protocol (SDP)", December 2002.
RFC3411	Harrington, D., Presuhn, R. and B. Wijnen, "An Architecture for Describing Simple Network Management Protocol (SNMP) Management Frameworks", STD 62, December 2002.
RFC3418	Presuhn, R., "Management Information Base (MIB) for the Simple Network Management Protocol (SNMP)", STD 62, December 2002.
RFC3428	Campbell, B. and J. Rosenberg, "Session Initiation Protocol Extension for Instant Messaging", September 2002.
RFC3455	Garcia-Martin, M., Henrikson, E. and D. Mills, "Private Header (P-Header) Extensions to the Session Initiation Protocol (SIP) for the 3rd-Generation Partnership Project (3GPP)", January 2003.
RFC3485	Garcia-Martin, M., Bormann, C., Ott, J., Price, R. and A.B. Roach, "The Session Initiation Protocol (SIP) and Session Description Protocol (SDP) Static Dictionary for Signaling Compression (SigComp)", February 2003.
RFC3486	G. Camarillo, "Compressing the Session Initiation Protocol (SIP)", February 2003.
RFC3515	R. Sparks, "The Session Initiation Protocol (SIP) REFER Method", April 2003.
RFC3520	Hamer, L-N., Gage, B., Kosinski, B. and H. Shieh, "Session Authorization Policy Element", April 2003.
RFC3524	Camarillo, G. and A. Monrad, "Mapping of Media Streams to Resource Reservation Flows", April 2003.
RFC3525	Groves, C., Pantaleo, M., Anderson, T. and T. Taylor, "Gateway Control Protocol Version 1", June 2003.
RFC3539	Aboba, B. and J. Wood, "Authentication, Authorization and Accounting (AAA) Transport Profile", June 2003.
RFC3550	Schulzrinne, H., Casner, S., Frederick, R. and V. Jacobson, "RTP: A Transport Protocol for Real-Time Applications", July 2003.
RFC3551	Schulzrinne, H. and S. Casner, "RTP Profile for Audio and Video Conferences with Minimal Control", July 2003.
RFC3588	Calhoun, P., Loughney, J., Guttman, E., Zorn, G. and J. Arkko, "Diameter Base Protocol", September 2003.
RFC3589	Loughney, J., "Diameter Command Codes for Third Generation Partnership Project (3GPP) Release 5", September 2003.
RFC3608	Willis, D. and B. Hoeneisen, "Session Initiation Protocol (SIP) Extension Header Field for Service Route Discovery During Registration", December 2002

Draft-hakala-diameter-credit-control-05.txt	Hakala, H., "Diameter Credit-control Application", November 2002 (Obsolete)
Draft-ietf-aaa-diameter-cc	Hakala, H., "Diameter Credit Control Application", November 2002 Work In Progress, October 2003.
Draft-ietf-aaa-diameter-mobileip	Calhoun, P., Johansson, T. and C. Perkins "Diameter Mobile IPv4 Application", Work In Progress, October 2003.
Draft-ietf-aaa-diameter-nasreq	Calhoun, P., Zorn, G., Spence, D. and D. Mitton, "Diameter Network Access Server Application", Work In Progress, October 2003.
Draft-ietf-aaa-diameter-sip-app	Garcia-Martin, M., "Diameter Session Initiation Protocol (SIP) Application", Work In Progress, October 2003.
Draft-ietf-aaa-eap	Eronen, P., Hiller, T. and G. Zorn, "Diameter Extensible Authentication Protocol (EAP) Application", Work In Progress, October 2003.
Draft-ietf-enum-rfc2916bis	Faltstrom, P. and M. Mealling, "The E.164 to URI DDDS Application (ENUM)", Work In Progress, November 2003.
Draft-ietf-enum-sip	Peterson, J., "enumservice registration for SIP Addresses-of-Record", Work In Progress, November.
Draft-ietf-geopriv-commom-policy	Schulzrinne, H., Morris, J., Tschofenig, H., Cuellar, J., Polk, J. and J. Rosenberg, "Common Policy", Work In Progress, February 2004.
Draft-ietf-impp-cpim-msgfmt	Atkins, D. and G. Klyne, "Common Presence and Instant Messaging: Message Format", Work In Progress, January 2003.
Draft-ietf-impp-cpim-pidf	Sugano, H. and S. Fujimoto, "Presence Information Data Format (PIDF)", Work In Progress, May 2003.
Draft-ietf-impp-im	Peterson, J., "Common Profile for Instant Messaging (CPIM)", Work In Progress, October 2003.
Draft-ietf-iptel-rfc2806bis	Schulzrinne, H. and A. Vaha-Sipila, "The tel URI for Telephone Numbers", Work In Progress, June 2003.
Draft-ietf-mmusic-sdp-new	Handley, M., Jacobson, V. and C. Perkins, "SDP: Session Description Protocol", Work In Progress, March 2003.
Draft-ietf-simple-event-list	Roach, A., Rosenberg, J. and B. Campbell, "A Session Initiation Protocol (SIP) Event Notification Extension for Resource Lists", Work In Progress, June 2003.
Draft-ietf-simple-presence	Rosenberg, J., "A Presence Event Package for the Session Initiation Protocol (SIP)", Work In Progress, January 2003.
Draft-ietf-simple-winfo-package	Rosenberg, J., "A Watcher Information Event Template-package for the Session Initiation Protocol (SIP)", Work In Progress, December 2002.
Draft-ietf-simple-xcap	Rosenberg, J., "The Extensible Markup Language (XML) Configuration Access Protocol (XCAP)", Work In Progress, October 2003.

Draft-ietf-simple-xcap-auth-usage	Rosenberg, J., "Extensible Markup Language (XML) Configuration Access Protocol (XCZP)", Work In Progress, October 2003.
Draft-ietf-simple-xcap-list-usage	Rosenberg, J., "An Extensible Markup Language (XML) Configuration Access Protocol (XCAP) Usage for Resource Lists", Work In Progress, February 2003.
Draft-ietf-sip-publish	Niemi, A., "Session Initiation Protocol (SIP) Extension for Event State Publication", Work In Progress, October 2003.
Draft-ietf-sipping-reg-event	Rosenberg, J., "A Session Initiation Protocol (SIP) Event Package for Registrations", October 2002.
Draft-ietf-sipping-conference-package	Rosenberg, J. and H. Schulzrinne, "A Session Initiation Protocol (SIP) Event Package for Conference State", Work In Progress, September 2003.
Draft-ietf-xcon-cpcp	Khartabil, H. and P. Koskelainen, "The Conference Policy Control Protocol", Work In Progress, April 2004.
Draft-rosenberg-simple-rules	Rosenberg, J., "Presence Authorization Rules", Work In Progress, February 2004.
Draft-rosenberg-simple-commom-policy-caps	Rosenberg, J., "An Extensible Markup Language (XML) Representation for Expressing Policy Capabilities".
Draft-rosenberg-simple-pres-policy-caps	Rosenberg, J., "An Extensible Markup Language (XML) Representation for Expressing Presence Policy Capabilities", Work In Progress, February 2004.

International Telecommunications Union (ITU)

ITU.H248.1	"Gateway Control Protocol: Version 1", ITU-T Recommendation H.248.1, March 2002.
ITU-T G.711	"Pulse code modulation (PCM) of voice frequencies".
ITU-T G.726	"40, 32, 24, 16 kbit/s adaptive differential pulse code modulation (ADPCM)".

Abbreviations

3GPP	Third Generation Partnership Project
3GPP2	Third Generation Partnership Project 2
A RR	IPv4 address resource record
AAAA RR	IPv6 address resource record
AAA	Authentication, authorization and accounting
AAL	ATM adaptation layer
ACA	Accounting-Answer
ACR	Accounting requests
ADSL	Asynchronous Digital Subscriber Line
AH	Authentication header
AKA	Authentication and key agreement
AMR	Adaptive multi-rate
AOR	Address of record
API	Application program interface
APN	Access point name
ARIB	Association of Radio Industries and Businesses (Japan)
AS	Application server
ATM	Asynchronous transfer mode
AUC	Authentication centre
AUID	Application usage ID
AUTN	Authentication token
AUTS	Synchronization token
AV	Authentication vector
AVP	Attribute value pair; audio video profile
B2BUA	Back to back UA
BCF	Bearer Charging Function
BER	Bit error ratio
BGCF	Breakout Gateway Control Function
BICC	Bearer Independent Call Control
BNF	Backus-Naur Form grammar
BS	Bearer service; billing system

BSF	Bootstrapping Server Function
BTS	Base Transceiver Station
CAMEL	Customized Applications for Mobile network Enhanced Logic
CAP	Camel Application Part
CCF	Charging Collection Function
CCSA	China Communications Standards Association
CDMA	Code Division Multiple Access
CDR	Charging Data Record
CGF	Charging Gateway Function
CK	Ciphering key
CN	Core Network
COPS	Common Open Policy Service
COPS-PR	Common Open Policy Service Usage for Policy Provisioning
CPCP	Conference Policy Control Protocol
CPIM	Common Presence and Instant Messaging
CPS	Conference policy server
CRLF	Carriage Return Line Feed
CS	Circuit-switched
CSCF	Call Session Control Function
CS CN	Circuit Switched Core Network
CSE	CAMEL Service Environment
CSRC	Contributing source
DDDS	Dynamic Delegation Discovery System
DES	Data Encryption Standard
DHCP	Dynamic Host Configuration Protocol
DL	Downlink
DOI	Domain of interpretation
DOS	Denial of service
DNS	Domain name system
DSL	Digital Subscriber Line
DTMF	Dual-tone multifrequency
EAP	Extensible Authentication Protocol
ECF	Event Charging Function
EDGE	Enhanced Data Rates for Global Evolution
ENUM	E.164 number
ESP	Encapsulation security payload
ETSI	European Telecommunications Standards Institute
FQDN	Fully qualified domain name
FSM	Finite state machine
GAA	Generic Authentication Architecture
GBA	Generic Bootstrapping Architecture

GBR	Guaranteed bit rate
G-CDR	GGSN-CDR
GCID	GPRS charging identifier
GGSN	Gateway GPRS Support Node
GPRS	General Packet Radio Service
GSM	Global System for Mobile Communications
HLR	Home location register
HSS	Home Subscriber Server
HTTP	Hyper Text Transfer Protocol
IAB	Internet Architecture Board
IANA	Internet Assigned Numbers Authority
ICID	IMS charging identifier
I-CSCF	Interrogating-CSCF
IESG	Internet Engineering Steering Group
IETF	Internet Engineering Task Force
IK	Integrity key
IKE	Internet Key Exchange
IMS-MGW	IP Multimedia Subsystem-Media Gateway Function
IMS	IP Multimedia Subsystem
IMSI	International Mobile Subscriber Identifier
IM-SSF	IP Multimedia Service Switching Function
IOI	Interoperator identifier
IP	Internet Protocol
IP-CAN	IP-Connectivity Access Network
IPsec	Internet Protocol security
IPv4	Internet Protocol Version 4
IPv6	Internet Protocol Version 6
IRC	Internet Relay Chat
ISAKMP	Internet Security Association and Key Management Protocol
ISC	IMS Service Control
ISDN	Integrated Services Digital Network
ISIM	IP Multimedia Services Identity Module
ISP	Internet Service Provider
ISUP	ISDN User Part
IV	Initialization vector
L1	Layer 1
LCS	Location services
LIA	Location-Info-Answer
LIR	Location-Info-Request
LPDP	Local policy decision point
M3UA	SS7 MTP3-user adaptation layer
MAA	Multimedia-Multimedia-Answer

MAC	Message Authentication Checksum
MAP	Mobile Application Part
MAR	Multimedia-Auth-Request
Mbone	Multicast backbone
MBR	Maximum bit rate
MCC	Mobile country code
MDS	Multimedia Delivery Service
MEGACO	Media Gateway Control Protocol
MGCF	Media Gateway Control Function
MGW	Media gateway function
MIB	Management information base
MID	Media stream identification
MITM	Man in the middle
MIME	Multipurpose Internet Mail Extension
MMS	Multimedia Messaging Service
MMSC	Multimedia Messaging Service Centre
MNC	Mobile network code
MOBILE IP	Mobile Internet Protocol
MPV	Music photo video
MRFC	Multimedia Resource Function Controller
MRFP	Media Resource Function Processor
MSC	Mobile switching centre
MSIN	Mobile Subscriber Identification Number
MSISDN	Mobile Subscriber International ISDN Number
MSRP	Message Session Relay Protocol
MTP	Message Transfer Part
MTPn	Message Transfer Part level n
MTU	Maximum transfer unit
NAF	Network Application Function
NAI	Network access identifier
NAPTR	Naming authority pointer
NAS	Network access server
NASREQ	Network Access Server Requirements
NDS	Network Domain Security
NTP	Network Time Protocol
OCS	Online Charging System
OMA	Open Mobile Alliance
OSA	Open Services Architecture
P2P	Peer to peer
PA	Presence agent
P-CSCF	Proxy-CSCF
PCMU	Pulse code modulation μ-law

PDF	Policy Decision Function
PDP	Packet Data Protocol; policy decision point
PEF	Policy Enforcement Function
PEP	Policy Enforcement Point
PIB	Policy information base
PKI	Public Key Infrastructure
PLMN	Public Land Mobile Network
PNA	Push-Notification-Answer
PoC	Push to talk over the cellular service
PNR	Push-Notification-Request
PPA	Push-Profile-Answer
PPR	Push-Profile-Request
PRACK	Provisional response acknowledgement
PRC	Provisioning class
PRI	Provisioning instance
PRID	Provisioning instance identifier
PS	Packet-switched; presence server
PSI	Public service identity
PSTN	Public Switched Telephone Network
PUA	Presence user agent; Profile-Update-Answer
PUR	Profile-Update-Request
QoS	Quality of service
RADIUS	Remote Authentication Dial In User Service
RAN	Radio access network
RAND	Random challenge
RES	Response
RFC	Requests For Comments
RLS	Resource list server
RNC	Radio network controller
ROAMOPS	Roaming operations
RSVP	Resource Reservation Setup Protocol
RTA	Registration-Termination-Answer
RTCP	RTP Control Protocol
RTP	Real-time Transport Protocol
RTP/AVP	RTP Audio and Video Profile
RTR	Registration-Termination-Request
S/MIME	Secure MIME
SA	Security association
SAA	Server-Assignment-Answer
SAD	Security Association Database
SAR	Server-Assignment-Request
SBLP	Service-based local policy

S-CDR	SGSN-CDR
SCF	Session Charging Function
SCS	Service Capability Server
S-CSCF	Serving-CSCF
SCTP	Stream Control Transmission Protocol
SDP	Session Description Protocol
SDU	Service Data Unit
SEG	Security Gateway
SGSN	Serving GPRS Support Node
SGW	Signalling Gateway
SHA	Secure Hash Algorithm
SigComp	Signalling Compression
SIM	Subscriber Identity Module
SIP	Session Initiation Protocol
SIPS	Secure SIP
SL	Subscriber locator
SLA	Service-level agreement
SLF	Subscription Locator Function
SMI	Structure for Management Information
S/MIME	Secure MIME
SMG	Special Mobile Group
SNA	Subscribe-Notifications-Answer
SNMP	Simple Network Management Protocol
SNR	Subscribe-Notifications-Request
SPD	Security Policy Database
SPI	Security Parameter Index
SPT	Service point trigger
SQN	Sequence number
SRF	Single reservation flow
SRV	Service records
SS7	Signaling System No. 7
SSF	Service Switching Function
SSRC	Synchronization source
TCP	Transmission Control Protocol
TCP/IP	TCP/IP stack
TD-CDMA	Time Division/Code Division Multiple Access
THIG	Topology Hiding Inter-network Gateway
TIA	Telecommunications Industry Association (North America)
TLS	Transport Layer Security
TTA	Telecommunications Technology Association (South Korea)
TTC	Telecommunications Technology Committee (Japan)
TTL	Time to live

TU	Transaction User
UA	User Agent
UAA	User-Authorization-Answer
UAC	User Agent Client
UAR	User-Authorization-Request
UAS	User agent server
UDA	User-Data-Answer
UDP	User Datagram Protocol
UDR	User-Data-Request
UDVM	Universal decompressor virtual machine
UE	User equipment
UICC	Universal Integrated Circuit Card
UL	Uplink
UMTS	Universal Mobile Telecommunications System
URI	Uniform resource identifier
URL	Universal resource locator
URN	Uniform resource name
USIM	Universal Subscriber Identity Module
UTRAN	UMTS terrestrial radio access network
VHE	Virtual home environment
VoIP	Voice over IP
WAP	Wireless Application Protocol
WB	Wideband
WCDMA	Wideband Code Division Multiple Access
WLAN	Wireless Local Area Network
XCAP	XML Configuration Access Protocol
XML	Extensible Markup Language
XRES	Expected response

Index

100rel 221, 290, 293
1xx (class) 266, 298
2xx (class) 266, 275–277, 288, 290–291, 295, 298–299
3xx (class) 266, 272, 278, 280
4xx (class) 267
5xx (class) 267
6xx (class) 266–267

AAA *see* authentication, authorization and accounting
AAAA 131, 283–284, 317
access point name 60, 82, 87, 324
ACK 164–165, 198, 202, 208, 210, 212, 217–222, 228, 230, 236, 238, 241–242, 266, 277, 290–291
activity-status 389
ad hoc conference 3, 387–388
address of record 176–180, 268, 271–274, 295, 320
AKA 22, 38, 61–62, 68–71, 139–143, 185, 280–281
Allow-Events header 288
AOR *see* address of record
application 3, 5, 9–10, 13, 15–18, 22, 25–27, 29, 42, 44–45, 52, 54–56, 59–61, 70–71, 77, 86, 91, 97–98, 103–105, 107, 109, 113–114, 126–127, 135, 137–139, 142, 144, 168–169, 171, 173, 175, 212, 219, 229, 255, 261–265, 270, 293, 301, 313–314, 317, 319, 324, 327–328, 330–340, 342, 344, 361–364, 369–370, 375, 377–379, 383, 385, 389
application server 22, 24–26, 29, 32, 39–44, 58, 67, 91, 94–98, 104–105, 107–109, 166, 182, 195, 212–215, 249–250, 253, 255–257, 264

architecture 1, 5–6, 8–11, 14–17, 25, 29, 32, 54–56, 60–62, 71, 91–92, 95, 97–99, 105, 119, 121, 139, 143, 155, 216, 261, 263, 340, 355, 361, 377, 379, 384, 387–388
AS *see* application server
asserted ID 67–69, 180, 198, 295
attribute value pair 74, 80–81, 84, 94, 98, 101, 223, 225–228, 233, 235–238, 241–242, 304, 314–315, 335–340, 344
audio video profile 74, 80–81, 84, 94, 98, 101, 223, 225–228, 233, 235–238, 241–242, 304, 314–315, 335–340, 344
AUID 369, 378
authentication 8, 22–23, 25, 30–36, 38, 40, 49–50, 54, 61–66, 68–72, 119, 125, 130, 133, 137–140, 142–144, 146, 148–149, 153, 156, 160–161, 168–169, 171, 176–177, 181–182, 195, 278–281, 327–332, 335, 338–340, 355–359
authentication, authorization and accounting 331–332, 334–336, 339, 343
authorization 8, 21, 23, 25, 32–33, 36, 38, 44–45, 48, 75, 77–78, 80, 82, 88–91, 95–96, 98–99, 101, 103, 112, 135, 140–143, 169, 185–186, 239–241, 245, 247, 278, 281, 298, 331–332, 338–339, 343, 354, 370, 379
AVP *see* attribute value pair; audio video profile

Backus–Naur Form grammar 270, 294
bearer charging function 97
BGCF *see* Breakout Gateway Control Function
binding 77, 88–89, 127, 129, 133, 176–178, 181, 186, 264, 271, 273–274, 279, 305

BNF *see* Backus–Naur Form grammar
body (message) 8, 106, 126, 175, 177, 180,
    186, 223, 227, 244, 267–268, 271, 276,
    289, 294, 376, 378–379, 389
bootstrapping 61, 71–72
Breakout Gateway Control Function 18,
    22, 26, 43–44, 67, 93, 95, 110
BYE 94, 164, 192, 198, 201, 217, 224, 253,
    266, 277–278

call control 25, 27, 109, 114, 121, 261, 388
Call-ID header 129–130, 132–133, 136,
    186, 201–202, 215, 221–222, 265, 267,
    271–274, 277
Call Session Control Protocol 15, 18–26,
    29–36, 38–40, 43–45, 48–52, 54, 57–58,
    64–67, 69–70, 72–75, 77–78, 83–86,
    89–91, 93–97, 99, 101–104, 106–115,
    119, 121, 123, 125–127, 129–135,
    137–144, 146, 148, 150–160, 162–168,
    170–175, 177–182, 184–186, 188,
    191–192, 194–199, 201–210, 212–219,
    239–242, 244–250, 253, 255–257, 325,
    340, 363
CAMEL *see* Customized Applications for
    Mobile network Enhanced Logic
CANCEL 165–166, 266, 277–278
carriage return line feed 266–267, 303
CCF *see* Charging Collection Function
CGF *see* Charging Gateway Function
charging 6, 8–9, 14–15, 17–19, 21–23, 25,
    28–29, 31, 34, 39–40, 45, 48, 54, 76–77,
    83, 91–99, 101, 167, 192, 245–250, 299,
    339, 354
Charging Collection Function 19, 21, 23–
    27, 92–94, 98, 101, 245, 249
charging correlation 14, 45, 76, 91, 99, 354
Charging Gateway Function 92–93
client (transaction) 6, 11, 55, 69, 114, 144,
    148, 150–154, 156–161, 186, 263–264,
    267, 270–271, 281, 289, 296, 308–309,
    319–321, 327–330, 332–334, 338–339,
    349–352, 365–366, 371, 378–379, 388
client error (response) 267
codec 20, 22, 77–78, 86, 103, 109, 114, 191,
    219–220, 223, 225–228, 236, 239–240,
    242, 244, 305, 307–308, 311, 314
Common Open Policy Service 349–354
compressing SIP 8, 300

compressor 361–364
conference factory URI 388–389
conference policy 371, 387–388
Conference Policy Control Protocol 371,
    388, 391
conference policy server 371
conference state 388, 390
confidentiality 20, 38, 63, 65, 70, 72,
    279–281, 327, 330, 355, 357, 359
Contact header 129–130, 132–133,
    136–137, 146, 150–152, 163–165,
    171–173, 175, 178, 186, 202, 204–210,
    212, 217–218, 222, 265, 267, 271–275,
    279–280, 293, 295, 300, 389
Content-Length header 129–130, 132–133,
    136, 171, 173, 175, 186, 265
Content-Type header 175, 265
COPS *see* Common Open Policy Service
CPCP *see* Conference Policy Control
    Protocol
cpim (message-cpim) 289, 377
CPS *see* conference policy server
CRLF *see* carriage return line feed
CSCF *see* Call Session Control Protocol
CSeq header 129–130, 132–133, 136, 186,
    201–202, 221–222, 265, 267, 271–273,
    275–277, 290
Customized Applications for Mobile
    network Enhanced Logic 5, 25–26, 42,
    97
Cx 30, 32, 34, 36, 38–39, 52, 75, 103, 106,
    133, 339–340

decompressor dispatcher 361, 363–364
deferred delivery messaging 383–385
design 8–10, 12, 17–18, 20, 25, 61, 75, 96,
    99, 155, 261–262, 314, 340
Dh 42–43, 339
DHCP *see* Dynamic Host Configuration
    Protocol
dialog 31, 109, 130, 134, 153, 163, 172,
    174, 185, 192, 195, 198–199, 201, 203,
    205, 207, 210, 212, 214–215, 217–219,
    222, 237, 239, 253, 263–264, 267,
    271–280, 285, 288, 291–294, 298, 362,
    384, 387
dialog-creating, or establishing,
    request 274, 288, 293–294
dialog-creating response 274, 279

Diameter credit control application 339–340, 344

Diameter SIP application 332, 339–340, 344

Diameter 6, 331–340, 342–344
Accounting-Answer 94, 96
Accounting-Request 94

Digest authentication 143, 281

discovery 25, 49–50, 72, 126, 137, 198, 273, 280, 283, 299

DNS *see* domain name system

DOI 357

domain name system 22, 58, 72–73, 126, 131, 174, 199, 204, 268, 283–284, 317, 319–321

Dx 30, 38, 339

Dynamic Host Configuration Protocol 72–73, 126, 365–366

dynamic payload 305, 314–315

ENUM 22, 110–111, 319–320

Event header 93, 95, 98, 171, 173, 175, 181, 184, 189, 245, 288–289, 294, 346, 376, 379, 388

event notification 130, 180, 184, 189, 287–289, 293–294, 378, 388

event notification package 388

Expires header 171–173, 288

expires parameter 33–34, 75, 129–130, 132–133, 136, 146, 148, 150, 164–165, 175, 177, 182–183, 186, 209, 274

expiry 333

fetching 274

focus 255–256, 371, 387–390, 392

forking proxy 48, 85, 89, 114, 264, 273, 285

format (message) 22, 34, 69, 105, 110, 170, 172, 225–226, 267, 280, 308, 311, 314, 318–319, 321, 335, 344, 351, 365–366

FQDN *see* fully qualified domain name

From header 5, 8, 14–15, 26, 68, 78, 81, 85, 99, 113, 130–132, 137, 142, 169–170, 174–175, 182, 194, 201–209, 213, 215, 224, 227, 253, 267, 271–277, 296, 325

fully qualified domain name 126, 268, 286, 319

gateway 12, 21–22, 24, 26–29, 65, 67, 72, 92–93, 97, 121, 123, 201, 263, 319–320, 323, 345–346, 355–356, 359

Gateway GPRS Support Node 12, 15, 21, 23, 28–29, 45, 48, 72, 76–78, 80, 82–83, 87, 89–91, 93, 97, 99, 121, 123, 125–126, 135, 201, 240, 245, 247, 323–325

GBA *see* Generic Bootstrapping Architecture

General Packet Radio Service 3, 5–6, 12–15, 18, 21, 28–29, 45, 48–50, 60, 72–73, 76, 90–91, 99, 101, 119, 121, 123, 125–126, 165, 201, 236–237, 240, 245, 250, 299, 323

Generic Bootstrapping Architecture 71–72

GGSN *see* Gateway GPRS Support Node

global failure (response) 267

global number 269

Gm 30–31, 51–52, 65–66, 68–69, 143–144, 155

Go 3, 5, 7–8, 11, 13–14, 18–19, 30–32, 35, 39, 45, 48, 58, 60, 62, 65–66, 69, 72, 74, 76–77, 83–85, 87, 89–91, 99, 103, 109–110, 113–114, 119, 129, 132, 140–144, 149, 151–152, 155, 157–158, 160, 162, 164–165, 168–169, 179, 185, 192, 205–207, 212, 216, 219, 221, 229–231, 238, 240, 242, 245, 253, 262, 264, 276–277, 296–297, 309, 311–312, 314, 324, 327–328, 330–332, 337, 339, 344, 349, 354, 356–359, 361, 363, 385

GPRS *see* General Packet Radio Service

Gq 45, 48, 76, 85

header 58, 66, 68–69, 106, 109, 114, 129–135, 137, 140–144, 146, 148, 150–154, 156–161, 163–172, 174–175, 178, 180, 183, 186, 192, 194–199, 201–210, 212–219, 221–222, 224, 227, 230, 232, 240, 246–250, 253, 265, 267–269, 271–277, 279–281, 285–291, 293–296, 298–300, 311, 314, 335–337, 344, 350, 355, 357–358, 376, 378–379, 389–390
Call-ID 129–130, 132–133, 136, 186, 201–202, 215, 221–222, 265, 267, 271–274, 277
Contact 129–130, 132–133, 136–137, 146,

header (*cont.*)
   150–152, 163–165, 171–173, 175, 178,
    186, 202, 204–210, 212, 217–218, 222,
    265, 267, 271–275, 279–280, 293, 295,
    300, 389
   Content-Type 175, 265
   CSeq 129–130, 132–133, 136, 186,
    201–202, 221–222, 265, 267, 271–273,
    275–277, 290
   Event 93, 95, 98, 171, 173, 175, 181, 184,
    189, 245, 288–289, 294, 346, 376, 379,
    388
   Expires 171–173, 288
   From 5, 8, 14–15, 26, 68, 78, 81, 85, 99,
    113, 130–132, 137, 142, 169–170,
    174–175, 182, 194, 201–209, 213, 215,
    224, 227, 253, 267, 271–277, 296, 325
   Max-Forwards 129–133, 136, 185, 265,
    267, 272
   P-Access-Network-Info 165–166, 299
   P-Associated-URI 168, 170–171, 300
   Path 129, 135, 199, 205, 294–295
   P-Called-Party-ID 197, 299–300
   P-Preferred-Identity 171–172, 194–195,
    197, 296
   Privacy 68, 194–198, 213, 279, 296
   P-Visited-Network-ID 166, 299
   RACK 164, 198, 202, 208, 210, 212,
    217–219, 221–222, 228, 230, 236, 238,
    241–242, 290–291
   Route 129–135, 137, 151–152, 163–165,
    171–172, 174, 185–186, 199, 202–210,
    212–218, 253, 272, 275–276, 279–280,
    285–286, 294–295, 299–300
   RSeq 222, 290
   Service-Route 129, 134–135, 172, 186,
    199, 202–203, 213, 299
   Subscription-State 175, 186, 288, 294
   Supported 221, 271, 273, 290, 293–295,
    378
   To 16, 19–23, 27, 36, 38, 43, 54, 57,
    62–63, 70, 93, 106, 108, 110, 112, 114,
    119, 129–131, 133–135, 137, 151,
    155–156, 168–169, 171, 175, 178, 180,
    182, 186, 194, 197, 201–203, 209, 214,
    240, 253, 267, 271–277, 279, 292–293,
    313, 329–330, 354, 359, 378, 389–390
   Via 13, 130–132, 134, 148, 151–154,
    163–165, 199, 202–210, 212, 214–215,

217–219, 267–268, 271–273, 280,
   286–287, 300
hijacking (registration) 278
history-status 389
Home Subscriber Server 18, 21–24, 32–36,
   38–43, 50, 52, 54, 57, 73–75, 101, 103,
   107, 111, 125, 127, 132–134, 137–142,
   144, 174, 184, 186, 199, 204–205, 212,
   255, 257
hop-by-hop (security mechanism) 65, 279,
   336
HSS *see* Home Subscriber Server
HTTP *see* Hyper Text Transfer Protocol
Hyper Text Transfer Protocol 44, 61, 70,
   139–140, 143, 155–157, 161, 262, 281,
   369

I-CSCF *see* Interrogating-Call Session
   Control Function
IETF *see* Internet Engineering Task Force
IKE *see* Internet Key Exchange
immediate messaging 383–384
impersonating a server 278
IMS identity module 54
IMS-MGW *see* Internet Protocol
   Multimedia Subsystem-Media
   Gateway Function
IMS Service Control 30, 32, 95–97
initial filter criteria 39, 101, 103–108, 127,
   212
instant messaging 289, 375, 383–384
integrity 19, 23, 38, 59, 61–66, 69–70, 72,
   135, 139–143, 166, 185–186, 208,
   279–281, 297, 327–328, 330, 355,
   357–359
intermediary 68, 263–264, 272–273, 275,
   284, 294, 298, 300, 343
Internet Engineering Task Force 6–9, 17,
   98, 233, 261–262, 287, 289, 305, 331,
   349, 354
Internet Key Exchange 65–66, 356–357,
   359
Internet Protocol Multimedia Subsystem-
   Media Gateway Function 27, 109–111
Internet Protocol Security 19, 63, 65–66,
   70, 125, 135, 142–144, 151–153,
   155–161, 163, 172, 186, 191, 195, 199,
   202–203, 205–210, 279–280, 297, 327,
   332–333, 355–357, 359

Internet Relay Chat 385
Internet Security Association and Key
    Management Protocol 65–66, 356–357
Interrogating-Call Session Control
    Function 19, 21–24, 30–36, 38–39, 43,
    50, 52, 73–75, 93–95, 111, 123, 127,
    129, 131–132, 134, 174, 186, 199,
    204–205, 210, 216, 250, 255–257, 340
INVITE 39, 51–52, 78, 83–84, 94, 106–109,
    111, 123, 125, 129, 134–135, 148, 153,
    166, 191–192, 194–199, 201–210,
    212–217, 219–224, 227, 229–230,
    232–233, 236–241, 246, 249–250, 256,
    264–266, 271, 273, 276–278, 290–291,
    293, 298, 384, 388–390
IPsec *see* Internet Protocol Security
IRC *see* Internet Relay Chat
ISAKMP *see* Internet Security Association
    and Key Management Protocol
ISC *see* IMS Service Control
isfocus 389

jitter 312

layering 17, 99
lr parameter 284, 285
local number 269
local sequence number 274–276
local tag 274
local URI 274–276
location server 264, 273
loose routing 272, 284
loose-route parameter 23, 129–133, 164,
    185, 218, 272, 285

MAC 143, 155, 158, 328, 330
Max-Forwards header 129–133, 136, 185,
    265, 267, 272
maximum transfer unit 286
media authorization 45, 48, 78, 90–91, 135,
    239–240, 245, 247, 298, 354
media description 78, 262, 276, 302
Media Gateway Control Function 18,
    26–27, 43, 67, 93–95, 109–111
message 8, 19–20, 23, 27, 30–32, 35, 45, 50,
    54–55, 57–58, 61–62, 68–70, 73, 77, 85,
    88, 90–91, 94, 97–98, 103, 106,
    109–111, 114–115, 119, 123, 126,
    129–131, 134–135, 144, 146, 148, 154,

156, 158, 162–164, 172, 180, 182, 184,
    191–192, 196, 201, 203, 208, 212, 230,
    239–241, 264–265, 267–268, 271,
    273–274, 276, 278–281, 284, 289–290,
    294, 298, 300–301, 303, 323, 328–330,
    332–337, 339, 350–351, 357, 361–366,
    375, 382–385, 391–392
    body 8, 106, 126, 175, 177, 180, 186, 223,
        227, 244, 267–268, 271, 276, 289, 294,
        376, 378–379, 389
    format 22, 34, 69, 105, 110, 170, 172,
        225–226, 267, 280, 308, 311, 314,
        318–319, 321, 335, 344, 351, 365–366
    payload 13, 19–20, 22, 45, 66, 68,
        225–227, 229, 268, 288, 304–305,
        311–312, 314–315, 355, 357–359
Messaging
    deferred delivery 383–385
    immediate 383–384
    session-based 383–385
Method
    CANCEL 165–166, 266, 277–278
    INVITE 39, 51–52, 78, 83–84, 94,
        106–109, 111, 123, 125, 129, 134–135,
        148, 153, 166, 191–192, 194–199,
        201–210, 212–217, 219–224, 227,
        229–230, 232–233, 236–241, 246,
        249–250, 256, 264–266, 271, 273,
        276–278, 290–291, 293, 298, 384,
        388–390
    PRACK 164, 198, 202, 208, 210, 212,
        217–219, 221–222, 228, 230, 236, 238,
        241–242, 290–291
    PUBLISH 289, 377, 379, 382
    REFER 293–294, 389
    REGISTER 19, 32–35, 40, 49–51, 73,
        112, 123, 126–127, 129–144, 146, 148,
        150–153, 156–161, 163–170, 172,
        176–177, 179, 181–182, 184–186, 194,
        264, 266, 269, 272–274, 294–295,
        299–300
    SUBSCRIBE 55, 96, 112–113, 134,
        171–174, 216, 287–288, 294, 378, 381,
        388–389
    UPDATE 164, 198, 202, 230, 237–239,
        247–249, 276, 291–292
Mg 43
MGCF *see* Media Gateway Control
    Function

Mi 43
Mj 43–44
Mk 29, 44
Mm 29–30, 43
MMS *see* Multimedia Messaging Service
mobility 3, 5, 28, 262, 273
Mp 44, 345
Mr 44
MRFC *see* Multimedia Resource Function
    Controller
MRFP *see* Multimedia Resource Function
    Processor
MTU *see* maximum transfer unit
multicast 261, 278, 365
multimedia 3–4, 6, 8, 11–13, 17, 24–27, 36,
    38, 43, 49, 59, 66–67, 83, 114, 119,
    121, 123, 126, 192, 201, 216, 223–224,
    255, 261, 263, 276, 301, 317, 323–324,
    344–345, 363, 377, 383–385, 387
Multimedia Messaging Service 60, 245,
    324–325, 385
Multimedia Resource Function
    Controller 18, 24, 44, 67, 93–98
Multimedia Resource Function
    Processor 18, 24, 44
Mw 29–32, 51–52

NAF *see* Network Application Function
NAI *see* network access identifier
naming authority pointer 73, 131, 283,
    317–320, 366
NAPTR *see* naming authority pointer
NDS 61, 63–66, 68, 155
network access identifier 54, 332
Network Application Function 71–72
non-adjacent contact 137, 294
non-repudiation 279

OCS *see* Online Charging System
offer/answer 25, 32, 38, 85, 105, 219–221,
    226–228, 231, 235–238, 241–242,
    276–277, 291–293, 298, 307–309,
    333–337, 365, 385
Online Charging System 23–25, 95, 98
Open Service Architecture 5, 25–26, 39
OPTIONS 164, 212, 266
OSA *see* Open Service Architecture
outbound proxy 123, 131, 151, 172, 198,
    202, 272

P-Access-Network-Info header 165–166,
    299
P-Associated-URI header 168, 170–171,
    300
P-Called-Party-ID header 197, 299–300
P-CSCF 15, 19–22, 29–32, 45, 48–52,
    64–67, 69–70, 72–73, 77–78, 83–86,
    89–91, 93–95, 99, 101, 114–115, 119,
    121, 123, 125–126, 129–131, 134–135,
    137, 139–142, 144, 146, 148, 150–160,
    162–168, 170, 172–175, 177, 180, 182,
    185–186, 191, 194–199, 201–203,
    205–210, 216–219, 239–242, 245–250,
    253, 325, 363
P-Preferred-Identity header 171–172,
    194–195, 197, 296
P-Visited-Network-ID header 166, 299
PA *see* presence agent
Packet Data Protocol 21, 29, 45, 50, 72,
    77–78, 82–91, 99, 123, 125–126, 129,
    192, 224, 229, 236–237, 240–242, 245,
    247, 249, 253, 323–325
page mode 289
path 23, 73, 101, 275, 285, 294–295, 382
Path header 129, 135, 199, 205, 294–295
payload 13, 19–20, 22, 45, 66, 68, 225–227,
    229, 268, 288, 304–305, 311–312,
    314–315, 355, 357–359
PDF *see* Policy Decision Function
PDP *see* Packet Data Protocol; policy
    decision point
PDP context 21, 29, 45, 50, 72, 77–78,
    82–91, 99, 125–126, 129, 192, 224, 229,
    236–237, 240–242, 245, 247, 249, 253,
    323–325
PIB *see* policy information base
Policy Decision Function 18, 20–21, 31,
    45, 48, 76–78, 80–83, 85–91, 99,
    239–240, 298
policy decision point 349–352
policy information base 352, 354
PRACK 164, 198, 202, 208, 210, 212,
    217–219, 221–222, 228, 230, 236, 238,
    241–242, 290–291
precondition 34, 207, 219, 230, 232–235,
    237–238, 292–293, 305
presence 3, 17, 26, 32, 44, 58, 106,
    135–137, 261, 264, 271, 289, 369,
    375–382

presence agent 36, 340, 376
presence server 135, 137, 377–37
presence user agent 41, 376–377, 381–382
presentity 376–381
privacy 14, 52, 68–69, 295–296
Privacy header 68, 194–198, 213, 279, 296
private user identity 23, 33–36, 38, 53–57, 59, 101, 113, 138, 140–142, 168–170
Protocol 3, 6–7, 11, 13, 19–21, 24, 49–50, 61, 65, 72, 81, 119, 121, 123, 126, 137, 139, 143, 155, 161, 165–166, 180–181, 184, 189, 191–192, 198, 201, 219, 224–225, 229, 239, 245, 250, 255, 259, 261–262, 266, 268, 279, 289, 301, 303–304, 307, 311–313, 317, 319, 323–324, 327–328, 331–332, 336–337, 345–346, 350, 354–356, 362–363, 365–366, 369, 371, 376–378, 383–384, 387–388
protocol stack 96, 208, 262, 361
provisional responses 221–222, 229, 273, 290, 298
proxy 19–20, 22–23, 32, 49, 72, 109, 119, 123, 131, 151, 156–157, 159–162, 172, 185, 191, 198, 202, 210, 262, 264, 272–273, 278–281, 284–285, 294–296, 298–300, 332, 334–335, 340, 363, 366, 377–378
proxy behaviour 284
proxy server 22, 264, 273, 285, 294–295, 298–300, 366
  statefull 20, 23, 210, 264, 284, 333, 338, 350
  stateless 264, 334–335, 338–339, 343
proxy-to-user (security mechanism) 280
PSI *see* public service identity
PUA *see* presence user agent
public service identity 44, 255–257
PUBLISH 289, 377, 379, 382

quality of service 13, 28, 45, 76–78, 81–83, 85–90, 94, 114, 233, 236–237, 292–293, 298, 311, 313, 323–324
QoS *see* quality of service

RACK header 164, 198, 202, 208, 210, 212, 217–219, 221–222, 228, 230, 236, 238, 241–242, 290–291
RADIUS 331, 335–336, 343

Real-time Control Protocol 9, 12–13, 27, 48, 78, 80–82, 84, 201, 223, 225–229, 233, 235–238, 241–242, 304–305, 311–315
reason phrase (responses) 266
receiving requests 83, 270, 287
receiving responses 286
redirect server 108, 264
REFER 293–294, 389
reference points
  Cx 30, 32, 34, 36, 38–39, 52, 75, 103, 106, 133, 339–340
  Dh 42–43, 339
  Gm 30–31, 51–52, 65–66, 68–69, 143–144, 155
  Go 3, 5, 7–8, 11, 13–14, 18–19, 30–32, 35, 39, 45, 48, 58, 60, 62, 65–66, 69, 72, 74, 76–77, 83–85, 87, 89–91, 99, 103, 109–110, 113–114, 119, 129, 132, 140–144, 149, 151–152, 155, 157–158, 160, 162, 164–165, 168–169, 179, 185, 192, 205–207, 212, 216, 219, 221, 229–231, 238, 240, 242, 245, 253, 262, 264, 276–277, 296–297, 309, 311–312, 314, 324, 327–328, 330–332, 337, 339, 344, 349, 354, 356–359, 361, 363, 385
  Gq 45, 48, 76, 85
  Mg 43
  Mi 43
  Mj 43–44
  Mk 29, 44
  Mm 29–30, 43
  Mp 44, 345
  Mr 44
  Rf 92–93, 97
  Ro 95–98, 339
  Sh 36, 39, 43, 67, 110, 113, 240, 250, 253, 339
  Si 42
  Ut 44, 70, 72, 379, 391
REGISTER 19, 32–35, 40, 49–51, 73, 112, 123, 126–127, 129–144, 146, 148, 150–153, 156–161, 163–170, 172, 176–177, 179, 181–182, 184–186, 194, 264, 266, 269, 272–274, 294–295, 299–300
registrar 20, 22, 123, 129–131, 135, 137, 168–169, 197, 199, 205, 217, 264, 269, 272–274, 294–295, 299–300

registration 6, 17, 19–20, 22–23, 25, 29–36, 38–39, 49–51, 54–55, 57, 59–60, 69–70, 73, 75, 98, 101–102, 104, 107, 110, 112–113, 119, 123, 125–127, 129–130, 132–135, 137–139, 143–144, 146, 148, 150–151, 153, 156, 159–161, 164, 167–189, 192, 194–195, 198–199, 202–203, 205–206, 212, 216, 245, 271, 273–274, 278, 288, 295, 299, 320, 340, 363

registration hijacking 278

reliability (of provisional responses) 221, 229, 290

remote sequence number 274–276

remote URI 274–276

request URI 35, 129, 131, 137, 164, 169, 171, 174, 192, 196–197, 202, 204–206, 210, 217–219, 299–300

request URI 389

Resource List Server 378, 382

response
   class
     1xx 266, 298
     2xx 266, 275–277, 288, 290–291, 295, 298–299
     3xx 266, 272, 278, 280
     4xx 267
     5xx 267
     6xx 266–267
   client error 267
   global failure 267
   provisional 221–222, 229, 273, 290, 298
   reason phrase 266
   server error 267
   success 3, 266, 272, 275, 340

retransmission (transaction) 270–271, 335

Rf 92–93, 97

RLS *see* Resource List Server

Ro 95–98, 339

roaming 6, 12–13, 15–17, 19–20, 23, 33, 64, 92, 119, 121, 166, 191, 250, 299, 331

Route header 129–135, 137, 151–152, 163–165, 171–172, 174, 185–186, 199, 202–210, 212–218, 253, 272, 275–276, 279–280, 285–286, 294–295, 299–300

route set 172, 272, 274–276, 285–286, 299

routing 9, 12, 18, 22, 29, 32–36, 38, 43, 54, 58, 110–111, 114, 119, 126, 131, 135, 150, 152–154, 164, 171, 174, 194, 196, 198–199, 202–203, 207–208, 210, 212, 214–216, 227, 255–257, 262, 268, 272, 274, 283–285, 300, 331, 333–337

RSeq header 222, 290

RTCP *see* RTP Control Protocol

RTP *see* Real-time Control Protocol

RTP Control Protocol 78, 81–82, 84, 304, 311, 313–314

RTP Payload Format 314

RTP Profile 314

RTP Profile for Audio and Video 314

S-CSCF *see* Server-Call Session Control Function

S/MIME *see* Secure-Multipurpose Internet Mail Extension

SBLP *see* session-based local policy

SCTP *see* Stream Control Transmission Protocol

SDP *see* Session Description Protocol
   line 303
     attribute 304–306
     format 303
     RTP map 305
   connection 303
   media 302, 304
   version 303

SDP message format 303

Secure-Multipurpose Internet Mail Extension 280–281

Secure Network Management Protocol 262, 354

security 6, 8–9, 14, 17, 19–20, 23, 25, 27–28, 30, 36, 44, 49, 59–70, 72, 114, 125, 135, 137–138, 140, 143–144, 146, 151–153, 155–161, 169, 186, 191, 194, 203, 268, 278–280, 296–297, 314, 319, 327–328, 330, 332–333, 335–337, 355–357, 359, 363, 371–372

security agreement 69, 296

security domain 27–28, 63–65, 68

security framework 278

security gateway 18, 27–29, 65, 68, 355–356, 359

security mechanism 30, 62–63, 66–67, 69, 125, 140, 144, 146, 153, 155–159, 161, 203, 296–297

SEG *see* security gateway

server 6, 18, 22–23, 25–26, 34, 43, 50, 52,
     58, 69, 71, 73–74, 91, 96, 101, 104–110,
     114, 125–127, 135, 137, 141, 144, 148,
     150–154, 156, 158–161, 172, 199,
     202–209, 212, 218, 255–256, 263–264,
     266–268, 270–271, 273, 278, 281, 283,
     285, 287, 289, 294–300, 317, 319–321,
     327–332, 334, 338–340, 343, 349–350,
     365–367, 369, 371, 377–379, 385,
     388–389, 391–392
   application 22, 24–26, 29, 32, 39–44, 58,
        67, 91, 94–98, 104–105, 107–109, 166,
        182, 195, 212–215, 249–250, 253,
        255–257, 264
   location server 264, 273
   proxy 22, 264, 273, 285, 294–295,
        298–300, 366
        outbound 123, 131, 151, 172, 198, 202,
             272
   redirect 108, 264
Server-Call Session Control Function
        19–26, 30–36, 38–40, 43–44, 49–50, 52,
        54, 57–58, 67, 73–75, 93–97, 101–104,
        106–114, 123, 125, 127, 129, 132–135,
        137–141, 143–144, 146, 156–160, 164,
        166–168, 170–172, 174–175, 177–182,
        184–186, 188, 191–192, 195–197, 199,
        202–206, 208, 210, 212–216, 218, 240,
        242, 245–250, 255–257, 340
server discovery 283
server error (response) 267
service records 73, 131, 268, 283–284, 321
service route discovery 299
Service-Route header 129, 134–135, 172,
        186, 199, 202–203, 213, 299
Serving GPRS Support Node 5, 12, 15, 23,
        28, 90, 93, 97, 99, 323–324
session description 7, 20, 78, 106, 119, 201,
        224, 229, 245, 268, 277–278, 291, 301–
        302, 307–308
session 3–4, 6, 9, 13–23, 25–32, 34, 43–45,
        49–52, 54–55, 57–58, 62, 66, 68,
        70–73, 75–78, 80, 82–84, 89–94,
        96–99, 101, 104, 106–111, 113–114,
        119, 121, 123, 137, 153, 161, 165–167,
        180–181, 184, 189, 191–192, 197–198,
        201, 207–210, 216–225, 227, 229–237,
        239–241, 245–247, 250, 253, 255,
        261, 266, 268, 276–278, 289, 291–292,

        298, 301–304, 307–308, 311, 313, 319,
        323, 327–330, 333–334, 338–340, 343,
        345, 362–363, 366, 375–376, 383–385,
        387
session-based local policy 20, 45, 76–78,
        85, 87
session-based messaging 383–385
Session Description Protocol 7, 9, 20, 22,
        29, 45, 68, 76–78, 80–81, 83–84, 86, 89,
        103, 106, 115, 119, 121, 201, 219–221,
        223–231, 233–238, 240–242, 244–245,
        268, 276–277, 289–292, 298, 301,
        303–305, 307–308, 315
SGSN *see* Serving GPRS Support Node
SGW *see* Signalling Gateway
Sh 36, 39, 43, 67, 110, 113, 240, 250, 253,
        339
Si 42
SigComp 217, 361–363
SigComp architecture 361
Signalling Gateway 18, 27
SIP 6, 8–9, 13–14, 19–27, 29–36, 39, 43–45,
        48–52, 54–59, 61, 65–66, 68–70, 73–75,
        77–78, 83, 89–91, 93–94, 96, 98,
        101–103, 105–106, 108–111, 114–115,
        119, 121, 123, 125–126, 129–137,
        139–144, 146, 148, 151–173, 175,
        179–181, 184–186, 189, 191–192,
        194–199, 201–210, 212–213, 215–222,
        224, 227–230, 232–233, 237, 239–241,
        245–250, 253, 261–281, 283–285,
        287–296, 298–300, 307, 319–321, 332,
        339–340, 344, 362–363, 366–367, 376,
        378–379, 381, 384–385, 387–388
   trapezoid 263
   URI 22, 34–36, 55–56, 58, 74, 102, 110,
        121, 123, 129, 167, 169–171, 179, 196,
        198–199, 202, 205, 268–269, 280,
        283–284, 300, 381, 385, 387
SIP compression 20, 30, 114, 125, 363
SIP extensions 287, 295, 298
SIP layers
   syntax and encoding 270
   transaction 270
   transaction user 226, 271, 283, 286–287,
        345
   transport 270
SIP URI 22, 34–36, 55–56, 58, 74, 102,
        110, 121, 123, 129, 167, 169–171, 179,

SIP URI (*cont.*)
196, 198–199, 202, 205, 268–269, 280, 283–284, 300, 381, 385, 387
sipfrag (message) 294
SLF *see* Subscription Locator Function
SNMP *see* Secure Network Management Protocol
soft state (registration) 274, 288–289
SRV *see* service records
start line 265–266, 294
state handler 362, 364
state machines (transmission) 26, 264, 271, 333
state publication 289
statefull (proxy server) 20, 23, 210, 264, 284, 333, 338, 350
stateless (proxy server) 264, 334–335, 338–339, 343
static payload 305, 314–315
stream 24–25, 27, 44, 78, 83–86, 88, 90, 109, 121, 191–192, 201, 220, 223–230, 233–234, 236–240, 242, 245, 247, 263–264, 276, 279, 292, 300, 307–309, 312, 325, 332, 337, 345–346, 389
Stream Control Transmission Protocol 27, 262, 279, 286, 332
strict route 284, 286
SUBSCRIBE 55, 96, 112–113, 134, 171–174, 216, 287–288, 294, 378, 381, 388–389
Subscription Locator Function 18, 24, 33, 38–39, 43, 204–205
Subscription-State header 175, 186, 288, 294
success (response) 3, 266, 272, 275, 340
Supported header 221, 271, 273, 290, 293–295, 378

tag 130, 132, 134, 157, 159, 171, 175, 179, 194, 201–202, 221, 233, 272–274, 289–290, 292–295, 378
TCP *see* Transmission Control Protocol
telephone uniform resource identifier 170, 269
tel URI *see* telephone uniform resource identifier
template package 288–289, 379–380
temporary public user identity 56–57, 102, 169–171

terminal 3–4, 16, 25, 43, 49, 57, 62, 64, 78, 109, 113–114, 121, 123, 168, 170–173, 175, 179, 188, 194, 221–222, 224–228, 230–237, 263, 331, 334, 343, 375, 377–378
text/plain 289
time description 302
timeout 270, 272, 288
TLS *see* Transport Layer Security
To header 16, 19–23, 27, 36, 38, 43, 54, 57, 62–63, 70, 93, 106, 108, 110, 112, 114, 119, 129–131, 133–135, 137, 151, 155–156, 168–169, 171, 175, 178, 180, 182, 186, 194, 197, 201–203, 209, 214, 240, 253, 267, 271–277, 279, 292–293, 313, 329–330, 354, 359, 378, 389–390
topology-hiding inter-network gateway 21
transaction 19, 30–31, 130, 201–202, 208, 212, 220, 264, 270–272, 277, 286–287, 290, 333, 366, 384
    client 6, 11, 55, 69, 114, 144, 148, 150–154, 156–161, 186, 263–264, 267, 270–271, 281, 289, 296, 308–309, 319–321, 327–330, 332–334, 338–339, 349–352, 365–366, 371, 378–379, 388
    retransmission 270–271, 335
    state machine 26, 264, 271, 333
    timeout 270, 272, 288
transaction user 226, 271, 283, 286–287, 345
Transmission Control Protocol 13, 78, 81–82, 84, 94, 126, 144, 148, 154, 262, 268, 279–280, 283, 286, 304, 311–314, 317, 327, 332, 335, 350, 356
transport 6, 9, 12–14, 17, 27, 30, 42, 45, 48, 72, 81, 84, 86, 99, 126, 140, 144, 154–155, 167, 170, 192, 201, 204, 225–226, 229, 233, 245, 262, 267–272, 279–281, 283, 286–287, 299, 301, 304–305, 311, 314, 319, 325, 327, 330, 332, 344, 350, 355, 357, 359, 361, 363, 365
    SCTP *see* Stream Control Transmission Protocol
    TCP *see* Transmission Control Protocol
    UDP *see* User Datagram Protocol
Transport Layer Security 72, 155, 158, 161, 268, 275, 279–280, 297, 319, 327–330, 332–333

trapezoid (SIP) 263
TU *see* transaction user

UA *see* user agent
UAC *see* user agent client
UAS *see* user agent server
UDP *see* User Datagram Protocol
UDVM *see* universal decompressor virtual
    machine
UE *see* user equipment
uniform resource identifier 22, 34–36,
    55–56, 58, 74, 102, 106, 108, 110, 121,
    123, 129, 131, 137, 140, 162, 164,
    167–172, 174–176, 178–179, 187, 192,
    195–199, 202, 204–206, 210, 215,
    217–219, 255, 261, 265–269, 271–272,
    274–277, 279–280, 283–286, 288–289,
    293, 295, 299–300, 319, 369, 371,
    378–379, 381, 384–385, 387–389
universal decompressor virtual
    machine 361–362, 364
Universal Subscriber Identity Module 56,
    60, 126, 169–170
UPDATE 164, 198, 202, 230, 237–239,
    247–249, 276, 291–292
URI *see* uniform resource identifier
user agent 19, 22, 32–35, 41, 68–69, 73, 75,
    108–109, 214–215, 263–264, 271–280,
    284, 286, 290–291, 293–300, 340,
    376–377, 381–382, 387–388
  B2BUA 214, 215
  client 263–264, 271–277, 279–280, 286,
    290–291, 300
  server 263–264, 271–275, 277, 279–280,
    284, 290–291
user agent client 263–264, 271–277,
    279–280, 286, 290–291, 300
user agent server 263–264, 271–275, 277,
    279–280, 284, 290–291
User Datagram Protocol 13, 78, 126,

129–133, 136, 154, 164–165, 171, 173,
    175, 185, 202–210, 212–213, 215–218,
    262, 265, 268, 280, 283, 304, 356, 365
user equipment 11–13, 17, 19–20, 22–23,
    28–31, 34, 36, 43–44, 49–52, 54–57,
    59–60, 62, 64–73, 75–78, 80, 82–87,
    89–91, 95, 99, 101, 108–115, 123,
    125–127, 129–135, 137–142, 144, 146,
    148, 150–160, 162–166, 168–172,
    174–175, 177, 179–187, 191–192,
    194–199, 201–203, 205–210, 212,
    216–227, 229–230, 232–233, 236–242,
    245–249, 253, 263, 323–325, 363, 381,
    384–385
user-to-user (security mechanism) 280
USIM *see* Universal Subscriber Identity
    Module
Ut 44, 70, 72, 379, 391

version 27, 98, 106, 110, 161, 164, 178,
    180, 182, 186, 188, 224, 227, 229,
    265–266, 301, 303, 311, 329, 345,
    350–351, 354, 365
Via header 13, 130–132, 134, 148, 151–154,
    163–165, 199, 202–210, 212, 214–215,
    217–219, 267–268, 271–273, 280,
    286–287, 300
video 15, 17, 48, 58, 78, 80, 83–84, 86, 89,
    93, 103, 113, 121, 201, 219, 223,
    225–229, 233, 241–242, 261, 304,
    314–315, 375, 385, 387
voice 3, 6, 12, 15, 78, 96, 106, 108–109,
    113–114, 197, 226, 261, 342, 375, 385

watcher 376–382
watcher information 379–380, 382
winfo 289, 379–380

XCAP 369–370, 378–379, 381